A Territorial Antelope:
The Uganda Waterbuck

A Territorial Antelope:
The Uganda Waterbuck

C. A. SPINAGE
UNDP,
B.P. 872, Bangui,
Central African Republic

1982

ACADEMIC PRESS

A Subsidiary of Harcourt Brace Jovanovich, Publishers

London   New York   Toronto   Sydney   San Francisco

Academic Press Inc. (London) Ltd.
24–28 Oval Road
London NW1

US edition published by
Academic Press Inc.
111 Fifth Avenue,
New York, New York 10003

British Library Cataloguing in Publication Data

Spinage, C. A.
  A territorial antelope.
  1. Uganda Kob 2. Territoriality (Zoology)
  I. Title
  599.73′58′096761   QL737.U5

ISBN 0–12–657720–X
LCCCN 81–66687

Printed in Great Britain by
St Edmundsbury Press, Bury St Edmunds, Suffolk

Contents

v

Preface

Territoriality, the name given to the social organization in which animals defend an area against competitors of their own kind, has been well known in birds since Eliot Howard defined it in 1920 in his book "Territory in Bird Life"; but in 1955 Bourlière was to write of mammals: "In general, and in so far as the paucity of observations permits judgement, it would seem that territorial behaviour is far from being as important in mammals as in birds." He was able to list only a half dozen inferences that suggested territoriality in mammals, and all of these related to smaller mammals. At the same time Bourlière quoted Hoier's observation of a buck defassa waterbuck which allegedly occupied the same area in the Congo Albert National Park from 1931 to 1939 (Hoier, 1950); and it was in 1955 that Verheyen postulated, from his observations in the same area, that the waterbuck was territorial in its behaviour (Verheyen, 1955).

It is difficult to understand today why the study of mammals, so much more closely related to the study of man than studies of birds, insects or fishes, was so neglected. Yet it was not until 1967 that a wide interest was really awakened in mammal behaviour by the popular writing of Robert Ardrey, whose book, "The Territorial Imperative", seems to have been prompted in no small measure by the observations of Buechner. In 1957 Buechner, an American Fulbright scholar working in northern Uganda, discovered, or rather his wife did, that the Uganda kob, a medium-sized antelope closely related to the waterbuck, exhibited a type of intensive territoriality, in the Semliki Valley of western Uganda, which had similarities with the "lek" behaviour of grouse on their breeding grounds in Scotland (Buechner, 1961).

But it was not these considerations which led me, in October 1964, to study the social organization of the waterbuck. This was prompted more by the growing interest in the natural regulation of animal populations stimulated by such works as Lack's "The Natural Regulation of Animal Numbers", and Andrewartha and Birch's "The Distribution and

Abundance of Animals", both of which appeared in 1954. Wynne-Edward's "Animal Dispersion in Relation to Social Behaviour", by the criticism which it attracted when it made its début in 1962, greatly stimulated my interest in the subject.

The concept of·territoriality was an important factor in the welter of speculation as to how natural population regulation might be achieved, and the apparent territorial organization of the waterbuck had been noted by the Director of the newly created Nuffield Institute of Tropical Animal Ecology, Dr R. M. Laws. This Institute was situated in the then Queen Elizabeth National Park of western Uganda, now the Rwenzori National Park and had been set up in response to the park authorities' concern at the great numbers of hippopotamuses inhabiting the area, which appeared to pose a threat of overgrazing and habitat deterioration to the detriment of the recently created park. I was invited to study the biology and social organization of the waterbuck, which was quite numerous there, as a part of an integrated overall study of the ecology of the park, which would relate to the presence of the hippopotamus.

My main objective was to try to establish whether territorial behaviour really existed in the waterbuck, and if so, what part this social system played in regulating the numbers of waterbuck, if any. However, at this period not only was virtually nothing known in detail of the social behaviour of almost all African ungulates, but also little was known of their biology. How long did they live for? How fast did they grow? When did they breed, and how often? These were just some of the questions to be answered, and this was why my study encompassed not only the behaviour of the waterbuck, but as much of its biology as I could learn in the three years which were available to me.

The result should not be considered as a book just about waterbuck. The interest of the waterbuck lies in its simple, basic type of territorial organization which the study revealed; but the book also presents the life history of an African antelope, detailing its anatomical, physiological and behavioural organization from birth to death, in which I hope that I have provided some insights into the biology of African antelopes as a whole. While, finally, I have allowed myself to speculate on the function and cause of territoriality.

The field work was carried out during 1964 to 1967, and this long gestation to the present has allowed me to put my findings into perspective in relation to studies both on the waterbuck, and to several other species on which studies have since been conducted, as well as to follow the vicissitudes of a waterbuck population for a period of some 44 years.

The original field work was financed by the Science Research Council of Great Britain, to whom I am eternally grateful. My thanks are also due to the Director and governing body of the former Nuffield Unit of Tropical Animal Ecology for providing me with facilities at the Mweya research station; and I am especially grateful to the former director, Dr R. M. Laws FRS, for his interest and encouragement. My thanks are also due to the former director of the Uganda National Parks, the late Francis Katete, who gave permission for the study to be carried out; also to Professor Sir Alan Parkes and Dr I. W. Rowlands for providing accommodation and facilities for the writing-up of the original field work at the Wellcome Institute of Comparative Physiology at the Zoological Society of London. This writing-up formed the subject of the degree of Doctor of Philosophy of the University of London.

My thanks are also due to Dr Roger Short FRS, who initiated me into the art of darting and the study of reproductive physiology; to Dr S. K. Eltringham, a former director of the Nuffield Unit of Tropical Animal Ecology for kindly providing me with some of his data; also to Dr G. Petrides for supplying me with his data, and to W. F. H. Ansell for generously allowing me to use his map of waterbuck distribution. Finally, but not least, my thanks are due to all those other many workers and colleagues, too numerous to mention, who assisted in countless ways.

July 1981   C. A. Spinage

List of Figures

List of Plates

1. Classification, Distribution and Origins of Waterbuck

Introduction

The antelope tribe Reduncini (Simpson, 1945) contains only two living genera: *Redunca* — the reedbucks, and *Kobus* — the kobs. The waterbuck is the largest of the kobs, and among the largest of the African antelopes. Of impressive appearance, the buck sports long, slightly curving horns, adorned with elegant chippendale ridges on the anterior faces. The body is solid, well proportioned and powerful, with a coarse but sleek coat. The hair of the coat gives the impression that it is thick, until examined closely, when it is found to be relatively sparse; but that on the neck is long, shaggy and wiry (Plates 1 and 2). The French call the waterbuck the *Cobe oncteux,* or "greasy kob", referring to the oily secretion that its skin exudes. When it is prolific this secretion imparts a dark, almost black look to the coat. The English name indicates its habit of always being found near to water; but despite its strong dependence on water it is one of the most widely distributed of the African antelopes, ranging from as far north as 14° to 29° in the south.

This wide range bears testimony to its success, wherever suitable permanent water exists; its habitat including such arid country as that surrounding the Webi and Schebeli rivers in Somalia, and the Awash in Ethiopia. Typically, however, it is a savanna and woodland species, localized near to permanent water and breeding well where not hunted by man. It seldom forms herds of larger than a hundred or so individuals, and commonly much less than this.

Ansell (1971) describes its former range in South Africa as: the north-eastern part of Natal, the eastern and northern Transvaal, the northern Cape Province along the upper Molopo River, and also along the upper Limpopo River. Today this range has been reduced in Natal to the

Plate 1. A 10-year-old defassa waterbuck, known as Y1, Mweya Peninsula.

Plate 2. Doe defassa on the Mweya Peninsula, overlooking the Kazinga Channel.

Hluhluwe and Umfolosi game reserves, and it is considered to be extinct within its former range in southern Botswana. Although its distribution has been fragmented in many areas to the north, it has fared much better there, and is found today to the north of South Africa throughout the southern savanna; while in the northern savanna it is distributed through Uganda and the southern Sudan, westwards as far as Senegal and the Gambia. In the east it ranges into the Somali arid zone and central Ethiopia, and as far north as the Atbara River in the Sudan (Fig. 1).

The type specimen on which the species is based, was collected in 1832 by a South African hunter-explorer named Andrew Steedman. He obtained his trophy "twenty-five days journey north of the Orange River

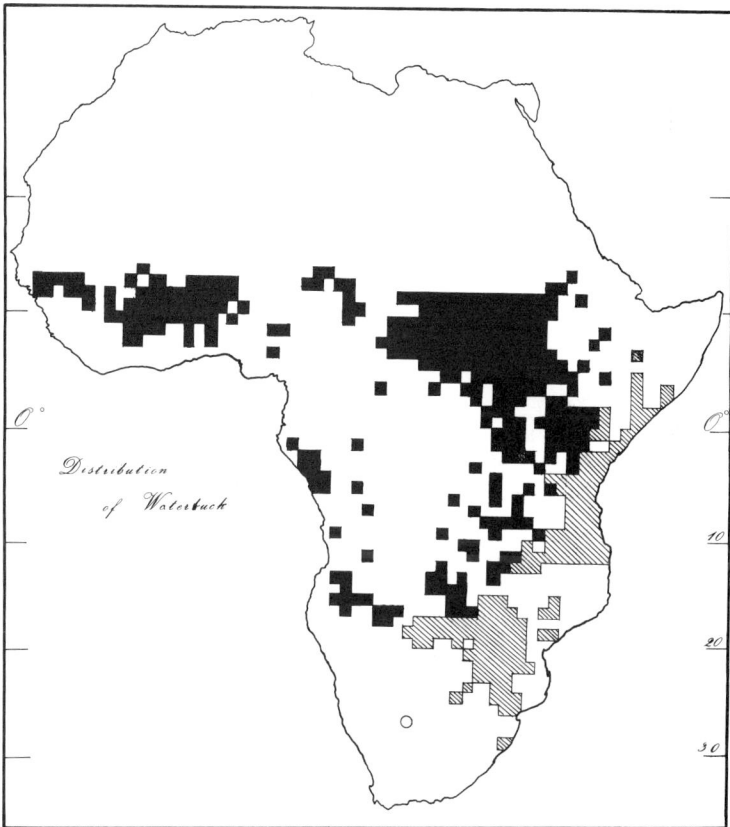

Fig. 1. The distribution of waterbuck in Africa. Blocked squares *Kobus ellipsiprymnus defassa*, hatched squares *K. e. ellipsiprymnus.* Circle shows where type specimen originally occurred. Based on a map by W.F.H. Ansell, with additions.

between Latakua and the west coast of Africa". This location has been placed as on the Molopo River, almost in the centre of South Africa, and a little to the east of the present-day Kalahari Gemsbok National Park, an area where the species is now extinct. Steedman's specimen was described by W. Ogilby in 1833 under the name of *Antilope ellipsiprymnus* Ogilby, later being referred to the genus *Kobus* as *Kobus ellipsiprymnus* Ogilby (Smith, 1840). The name *ellipsiprymnus* referred to the distinctive white ellipse which characterizes the animal's rump.

Some years before Steedman collected his specimen, a horn had been brought to Cape Town. The Dutch Government offered a large reward for a complete specimen of the animal, but without success. Steedman subsequently discovered that the Governor of the Cape had, in fact, possessed an imperfect specimen, which he gave to the French naturalist Jules Verreaux; but the latter did not apparently describe it.

Two years after Steedman had collected his specimen another was secured, this time in Abyssinia by the German naturalist Dr Edward Rüppell. The new animal differed principally in that the rump was marked with a white blaze, similar to that of the red deer. Rüppell claimed this to be a new species and called the animal after its Amharic name "defassa", the defassa waterbuck *Antilope defassa* Rüppell (Rüppell, 1835).

The Races of Waterbuck

Steedman's waterbuck, which has come to be known as the "common waterbuck", although the defassa is by far the more widely distributed and the more numerous of the two, is described as grizzled grey or brown in colour. The typical defassa, as found by Rüppell "about Dembea lake and in the Kulla", was bright rufous with much white on the face and throat. But coat colour is extremely variable in the defassa, such that 29 races have been described, compared with only 8 for the common waterbuck (Table 1). This ebullience of the early taxonomists has now been reduced in the latest taxonomic authority (Ansell, 1971) to four and nine races of the common and defassa waterbuck respectively (Table 2). Even so, I am unable to identify any obvious differences between waterbuck from the Rwenzori National Park of Uganda, and those occurring in the Saint Floris National Park in the north of the Central African Republic, 1600 km to the north-west.

The various races have all been attributed to differences in horn shape and coat colour and pattern; the former varying in the amount of curvature, and the coat colour varying from light rufous, through deep rufous, grey, silver grey, grizzled and very dark brown. Differences are

Table 1 The races of waterbuck (after Allen, 1939).

Kobus ellipsiprymnus ellipsiprymnus (Ogilby, 1833)
                    kondensis (Matschie, 1911)
                    lipuwa (Matschie, 1911)
                    kulu (Matschie, 1911)
                    kuru (Heller, 1913)
                    pallidus (Matschie, 1910)
                    thikae (Matschie, 1910)
                    canescens (Lönnberg, 1912)

Kobus defassa defassa (Rüppell, 1835)
                    albertensis (Matschie, 1910)
                    annectens (Schwarz, 1913)
                    crawshayi (P.L. Sclater, 1894)
                    harnieri (Murie, 1867)
                    breviceps (Matschie, 1910)
                    ladoensis (Matschie, 1910)
                    griseotinctus (Matschie, 1910)
                    hawashensis (Matschie, 1910)
                    matschiei (Neumann, 1905)
                    nzoiae (Matschie, 1910)
                    fulvifrons (Matschie, 1910)
                    penricei (W. Rothschild, 1895)
                    frommi (Matschie, 1911)
                    münzneri (Matschie, 1911)
                    raineyi (Heller, 1913)
                    schubotzi (Schwarz, 1913)
                    tjäderi (Lönnberg, 1907)
                    powelli (Matschie, 1910)
                    angusticeps (Matschie, 1910)
                    togoensis (Schwarz, 1914)
                    tschadensis (Schwarz, 1913)
                    ugandae (Neumann, 1905)
                    unctuosus (Laurillard, 1842)
                    adolfi-friderici (Matschie, 1906)
                    avellanifrons (Matschie, 1910)
                    cottoni (Matschie, 1910)
                    dianae (Matschie, 1910)
                    uwendensis (Matschie, 1911)

---

also said to exist in the extent of the white facial markings. But in the main area in which I studied the defassa waterbuck, the Mweya Peninsula of the Rwenzori Park, at least three "races" could be identified in less than 5 km^2.

There was a big rufous buck with long horns, a short, stocky buck, silver-grey in colour, and the more grizzled form of intermediate size. All

Table 2   The races of Waterbuck (after Ansell, 1971).

---

ellipsiprymnus group

Kobus ellipsiprymnus ellipsiprymnus (Ogilby, 1833)
     kondensis (Matschie, 1911), includes *lipuwa* and *kulu*
     thikae (Matschie, 1910), includes *canescens* and *kuru*
     pallidus (Matschie, 1910)

defassa group

Kobus ellipsiprymnus defassa (Rüppell, 1835), includes *matschiei* and *hawashensis*
     unctuosus (Laurillard, 1842), includes *togoensis*
     harnieri (Murie, 1867), includes *ugandae*, *avellanifrons*, *cottoni*,
       dianae, *breviceps*, *ladoensis*, *albertensis* and *griseotinctus*
     crawshayi (P. L. Sclater, 1894), includes *frommi*, *münzneri* and
       uwendensis
     penricei (Rothschild, 1895)
     adolfi-friderici (Matschie, 1906), includes
       nzoiae, *fulvifrons* and *raineyi*
     tjäderi (Lönnberg, 1907), includes *powelli* and *angusticeps*
     annectens (Schwarz, 1913), includes *schubotzi*
     tschadensis (Schwarz, 1913)

---

three types could be found elsewhere in the Park, in addition to possibly, a fourth, very dark form, although this dark colour may have resulted from the skin secretion. The horns tend to be long and sharply pointed, with a slight forward curvature, and ridged on the leading side. Among the Peninsula bucks, and others in the Park, I could distinguish three quite separate horn shapes. These I classified as straight, intermediate, and bowed, depending upon the amount of curvature and inward rotation (Fig. 2).

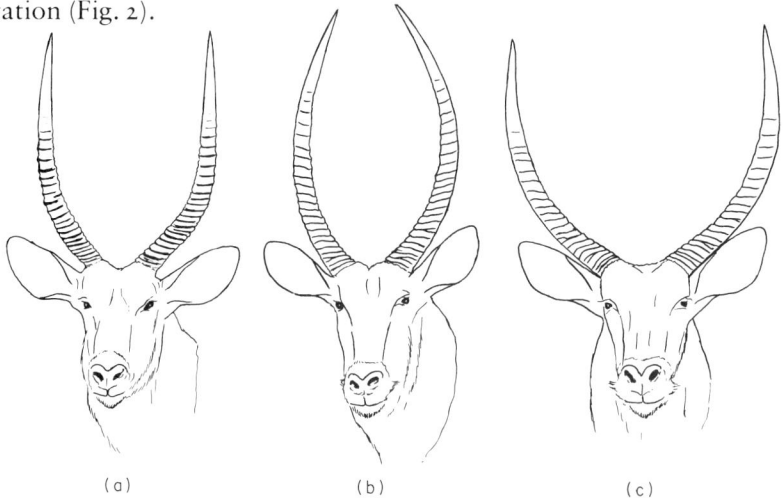

   (a)   (b)   (c)

Fig. 2. Horn shapes in waterbuck: (a) "straight", (b) "intermediate", (c) 'bowed".

Does were more uniform in appearance, but even so there were light coloured ones and dark coloured ones, and a distinct separation into long-legged and short-legged forms, the latter the more common of the two.

Using the measurements of trophy heads given in Rowland Ward's "Records of Big Game" (Best *et al.*, 1962), I analysed the horn shape of seven main groups: the common, typical defassa, Uganda defassa, Rhodesian defassa, Angolan defassa and West African defassa. These groups are defined by Best *et al.* (1962) as follows:

> "the common waterbuck inhabits the area from the Limpopo River through Zululand, the Transvaal, near Lake Ngami, Rhodesia, Malawi, Zambia, Tanzania and Kenya to the Webi Shebeli river in Somalia. The typical defassa is found in the highlands of Ethiopia, south through Kenya into northern and central Tanzania. West of the typical defassa's range the Uganda defassa spreads from the foot of the Ethiopian highlands, through the southern Sudan, the Bahr el Ghazal basin, and south through Uganda to western Tanzania. The Rhodesia defassa is found in Zambia, Malawi and Zimbabwe Rhodesia; while the Angola defassa is found from Congo Brazzaville and southern Gabon, through Angola to the Upper Okavango and the eastern end of the Caprivi Strip in Namibia. The greatest range is shown by the West African defassa, or Sing-sing, which occurs in Zaire, through the Central African Republic, west to Senegal and Gambia."

To these major groups I added a random collection of adult specimens from the Rwenzori Park (Table 3).

With these measurements I examined the relationship of width between the tips of the horns within groups, and among groups, to determine whether a separation of horn types could be made on a geographical basis. Although the Rowland Ward measurements represent a biased sample with regard to length, they nevertheless cover a wide range of lengths, and unless shape is a reflection of length, then the bias should not significantly influence the analysis of horn shape. Horn shape does not, in fact, seem to be a function of horn length. Width between the tips provides an indication as to whether the horns are straight, intermediate, or bowed in shape.

The analysis showed that none of these distributions of degree of "bow" approximated to that of the normal distribution at the 95% level of probability. The widest departures from normality were shown by the Rwenzori group, followed by the Angolan, with distributions skewed sharply to the right by a preponderance of intermediate forms. In general, "straight" horns were the most uncommon type, and "intermediate" tending towards the "bow" shape were the most common (Table 4).

Using Bartlett's test for homogeneity of variance it was found that the residual variance was the same for all groups ($P = > 0.05$), and thus the

Table 3 The Rowland Ward (1962) grouping of races.

Race	Origin
Typical defassa *Kobus defassa defassa*	Ethiopia
angusticeps	North Kenya
fulvifrons	West Kenya
hawashensis	Ethiopia
matschie	Ethiopia
nzoiae	West Kenya
powelli	Central Kenya
raineyi	West Kenya
tjäderi	West Kenya
Uganda defassa *Kobus defassa ugandae*	West Uganda
adolfi-friderici	North Tanzania
albertensis	West Uganda
breviceps	South Sudan
cottoni	West Zaire
dianae	West Zaire
griseotinctus	South Sudan
harnieri	South Sudan
ladoensis	South Sudan
uwendensis	South-west Tanzania
Rhodesian defassa *Kobus defassa crawshayi*	Zambia
Angola defassa *Kobus defassa penricei*	Angola
frommi	South-west Tanzania
münzneri	South-west Tanzania
West African defassa (Sing-sing) *Kobus defassa unctuosus*	Senegal
annectens	South-west Sudan
avellanifrons	West Zaire
togoensis	Togo
schubotzi	North Zaire
tschadensis	South-west Sudan

Table 4 A comparison of waterbuck groups by means of horn width between the tips with the normal distribution.

Group	Chi X^2	df	P
Rwenzori	46·22	1	$\ll 0.005$
Angola	4·38	1	$< 0.05 > 0.025$
Uganda	4·47	3	$< 0.5 > 0.25$
Typical	1.13	2	$< 0.75 > 0.5$
Rhodesia	0·36	1	$< 0.75 > 0.5$
West African	0·37	2	$< 0.9 > 0.75$
Common	1·12	1	$< 0.9 > 0.75$

groups were not significantly different from one another. This suggests that all of the groups intergrade, but using the "*t*" test between groups it was found that significant differences existed (Table 5). The common, typical defassa, Uganda and Rwenzori animals seem to form a related unit, but surprisingly enough significantly separate from the Uganda group. The Rhodesian and Angola groups unite in a distinct separation, but the Rhodesia group is united with the typical. The greatest difference exists between the Angola and Rwenzori groups, and the least difference between the Rwenzori and common groups (Fig. 3 and Table 5). This suggests that distinct geographical races do exist, the only anomaly being the separation of the West African group from the Uganda group, but we will see later that this can be explained.

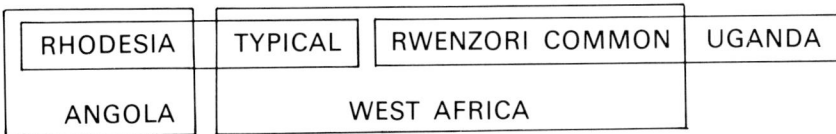

RHODESIA	TYPICAL	RWENZORI COMMON	UGANDA
ANGOLA		WEST AFRICA	

Fig. 3. The grouping of Rowland Ward's waterbuck divisions.

Table 5 The separation of waterbuck groups according to horn shape.

Group	t	df	p
Typical × Uganda	− 1·62	64	< 0·2 > 0·1
× Rhodesia	+ 1·95	34	< 0·1 > 0·05
× Angola	+ 2·76	37	< 0·01 > 0·001
× West African	+ 0·96	68	< 0·4 > 0·3
× Common	− 0·51	88	< 0·7 > 0·6
× Rwenzori	+ 0·37	79	< 0·8 > 0·7
Uganda × Rhodesia	+ 3·70	50	> 0·001
× Angola	+ 4·80	53	> 0·001
× West African	+ 3·46	84	> 0·001
× Common	+ 1·53	104	< 0·2 > 0·1
× Rwenzori	− 1·67	95	< 0·1 > 0·05
Rhodesia × Angola	+ 0·74	23	< 0·5 > 0·7
× West African	− 2·03	54	< 0·05 > 0·02
× Common	− 2·71	74	< 0·01 > 0·001
× Rwenzori	+ 2·70	65	< 0·01 > 0·001
Angola × West African	− 3·12	57	< 0·01 > 0·001
× Common	− 3·76	77	> 0·001
× Rwenzori	+ 3·69	68	> 0·001
West African × Common	− 1·95	108	< 0·1 > 0·05
× Rwenzori	+ 1·75	99	< 0·1 > 0·05
Common × Rwenzori	+ 0·18	119	< 0·9 > 0·8

t = value of Student's t test.
df = degrees of freedom.
p = probability that the populations differ.

Although the analysis showed that the Rwenzori animals apparently have the greatest affinity to the common waterbuck, followed by the typical defassa, and only a relatively weak link with the Uganda group, when I carried out my study in the field I was unaware of this fact. I had assumed the Rwenzori animals to represent the Uganda race, which was based on a specimen collected in 1894 by Oscar Neumann on the Mayanja River, some 435 km to the north-east of the park. This was named *Kobus defassa ugandae* Neumann, the Uganda defassa water-buck, but is now regarded by Ansell (1971) as a synonym of *Kobus defassa harnieri* Murie, a race described from the White Nile in 1867 (Murie, 1867).

No comparable descriptive work on African mammals has appeared since that of Roosevelt and Heller (1915), who gave the range of the Uganda race as from the western base of Mount Elgon, westward throughout Uganda to the Semliki Valley, north as far as the limits of the Victoria Nile drainage, and south to Lake Kivu. For the Nile defassa they gave the range as extending from the White Nile district east to the foot of the Abyssinian highlands, south as far as Lake Albert, and westward throughout the Bahr-el-Ghazal watershed.

The Species of Waterbuck

Despite the fact that there is doubt about the status of the races of waterbuck, one would have thought that the two "species" were distinct enough: the defassa with its white blaze on the rump and predominantly rufous colouring, and the common with its white ellipse and rather uniformly grizzled coat (Plates 3 and 4). The former, as we have seen, is the more widely distributed of the two, ranging from central Ethiopia in the east to Senegal in the west, and south to the Tropic of Capricorn. The common waterbuck is restricted to the east of the Rift Valley for the greater part of its range, from about 6° north of the Equator, then extending much further south than its congener to about latitude 29°, and reaching westwards to the north-eastern tip of Namibia.

In three places in East Africa, namely on the Uaso Nyiro River in northern Kenya, on a small stretch of the Athi River in the Nairobi National Park of Kenya, and in the Ngorongoro Crater of northern Tanzania, the two species meet and apparently interbreed (Fig. 4). Thus, in the Nairobi Park, one can meet with animals displaying all degrees of pattern on the rump, from the ellipse to the blaze, the intermediate patterns being the only feature which distinguishes them as possible hybrids (Plate 5). These animals were described by Kiley-Worthington (1965), and I was able to observe them for myself in 1966. Their existence

Plate 3. Doe common waterbuck, showing the difference in rump patterns; Tsavo East National Park, Kenya.

Plate 4. Hybrid common × defassa waterbuck in the Nairobi National Park, Kenya.

Plate 5. The glacier-capped Rwenzori Mountains from Kayanja, with waterbuck bachelor herd and does in the foreground.

had been known of for some time and, according to Cowie (1956), the population was studied by a scientist "many years ago". Presumably this study was some time after 1946, the year that the Park was gazetted, but I have been unable to trace any further reference to it.

This population, numbering some 40 to 50 head, does not appear to increase either its range, or its population size, and would thus seem to represent what Mayr (1942) termed "allopatric hybridization". This results from the meeting of two incompletely separated species in a border zone, consequent upon the premature breakdown of a geographical barrier. The two populations, having been isolated in this case perhaps by the tectonics of the Rift Valley, but more probably by the climate of the Rift Valley, have developed to full species in reproductive isolation, but are seemingly closely competitive owing to a lack of development of significant differences in their ecological requirements. This results in a narrow overlap zone, there being some measure of reduced viability in the hybrids preventing the exchange of genes over a wider area. However, in the Tsavo West National Park of Kenya, I have seen similarly varied rump patterns among the common waterbuck, and yet there is no known population of defassa within several hundred kilometres of the area.

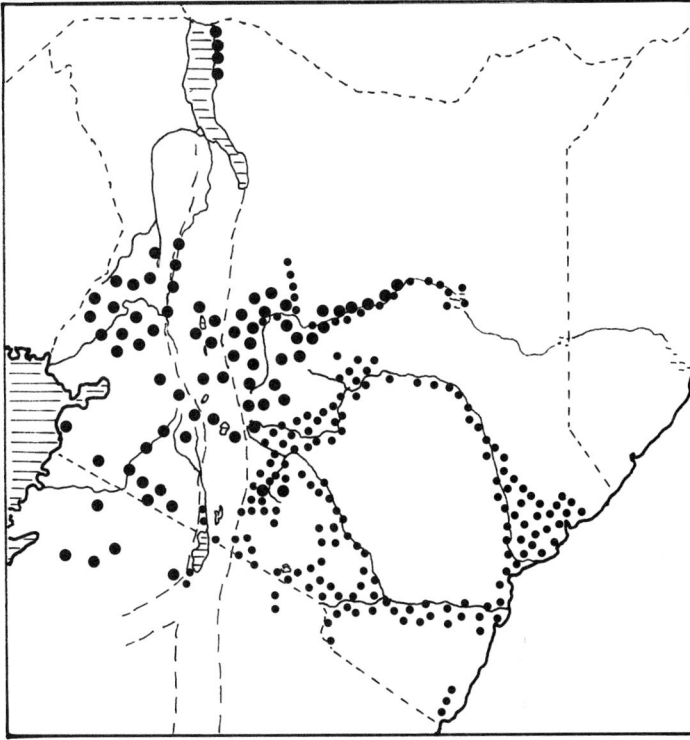

Fig. 4. The overlapping of *Kobus ellipsiprymnus defassa*, large dots, and *K. e. ellipsiprymnus*, small dots, in East Africa. Based on a map by Stewart and Stewart (1963).

Ansell (1969) has described how a common waterbuck doe appeared in the midst of a defassa population in southern Zambia in 1968, some 100 km outside of the common waterbuck's normal range. The doe bred with a defassa waterbuck and produced a hybrid offspring, subsequently followed by at least two other births (Ansell, 1978). Two other sites of overlap have since been identified. This, and the other evidence of hybridization in East Africa, has led Ansell and other taxonomists to consider the common and defassa waterbucks as conspecific (Haltenorth, 1963; Heyden, 1969; Ansell, 1971). Ansell assigns them the status of superspecies, which Mayr (1942) defines as "a monophyletic group of geographically representative (allopatric) species which are morphologically too distinct to be included in one species". Thus Ansell divides the waterbucks into the *ellipsiprymnus* group and the *defassa* group; *Kobus defassa* now becoming *Kobus ellipsiprymnus* of the *defassa* group (Table 2).

As if this confusion over races and species was not sufficient to contend with, there is indecision even as to the status of genera and species. Thus, some would have us consider all waterbucks, lechwes and kobs, as one species, for all have been found to interbreed and to produce apparently viable offspring. This fits the taxonomist's rule of thumb that interbreeding animals should be considered as of the same species. Originally the waterbucks, lechwes and kobs were assigned to three separate genera, *Kobus, Hydrotragus* (= *Onotragus*) and *Adenota* respectively, based upon their distinctive behavioural and anatomical attributes. The lechwes differ from the waterbuck in their much smaller size, large bucks reaching only some 136 kg in weight, the shorter horns with a double curvature, and the presence of inguinal pouches. The colour ranges from almost black in the Nile lechwe, to rufous in the red lechwe of Zambia and further south. It is much more aquatic than the waterbuck and, next to the sitatunga and water chevrotain, is the most semi-amphibious of all African antelopes, particularly favouring inundated floodplains. A very gregarious species, in parts of Zambia its herds number many thousands of individuals. The distribution is markedly discontinuous; it occurs in the southern Sudan, and again in Zambia, being absent from the area in between; thence it extends in a patchy distribution as far south as Lake Ngami in Botswana.

The kob, on the other hand, is much more widely distributed, occurring throughout the Sudanian Domain, or, as it used to be termed, the Guinea savanna zone. It has some affinity for the presence of water, but is by no means an aquatically inclined species, preferring to enter the floodplains after they have dried out. It appears to favour those parts of Africa where a high rainfall results in tall grass. Smaller still than the lechwes, Uganda kob bucks reach only some 97 kg in weight. The horns are shorter, but are possessed of the double curvature. Inguinal pouches are present. Unlike the waterbuck or lechwes, the coat colour is uniformly sandy-coloured throughout its range, with no tendency to darkening. In suitable localities it is very gregarious, occurring in large herds, but not attaining the numbers found in lechwe herds.

Although interbreeding has not been demonstrated, another closely related animal, assigned to the same genus, is the puku *Kobus vardoni*; considered by some to be intermediate between the lechwes and the kobs. It is the smallest of the group, attaining only some 45 kg in weight, and resembling the kob in general appearance, but, unlike the kob, it possesses antorbital glands in addition to its inguinal pouches. The puku occurs from the Chobi and Zambesi Rivers, north to southern Tanzania. It is gregarious in nature and favours a similar habitat to the kob.

Hindle (1951) drew attention to the fact that, at the Khartoum Zoo, in 1936, a doe lechwe was crossed with a defassa waterbuck, giving birth to a hybrid doe. This hybrid was subsequently crossed with a kob, and gave birth to an offspring buck which resembled the sire. In 1951 this second generation hybrid buck was sire to a defassa waterbuck doe, which allegedly became pregnant (Daniell, 1951). Unfortunately this is the last that we hear of the experiment.

Based on these experiments, which are but poorly documented, all waterbuck, lechwes and kobs are now considered as congeneric, of the genus *Kobus*. But opinion has baulked at considering them as being all of the same species, despite their apparent ready hybridization. To pose the riddle: "When is a waterbuck not a waterbuck?", and to answer, "When it's a kob", would be to ignore the distinctive anatomical characters of these animals, as well as their differing behavioural and ecological requirements. In the wild they do not interbreed, although they often share the same general habitat. Wilson (1975) has defined the issue in relation to species:

"The existence of natural conditions is a basic part of the definition of the species. In establishing the limits of a species it is not enough merely to prove that the genes of two or more populations can be exchanged under experimental conditions. The populations must be demonstrated to interbreed fully in the free state."

De Vos and Dowsett (1966) attempted to justify the congeneric status of the waterbuck, lechwes and kobs, by a behavioural study of the three in Zambia. Their resulting data were too generalized to warrant any conclusions in this respect, for all bovid behaviour patterns are basically similar, and the authors did not distinguish any specific patterns which would serve to separate the three species within the general bovid encompass of behaviour. But following the experiences of the Khartoum Zoo, which were more within the realms of teratological experimentation, rather than the pattern of nature, the most recent classification adopts the consensus of taxonomic opinion in relation to the waterbuck, lechwe, kob and puku by, in fact, reverting to Flower and Lydekker (1891) who had considered them to be congeneric 80 years before. The complete classification (Ansell, 1971) thus stands as follows:

Family: Bovidae Gray
Subfamily: Hippotraginae Brooke
Tribe: Reduncini Simpson

Genus: *Redunca* H. Smith, 1827
Redunca arundinum (Boddaert, 1785) reedbuck
Redunca redunca (Pallas, 1777) bohor reedbuck
Redunca fulvorufula (Afzelius, 1815) mountain reedbuck

Genus: *Kobus* A. Smith, 1840
Kobus kob (Erxleben, 1777) kob
Kobus ellipsiprymnus (Ogilby, 1833) waterbuck
Kobus leche (Gray, 1850) lechwe
Kobus megaceros (Fitzinger, 1855) Nile lechwe
Kobus vardoni (Livingstone, 1857) puku

Origins of Waterbuck

Unfortunately the fossil record tells us little of the origins of the waterbuck. Fossils of kobs are rare, but some have been found which could be ancestral to, or related to, all living species of *Kobus*; except for the Nile lechwe, for which no lineage has been identified (Gentry, 1978). These fossils consist of a reduncine horn core base of late Pleistocene date from the Sudan, which resembles that of a very large kob; while Upper Pleistocene beds in South Africa have produced horn cores and a cranium named *Kobus venterae* Broom, indistinguishable from the living *Kobus leche*. The famous Lower Pleistocene beds of Olduvai Gorge in Tanzania have produced the skull and horn cores of *Kobus sigmoidalis* Arambourg, an animal which Gentry considers to have been evolving into *K. ellipsiprymnus*, and which was perhaps also an ancestor of *K. leche*.

Reduncines of Pliocene date with kob-like characters have been known for a long time from the Siwalik Hills of India. These might be ancestral to both kobs and to *K. sigmoidalis*, or, may be merely primitive Indian off-shoots already isolated from African forms. They show some characters of the *Boselaphini*, a mainly fossil group with only two living species, the Indian nilgai *Boselaphus tragocamelus*, and the Indian four-horned antelope *Tetracerus quadricornis*. Other fossils resemble those of reduncines from South Africa and Kenya, suggesting the near contemporaneous occurrence of similar primitive reduncine forms in both Asia and Africa during the earliest Pliocene.

Gentry points out that the involvement of the Siwaliks in the history of the African bovids does not necessitate the opinion that the African bovids originated in India. All we can say is, that the two areas have shared many of their antelopes, at least at the generic level. Gentry provides us with a tentative phylogeny (Fig. 5) but, interesting as these speculations may be on the lineage of the waterbuck, we must remember that the postulated ancestor, *K. sigmoidalis*, is based upon only one skull and two horn core fragments.

As to what sort of a beast the waterbuck may have derived from, its dependence on water, which we will consider later, and its ambulatory nature, suggest an ancestor closely linked to an aquatic habitat. We do not know enough of the physiology of the lechwes to know whether, or not, they are as dependent on water as their relative, but their closer affinity for an aquatic medium would suggest this to be so. Like the waterbuck, lechwe are ambulatory species, not cursorial, and hence are not able to run fast with ease over open plains. This latter state has its highest attainment in the kob; which has also become emancipated, to some extent, from a close dependence on water.

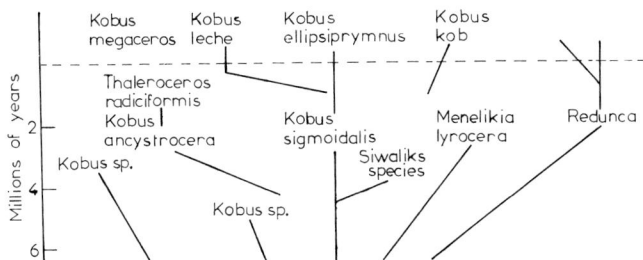

Fig. 5. The phylogeny of the waterbuck, after Gentry (1978).

Following the "dispersal theory" proposed by Geist (1974), in which new species are characterized by individuals of larger body size, larger horns, and more distinctive coat patterning than the parent species, the common waterbuck, with its smaller size and smaller rump pattern would represent the parental origin of the defassa waterbuck. The only contradiction to Geist's theory is that he attributes a wider geographical range to the parental species, which is more generalized in its ecological requirements, whereas, in the waterbuck, it is the defassa which has the wider range. However, the theory also postulates that the evolving species is ecologically more specialized, and we must remember that the defassa waterbuck is very specialized, as we shall see, requiring a high protein diet and having an acute dependence on water, whereas there is some indication that the common waterbuck may have a less specialized diet. Should more arid conditions prevail, such as those experienced in Last Glacial times around 12 500 B.P., undoubtedly the defassa waterbuck's distribution would be greatly curtailed. Comparable studies upon diet and water dependence have not been conducted on the common waterbuck so whether it really has more generalized requirements is not known. We must remember, however, that the defassa waterbuck is in a very early stage of subspeciation, so early in fact that reproductive barriers do not yet exist between the two forms.

Assuming Geist's theory to offer the correct explanation with regard to

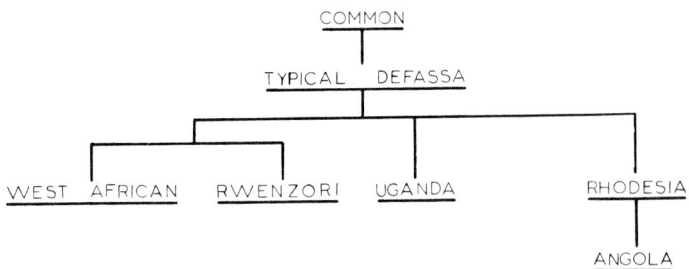

Fig. 6. The phylogeny of the main groups of waterbuck.

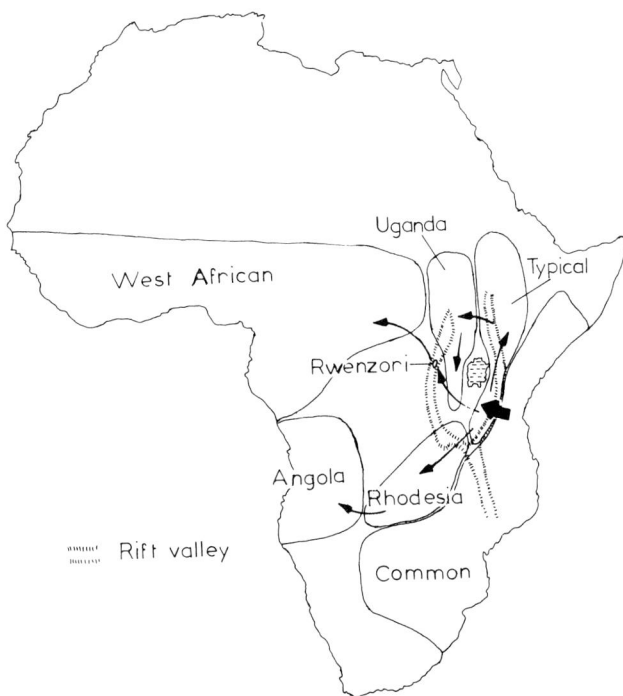

Fig. 7. Hypothetical dispersal of the groups of waterbuck from a common waterbuck origin.

parent species and subspecies, the ancestral home of the waterbuck appears to have been the eastern coast of Africa, bounded by the arid Horn of Africa to the north and the East African Rift Valley to the west; its southern extension being limited by the cold winters of the Cape. Sometime, probably as recently as the last East African pluvial of about 10 000 B.P., the Rift Valley barrier was breached by a common waterbuck population, which was to give rise to the defassa waterbuck. The analysis of horn shape suggests that, probably from an origin in, or near to, Kenya, the typical defassa group was the first to emerge. From this population one offshoot could have gone north to give rise to the Uganda group, while another could have expanded south round Lake Victoria (much greater in extent at the time) to give rise to the Rwenzori type and the West African group; a third offshoot expanding southwards to form the Rhodesia group, which in turn gave rise to the Angola group (Figs 6 and 7).

2. The Study Area

History

What has history to do with the ecology of an animal species? My opinion, shared more and more in recent years by ecologists who are witness to the rapid habitat changes taking place today, is that it has a great deal to do with the understanding of animal/habitat relationships. There are very few habitats in Africa which have not been modified by the presence of man; whether by fire, by settlement, or by agriculture and husbandry. Many of the vegetation associations are secondary, and the animal populations are in a state of flux with the disturbed primary components. The history, such as we know it, of the Rwenzori National Park, shows it to be a salient example of this type of change, and not, as many would suppose, an island of pristine Africa, selected and set aside by man as an example of virgin, untouched habitat, of the Africa that was. On the contrary, it is an area where circumstances have ousted man in very recent times, and animals have taken advantage of the situation by colonizing the vacated area.

In the last decade of the nineteenth century, a squabble ensued between Great Britain, Germany and Belgium, concerning which power should administer what, in the vast hinterland of an unknown Africa. Stanley's treaties had already given Britain control of Uganda, and the boundary with Leopold's Congo was staked along the $30°$ meridian, through a place called Katunguru. This meant that the whole of the area north of the Kazinga Channel, which we know today as part of the Rwenzori National Park, was placed in Belgian hands. The Colonial Office then discovered that a "mistake" had been made. The $30°$ meridian was in the wrong place. It ought to have run west of a village called Katwe. This "discovery" was caustically described by the German explorer Von Mecklenberg as nothing more than a British trick; for the British had discovered the commercial importance of Katwe's salt deposits. There was probably a lot of truth in Von Mecklenberg's assumption, for the meridian lies about 3·2 km west of Katunguru, but 16 km *east* of Katwe.

Finally, the Lubilia River, much further west, was agreed as the boundary. So today much of this formerly disputed region forms part of the Rwenzori National Park. Whether its future would have been different had the Belgians retained it, is doubtful. They made the country on their side of the boundary into a national park 20 years before Uganda followed suit.

The Rwenzori National Park, gazetted in April 1952 as the Queen Elizabeth National Park, lies at the southernmost tip of the Rwenzori Mountains. Its 1980 km² of hills, plains, forest and swamp, straddle two lakes: Lake Edward of 2150 km² in the west, almost half of which is in neighbouring Zaire where it reaches over 115 m in depth, and Lake George to the east. A shallow puddle in comparison of only some 3 m in depth, and 389 km² in area. Linking these two lakes is the 32 km long, 0·8 km wide and 3m deep, Kazinga Channel. Its original name was Kafuru: *kazinga* meaning "an island", and referred to the land on the opposite bank (Fig. 8).

The lake's gently sloping beaches, the water's profusion of fish, and the animals coming to drink at the shore lines, probably resulted in the area being inhabited by man since prehistoric times. Kanyesewa, a narrow ridge connecting Mweya Peninsula with the mainland of Busongora, is believed to have been inhabited since about 50 000 B.C. After the rains, stone balls about the size of cricket balls, probably used for grinding grain (and not, as has been suggested, to make bolas), and stone handaxes, come to light on the surface. These are remnants of a Middle-Pleistocene, fire-using, stone-age culture, known as the Acheulean (Posnansky, 1965). A much later settlement, on the far side of Lake Edward, in Zaire, was covered with ash from the explosion of craters in the north of the Park, and the same pompeian fate probably befell the Mweya settlement. But the area seems always to have been popular, judging from the profusion of waste quartz flakes from handaxe construction. These are found all along the Kazinga Channel. Burial remains near Mweya sometime after 1500 A.D., show its persistence as a favourable site of occupation.

The date of the advent of agriculture to this part is unknown, although one guess puts it at about a thousand years ago. In recent times, a century ago, it was cattle country, ruled over by the Wasongora, a people who had become rich and powerful from the Katwe salt deposits. The largest salt-producing district in central Africa, Katwe's trade extended both east and west across the continent. In about 1871, the Warasura, Waganda warrior hordes of King Mtesa of Buganda, swept through the

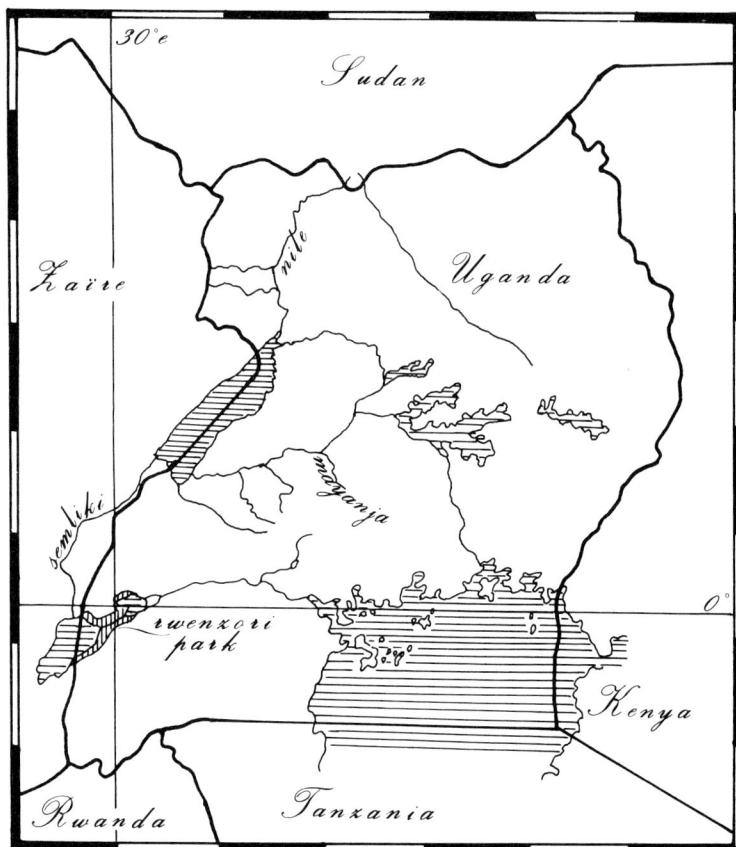

Fig. 8. Location map of the Rwenzori National Park in East Africa.

area under the command of the Katekiro. Only one man remained unconquered under the onslaught, and that was Kaiyura, the chief of Mweya Peninsula. Kakuri, the chief of Katwe, fled to the small islands in Katwe Bay. The raiders despoiled the immense herds of Wasongora cattle, driving off all that they were able. Some of the raiders remained behind in occupation with a portion of the cattle.

It was 18 years after this raid when H.M. Stanley visited Katwe. He recorded signs of numerous cattle, but the Warasura who now owned them, fearing a raid, had apparently driven them off into hiding. Stanley recorded the area as quiet and almost deserted, although Katwe had a population of about 2000 people. The plain to the west of the Nyamagasani River was described as: "remarkable for its growth of euphorbia . . . planted by generations of Wasongora to form zeribas to protect their herds from beasts of prey . . ." (Stanley, 1890).

In general, Stanley's description fits the area as it is today, except that the clustered huts have gone from the plain, and such of the zeribas as remain are now disordered thickets. His reference to three-feet high grass, with spikelets "which pierced the thickest clothing", shows the plain to have been covered with the fire-encouraged *Themeda triandra*, as it is today. Stanley sailed from Katwe across Lake Edward, then known as Lake Muta Nzige "the killer of locusts", to Habibale at the southern tip of Mweya Peninsula. He recorded that he could see nothing from there but a "formless void", an indication, as this was in July, that the dense smoke haze from country-wide burning must have been as prevalent then as it is now.

Leaving Katwe, one of Stanley's party, Jephson, wrote: ". . . everything was dried up and sere, and nothing in the shape of trees except euphorbias were to be seen, the whole plain had a peculiarly desolate and dead look" (Middleton, 1969). There was little room for wildlife at this time. In June 1889 the Mweya Peninsula had 81 huts and was "rich in sheep and goats". But although it had managed to escape the ravages of the Warasura, its circumstances changed rapidly.

Two years after Stanley's visit, Lugard, who camped for one night on the Mweya plateau, was to describe it as: "deserted villages surrounded by hedges of cactus (sic) marked the dwellings of the people before the advent of Kabba-Rega and his guns" (Lugard, 1893). But the advent of the Warasura had been some 20 years before, and it was probably Lugard's arrival which had caused the inhabitants to desert the area. Three years after Lugard's visit, Elliot (1896) described Mweya as a "most miserable collection of huts, and his [Kaiyura's] people to be a puny, half-starved race . . ."

In 1890 the area had been visited by the great rinderpest plague which had devastated almost the whole of Africa, and Lugard had witnessed its effects. He described the plain between the Kichwamba escarpment and the Kazinga Channel in dismal terms:

"The great plain spread out before us must have swarmed at one time with elephant and buffalo, for their tracks were everywhere; but the former had left, and the latter were dead. Shukri bagged two waterbuck, and I got an nsunu [Uganda kob] but game was very scarce" (Lugard, 1893).

Later, he recorded seeing 13 waterbuck near to Katwe (Perham and Bull, 1959).

It was probably at about this time that people started to die of sleeping-sickness, although it is not generally accepted to have been present in the area before the year 1910. But in the Lake Victoria area an estimated 30 000 people had died from the disease in 1901. To avert a similar catastrophe here, it was decided to evacuate most of the inhabitants, leaving some to maintain the salt and fishing industries. To protect these people extensive bush clearings were to be made around Lake Katwe and the village, to remove the habitat of the tsetse fly vector.

One of the first places to be cleared was the shore of the salt lake at Katwe; but although Stanley had described it as "ringed around with Ukindu palms, scrubby bush, reedy cane, euphorbia, aloetic plants . . ." (Stanley, 1890), his companion Jephson wrote in his private diary: "One side and end of this lake is utterly barren . . . the other side and end is clothed all round by a narrow belt of rank vegetation amongst which a few palm trees may be seen growing . . ." (Middleton, 1969).

Kazinga village, today a corrugated-iron eyesore which all park visitors taking the launch trip pass by, described by Lugard as "canoes in great number . . . huts miserable, small and temporary surrounded by cactus hedge", was one of the evacuated places. In 1921 its inhabitants were allowed to return, and a government entomologist, Dr Hale-Carpenter, was sent to survey the whole area to see if it was now free of trypanosomiasis. In that year 1226 people were examined for sleeping-sickness at Katwe, and only three cases were found. From this figure it can be inferred that most of the population had remained there.

The Mweya Peninsula seemed to be a focal point for the area, and of it Hale-Carpenter wrote:

> "I propose to allow the re-population of Mwaya (sic) Peninsula . . . a very few *palpalis* occur on part of the western coast, but as this is simply a dense thicket of thorny euphorbia and creepers growing on precipitous slopes, and there are many other places where access to water is easy and open, I do not think any native would go into it. The rest of the peninsula is open grassland with a few euphorbia trees, and there is no other shelter on the shore for *Glossina* except on the eastern face . . ."

A permit for settlement, and to operate a ferry to Kazinga, carried the following conditions:

> "The whole of the shore of the Mwaya Peninsula facing Kazinga village must be kept clear of all bush between the following points: (a) on the west, the point known as Habibale where the coast turns northwards, (b) on the east, a point directly opposite the end of Kayumbura ridge on the Ankole side, to the west of the angle where the coast turns northwards to join Kanyeseswa, (c) on the western side of the isthmus of Kanyeseswa all bush between the road and the crest on the ridge shall be cleared."

Kazinga village was open for 'occupation, grazing cattle and cultivation' in May 1921, as long as the bank facing the Peninsula was kept cleared (Hale-Carpenter, 1921). In 1922, the village, which then had 368 inhabitants, requested an extension southwards for 2 miles to a sandy beach known as Olwambo lwente, today called Lion Bay. Bush was to be cut back for 400 yards from the water, and settlement was allowed up to one mile back, bounded by Luhembo ridge and Ogwempara (rwempara = the place of the kob) (Hale-Carpenter, 1922).

The shore of Lake Edward, from Bwenda Bay to Rwenjubu Bay (rwenjubu = the place of the hippopotamus) was completely cleared of bush, and so maintained until at least 1933. The area at Kayanja was cleared, for here there was a traditional cattle-watering point, Kabahango. Further west, in the Mpondwe area of the Congo border, there was open grassland, with settlement and cultivation along the banks of the Lubilia River. Bush in the ravines here had to be cleared. A note states that no game, except kob, was seen here. The road from Katwe to Katunguru, thence to Kasenyi salt lake in the east of the Park, was cleared for half a mile on either side, as was also the road from Katunguru to Kikorongo (Fig. 9).

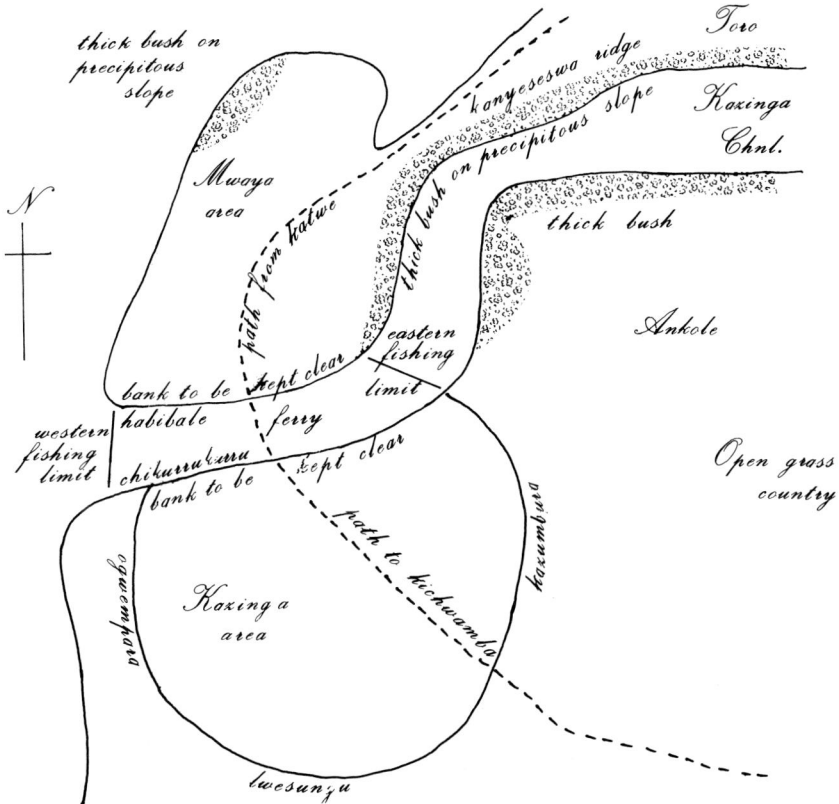

Fig. 9. The clearing of bush for tsetse fly control on and around Mweya Peninsula, based on a copy of G.D. Hale-Carpenter's 1921 sketch map.

These details of bush clearing have been related at some length, because I believe that the clearing may have had a profound and long-lasting effect upon the ecology of the area, by opening-up easily accessible lake-side grazing for the hippopotamus. This led to an increase in hippopotamus numbers which subsequently affected the other herbivore populations of the area.

In spite of these measures, however, there was an increase in sleeping-sickness, and the whole area was again closed in 1924. Inhabitants, other than those left to maintain the salt and fishing industries, were once more evacuated to Katunguru and Muhokya. Some 15 000 head of cattle are reported as having been moved from Katwe and the surrounding country. Officially the area remained closed to human inhabitants until it was declared a national park in 1952, although many people had gradually found their way back again.

The rich animal populations, which resulted in the area being made into a national park, would seem to have stemmed mainly from the time of this last evacuation. Game was probably plentiful in some parts even before this date, for in 1925 it was reported that in the Chikarara [Kikeri] area game was "plentiful, large herds of kob, many topi and waterbuck, and some reedbuck" (Anon., 1925).

In 1931 an expedition from Cambridge University, studying the lakes, passed this way. By this time game was reported in "enormous numbers" in the Katwe area, and animals had already invaded the de-peopled Mweya Peninsula. The Worthingtons wrote:

> "besides the wealth of bird life abundant game shared the small peninsula . . . four elephants . . . in addition to these the resident population consisted of a waterbuck, two doe reedbuck, a Uganda kob, a bushbuck, a herd of warthogs . . . two buffaloes . . . a lion and plenty of hippos and hyaena. Every day different animals strayed in across the narrow neck of the peninsula, but they were only visitors" (Worthington and Worthington, 1933).

By today's standards this enumeration would not be "abundant", so we must also consider the "enormous numbers" of the Katwe area to represent far less game than we would expect to meet there today. Nevertheless the Worthingtons' observations signify a marked change in the status of wildlife in the area.

This visit was followed 2 years later by that of a government official who came to see if the area could be re-settled. He found large numbers of hippopotamuses along the shores of Lake Edward, but considered it unlikely that the flies (tsetse flies) "come into much contact with waterbuck or impala [kob] but it may possibly feed on buffalo or elephant . . . The protection of hippopotamus might well be removed for such a period as is required to reduce these animals to reasonable numbers." Of elephant he wrote: "Elephant will require driving down to the Nyamagasani as they have made the growing of bananas, the chief local food, a matter of considerable difficulty (in the Rwempyo valley)" (Hancock, 1933).

In 1937 a game department report (Pitman, 1937) stated that waterbuck were plentiful in the Lake George area, and in the "adjoining sleeping-sickness areas between Katwe and the Lubilia River". The clearings seem to have been maintained until 1939, when it was considered that their maintenance was no longer necessary as sleeping-sickness had abated. No doubt the war prevented effective supervision, and also any organized re-settlement plan, such that the latter was never put into effect.

Thus circumstances favoured the wildlife of this area to the extent that, by the time the first estimates of numbers were made in 1956, the biomass of wild ungulates was found to be one of the highest in Africa (Petrides and Swank, 1965).

Topography and Geology

Most of the area comprising the Rwenzori National Park (Fig. 10) lies at an altitude of 900 to 1000 m above sea level; dominated to the north by the great horst of the Rwenzori Mountains, after which the Park now takes its name (Plate 5). This massive, glacier-capped upthrust of the earth's crust, the highest point of which reaches to 5119 m, is considered to be of early Pleistocene date; the result of a gigantic see-sawing of the Rift Valley floor. The southern foothills, which comprise a part of the Park, express themselves in a remarkable moon-like landscape, peppered with a multitude of small craters; at least 78 occur in 155 km². Another series occurs to the east, in the Chambura Game Reserve, outside of the Park (Fig. 11). These craters are thought to be geologically recent, their formation culminating about 8000 B.C.; and representing a tardy protest at the Rwenzoris' uplift. As explosion craters, formed by the expulsion of gas and steam, no lava flow is associated with them, but their formation was accompanied by a vast belching of volcanic ash and dust, which settled over the whole area as a rich, alkaline volcanic tuff. Such craters are considered by volcanologists to usually precede volcanic activity proper, but this second phase has yet to take place, and perhaps it never will. But the area is by no means volcanically inert, and in 1898 when the explorers Grogan and Sharp travelled along the east shore of Lake Edward, at its southern end near to the Ntungwe River, they noted several geysers, shooting "vast jets of steam into the air" (Grogan and Sharp, 1900). In 1925 a government

Plate 6. A view looking west over the Peninsula from the high ground to the lower plateau, with Katwe in the distance, showing *Capparis* clumps and the open patches of grassland.

Plate 7. Hippopotamus on land; note the closely-cropped grass cover.

official passing the same area, recorded four active "ground volcanoes" 3 to 4 miles north-west of Kabwema (Anon., 1925), which were probably a more subdued phase of Grogan and Sharp's geysers. Today perhaps they have been covered by a rise in water level, for there is no recorded trace.

Before the period of tectonic grandeur signified by the craters' eruptions, the lakes, which form the focal point of the Park, were much more extensive. They reached their maximum during one of East Africa's pluvials, perhaps about 60 000 B.C., producing a bedded sequence of sands, clays and gravels. During a particularly arid interpluvial the lakes appear to have dried up altogether, wiping out the aquatic fauna and producing a fossiliferous ferruginized bed known as the Kaiso bed. This bed is of interest as it contains the remains of crocodile, nile perch, and other aquatic fauna typical of the Nile system, suggesting a one-time link with Lake Albert to the north of the Rwenzori range. This fauna has never returned to the area, in spite of the fact that the lakes were restored in a following pluvial, and especially surprising in view of the ubiquitous occurrence of the crocodile in Africa.

When the lakes were restored there seems to have been a sinking of the Rift Valley floor, enabling Lake Edward to capture much of the water which drained from a previously larger Lake George, through the Kazinga Channel. Today the two lake levels are almost equal, and the flow between them is negligible. The Semliki River, which provided the link with Lake Albert to the north, was blocked by movements associated with the uplifting of the Rwenzoris, but then cut back until it was captured again by the Lake Albert watershed; today providing the overflow for Lake Edward into Lake Albert (Beadle, 1965). Retreat of the lakes from their high levels was not a continuous process, many terraces being formed during pauses in the retreat. Closer to the Kazinga Channel, however, the terrace-like features have been shown to be due to faulting of the Kaiso beds (Harrop, 1960). Such a fault separates the Mweya Peninsula into two halves: a raised plateau in the east, and a low-flying, flat area, to the west.

Soils

The volcanic tuff from the explosion craters, which settled over the area, has produced a eutrophic brown soil, generally rich in plant nutrients, and with a high reserve of weatherable minerals. It is sometimes saturated with bases to more than 50% exchange capacity, and characterized by a medium to high productivity (Harrop, 1960). The dry climate, however, limits their agricultural potential.

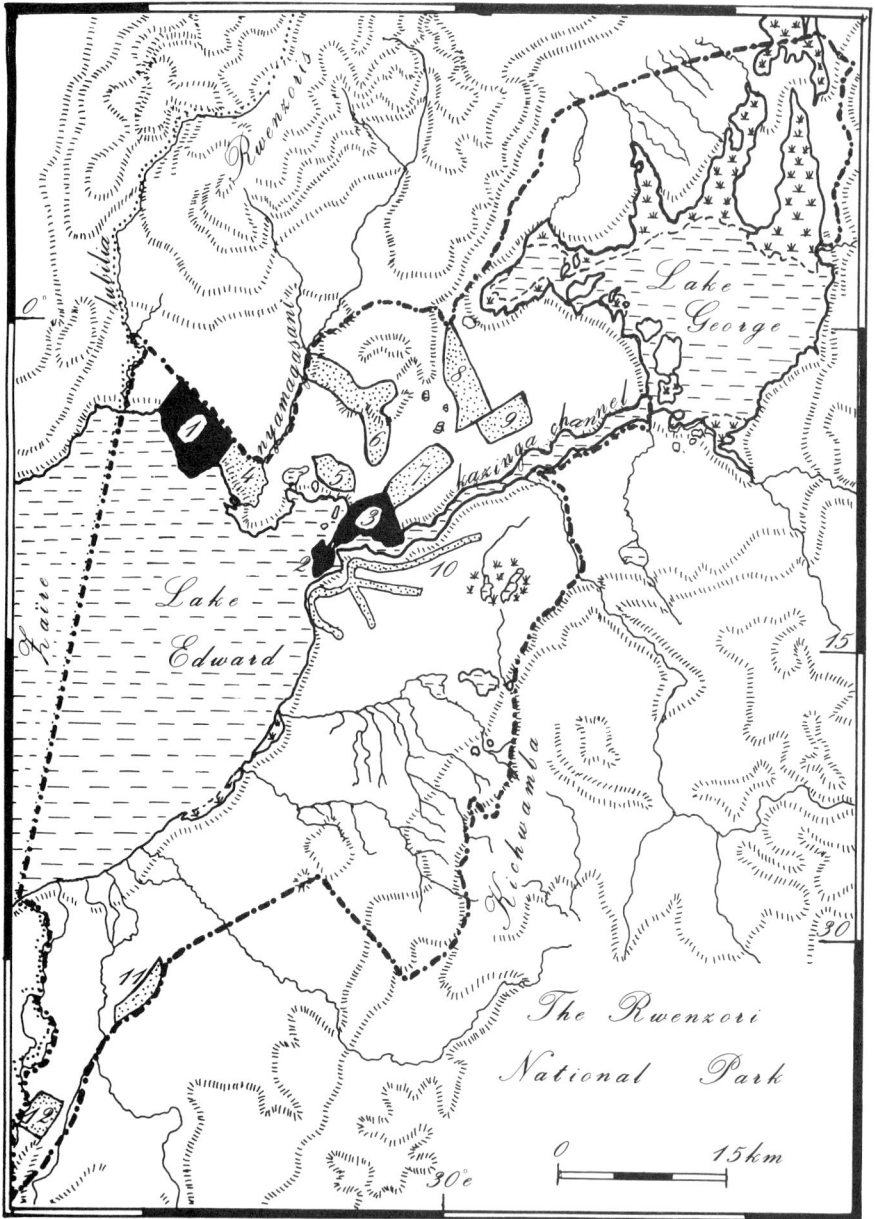

Fig. 10. Map of the Rwenzori National Park; showing the main study areas, blocked, and the game count areas, stippled.

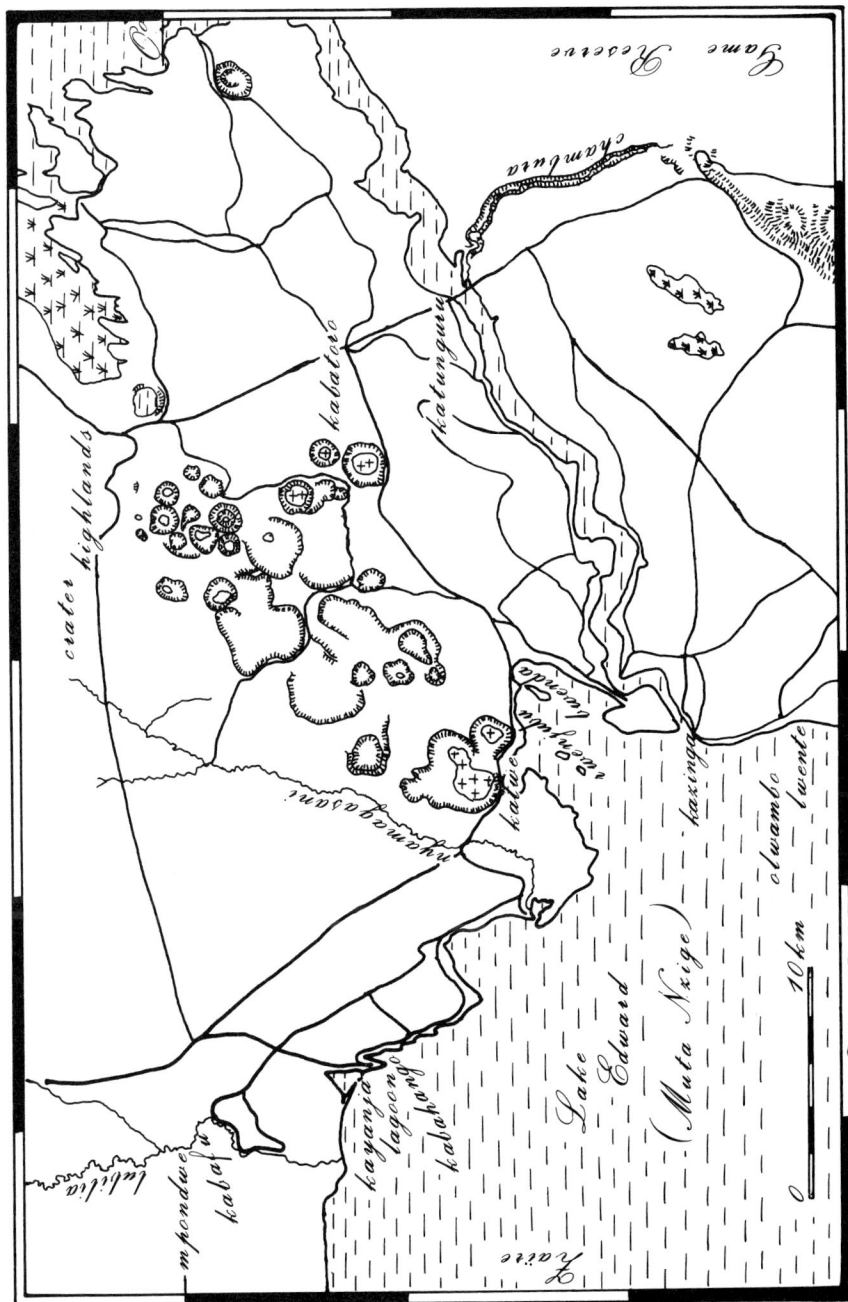

Fig. 11. Map of the north-east part of the Rwenzori Park, showing the Crater Highlands.

Climate

Rainfall in the Park ranges from 1200 mm per year to less than 800 mm; the central area lying in a rain shadow which probably results from interference with circulation patterns by the Rwenzoris to the north, and the Kichwamba escarpment to the south-east (Fig. 12). Thus the 10-year average for 1954–1963, for the Mweya Peninsula, showed a rainfall of only 669·5 mm per year. Dry periods occur in December to February, and June to July. There was an annual evaporative demand, during the period 1954–1963, of 1880 mm (E_o). There was little seasonal variation in this, and in no single month did the rainfall exceed evaporation (Table 6). Climatically, using the Thornthwaite classification (Thornthwaite, 1948), the climatic index was -38, where I_m (the moisture index) equals:

$$\frac{(100 \times 0) - (60 \times 1189)}{1180}.$$

Under this classification -40 is arid, thus Mweya and its environs are distinctly semi-arid, and the water deficiency is large compared to the need, so that the growing period is short. The years since 1961 have, on average, been wetter than those of the preceding half-century, with more rain falling in the short rains of February to April (Fig. 13). One can suppose that this will have had some effect upon the vegetation and animal populations.

Table 6 Climatic data for Mweya 1954–1963

Month	Rainfall (mm) 1954–1963	Approximate E_o (mm)	Deficit $E_o - R$
Jan.	26·16	160	134
Feb.	24·64	160	135
Mar.	64·52	175	110
Apr.	75·69	150	74
May	62·23	160	78
Jun.	29·46	145	115
Jul.	25·4	150	125
Aug.	71·12	145	74
Sep.	84.58	160	75
Oct.	82·8	165	82
Nov.	71·12	150	79
Dec.	51·82	160	108
Total	669·54	1880	1189

There are no E_o figures for Mweya, those above being calculated from the E_o figures for Nyakatonzi and Kasese, given in Rijks and Owen (1965). These places lie 24 and 48 km from Mweya respectively.

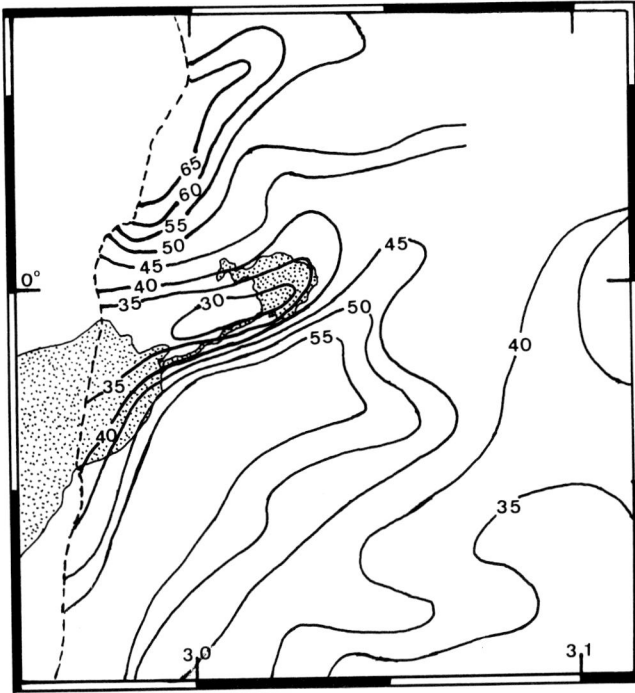

Fig. 12. Isohyets (ins.) of the Rwenzori National Park. Lakes are shown stippled.

The mean maximum annual temperature at Mweya, from 1955 to 1961, was 28·7°C, and the minimum was 18·3°C. The highest temperature recorded during this period was 35·8°C, and the lowest 12·3°C. The dry seasons would undoubtedly have a much greater desiccating effect than they do, both upon soil and vegetation alike, if it was not for the dense pall of smoke created by the country-wide grass fires, which at times completely blocks out insolation. Apparently becoming trapped in this depression the smoke can reduce visibility to as little as half a km. The haze was reported by Stanley (1890), and provides an interesting example of how man modifies the environment; in this case probably benefiting both plant and animal populations by ameliorating the dry season's intensity.

Vegetation

I have never found the Rwenzori Park to be the inhospitable-looking place that the early travellers recorded: its stately cactus-like *Euphorbia*

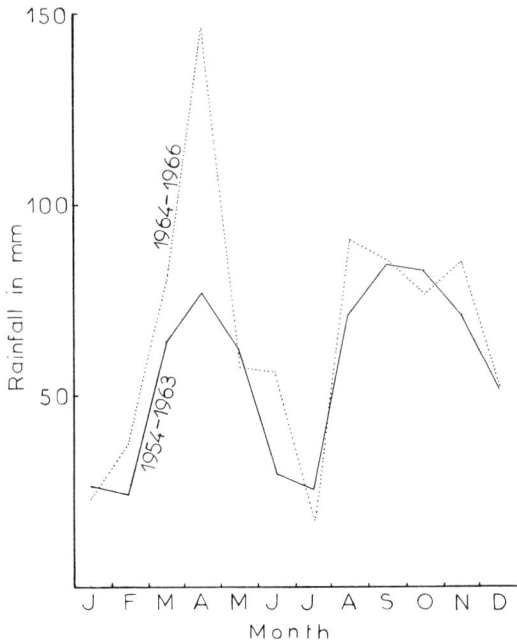

Fig. 13. Graph of the mean monthly rainfall in mm, recorded at Mweya, for 1954–1963 and 1964–1966.

candelabra trees (true cacti do not occur in Africa) with their erect, spiky fronds, impart a pleasing, never-to-be-forgotten aspect to the rolling grasslands with their dotted thickets. According to Osmaston (1965) this present appearance is due almost entirely to fire; annual or biennial grass fires maintaining open grassland which would otherwise develop into thorny thicket, woodland, and forest in the wetter areas. In one of the earliest references to the modification of vegetation by elephants, Dale and Osmaston (1954) noted that the ground-water forest at the mouth of the Nyamagasani River, cut extensively in the past for fish smoking, had reached a "game climax"; elephants inhibiting the forest re-growth. There is no other evidence of elephants having exerted any significant impact on the vegetation of the Park in the past, although Field (1971) claims that some elephant-induced change may now be taking place. He reports losses of up to 99·5% of trees in the Crater Highlands having taken place since 1954, and only 21% of standing trees showing no evidence of barking by elephants. But in studies conducted in the adjacent Kibale Forest Reserve, Wing and Buss (1970) found that only 20% of all trees and shrubs showed some sign of consumption by elephants, and that this did not represent over-use.

Four main vegetation communities can be found in the Park: bushed grasslands, thicket, *Acacia* woodland, and semi-deciduous forest. The bushed grasslands occur on the slopes of the Crater Highlands, in the central region of the Park around the Kazinga Channel and the two lakes, and in the southern extremity. Thicket is associated with grassland areas; the *Acacia* woodland is scattered in distribution, and the semi-deciduous forest cuts across the centre of the Park.

Bushed grasslands

Due to the short growing season the grass tends to be short in height compared with that found in higher rainfall areas. For the most part, communities comprise the coarse, fire-encouraged species: *Themeda triandra* (red-oat grass), *Heteropogon contortus* and *Hyparrhenia filipendula*. The rolling slopes of the Crater Highlands are clothed with much tussock grass, *Cymbopogon afronardus*, and, in the depressions, Indian sword grass, *Imperata cylindrica*. Both of these tussock grasses are evidence of former land use; the first probably associated with overgrazing by domestic stock, and the second associated with abandoned cultivation sites.

I conducted a brief analysis of the grass composition on the Mweya Peninsula and in the Kayanja area; while Dr C.R. Field kindly supplied me with figures for the mainland area, adjacent to the Peninsula. Using the technique of recording presence or absence of a species within a 50 × 50 cm quadrat, thrown down at random intervals along selected transect lines, an estimation of species availability could be made; an important factor in relation to the feeding preference of the waterbuck.

The Mweya Peninsula was found to be rich in grass species: 21 were recorded, discounting those shade-loving species found in thickets. *Sporobolus pyramidalis* was the dominant, with the fire-induced *Themeda triandra,* common throughout much of the Park, conspicuously absent. The adjoining mainland area showed identical dominants. Kayanja was less rich in species, with only 14 recorded, and here, fire-loving species such as *Heteropogon contortus* and *Hyparrhenia filipendula* were dominant in association with *Themeda triandra*. The latter species increased in frequency the further one moved from the lakeshore, and the closer one approached to the main road, where the fires usually start from. Fire was generally prevented on the Peninsula, but it must be noted that this was the grassland composition at the time of the study. Considerable changes can take place here in the grassland composition, related to the hippopotamus control experiments, for the numbers of hippopotamuses are sufficient to modify the grassland composition in certain areas, as we shall see later.

The grasslands to the north of the Park are characterized by the *Euphorbia candelabra* trees, which dot the landscape, and whose place is taken in the southern extremity of the Park by the gnarled, large wild fig trees *Ficus gnaphalacarpa*, whose branches form a favourite resting place for lions.

Thicket

Thicket, comprising almost impenetrable associations of the thorny scrambler *Capparis tomentosa*, and the even more thorny *Azima tetracantha*, was undoubtedly once much more extensive along the low-lying lakeshore regions, until the clearing operations commenced in 1912. Away from the water the plains are dotted with clumps of *Capparis*, which bears a galaxy of heavily scented, attractive white flowers in the dry season, much favoured by waterbuck. These clumps are usually associated with termite mounds, but the termitaria flora is generally impoverished in comparison with that found in some areas of East Africa; such as in the Akagera Park of Rwanda. This impoverishment of the mounds may be the result of elephant feeding; elephants are absent from the Akagera Park.

Some observers have claimed that the thicket is increasing in extent in such areas as the Mweya Peninsula, due to overgrazing by hippopotamuses. The theory is that the hippopotamus eats the grass but not the shrubs, and thus removes competition for the establishment of woody seedlings. Although higher rainfall in recent years may well have resulted in a more vigorous thicket growth, regardless of other factors, photographs of the northern tip of the Mweya Peninsula taken by me in December 1959, and November 1968, show little change to have taken place (Plates 8 and 9). What may be more important is a rise in water level in 1961 which flooded much of the *Panicum repens* and *Cynodon dactylon* grasses along the water's edge; grasses which would have remained green in the dry season (Plates 10 and 11).

True relic thicket of the climatic climax clothes the steeper sides of several of the small craters of the Crater Highlands, where fire has never had access (fire burns more slowly downhill than uphill). Unfortunately the composition of this thicket has not been studied.

Acacia *woodland*

Acacia woodland, so characteristic of the African landscape further east, is not typical of this area. Some mature woodland of *A. xanthophloea*, the "yellow-fever tree", which favours a high water table, is found near Lake George; and a woodland of *A. sieberiana*, another water lover,

Plate 8. The northern tip of the Mweya Peninsula, December 1959.

Plate 9. The same view, November 1968. Some change in the grass cover is evident, but there is little perceptible change in the bush.

Plate 10. The east side of the Mweya Peninsula, December 1959, with a lone buck on the skyline.

Plate 11. A further view of the east side of the Mweya Peninsula in December 1959, showing the exposed grassy shorelines which were later covered by a rise in the water level.

occurs in the Kayanja area, where a particularly vigorous regeneration was taking place. The stunted, fire resistant, *A. gerrardii,* is found in parts of the Crater Highlands. It is possible that the general paucity of *Acacia* is due to past overcutting for fish smoking.

Semi-deciduous forest

Semi-deciduous closed forest occupies an area totalling some 450 km² south of the Kazinga Channel, known as the Maramagambo Forest. This forest occupies the higher ground to the east, with a relatively broad belt reaching down to the lake. Its species composition has not been studied, but it is said to contain much Uganda ironwood, *Cynometra alexandrii.*

Fauna

The larger mammalian fauna of the Park is characterized by a few large species furnishing an exceptionally high biomass. The paucity of species (there are no rhinoceros, zebra, giraffe, hartebeest, eland, roan, oribi or impala, although several of these species exist within 160 km to the east) is probably, in part at least, the result of exterminations from the previous heavy settlement of the area, perhaps coupled with the effects of the rinderpest plague. Suffice to say that Parke, Stanley's surgeon, casually noted "a few giraffes" in the region between Kikorongo and Muhokya (Parke, 1891). Although it is easy to postulate faunal barriers, presented by escarpments or forest, one is nevertheless left wondering why there are no hartebeest, eland, roan or zebra there. Kikorongo lake is rich in fossils, but a high water table in recent years has prevented their study. Perhaps one day some of the questions regarding mammal distributions here may be answered.

Aerial counts conducted by NUTAE in 1966 to 1967 revealed the presence of some 3000 elephants, 20 000 buffalo and 12 000 hippopotamus (Field, 1966, 1967). In addition there were an estimated 20 000 Uganda kob, 5000 topi, 3000 waterbuck, and numerous warthog; with lesser numbers of bushbuck and Chanler's reedbuck, to name but the commonest ungulates.

Topi are restricted to the southern sector of the Park, south of the Maramagambo Forest; reedbuck are localized in distribution, and kob occur only in numbers in open grassland where there is little thicket. Both elephant and buffalo are widely distributed, although dispersion is seasonal, the majority remaining in the forest during the dry seasons. At a rough estimate, since a true estimate cannot be made without an accurate knowledge of the population structure of each species, biomass must be in excess of 12 290 kg/km². This is probably similar to the adjacent Kivu National Park in Zaire, formerly the Albert National Park; although higher figures have been given.

More detailed aerial counts were conducted during 1971 and 1972, and estimates for the smaller ungulate species gave figures (with their 95% confidence limits) of 12 000 ± 3545 kob, 4932 ± 4599 topi, 1530 ± 304 warthog, and 3563 ± 1228 waterbuck. By using a ground-to-air correction factor it was suggested that the total waterbuck population was probably of the order of 4454 animals (Eltringham and Din, 1977). It is difficult to make any comparisons between these two sets of estimates made with a 5-year interval between them; unless there are grounds for supposing the populations to have remained stationary over this period. Increased rainfall in recent years might be expected to have resulted in an increase in population sizes. Nevertheless, it is thought that the kob figure does not represent a decline, and the more recent waterbuck estimate may approach more closely to the true total. Due to their fossorial habit, the numbers of warthog could well be more than twice the number recorded.

Large carnivores within the area comprise lion, leopard and spotted hyaena. The hunting dog has not been seen in the Park since 1952, previous policy having been to shoot it on sight. Reports of the Uganda Game Department show that at one time they were not uncommon (Pitman, 1948–1952):

> 1948 — 12 shot near the Nyamagasani River and poison baits laid;
> 1950 — 2 seen, one shot, near Lake George;
> 1951 — 20 killed near Lake George;
> 1952 — 12 seen in the Queen Elizabeth Park.

In 1931 the Worthingtons reported seeing a pack of 20 chase a kob across the Mweya Peninsula (Worthington and Worthington, 1933).

There is a variety of smaller carnivores, but none is numerous; although high densities of rodents exist in the long-grass areas of the Crater Highlands.

The Park is renowned for its profusion of bird life, particularly aquatic species. Cormorants and pelicans are plentiful along the Kazinga Channel, the rare shoe-bill stork hides in the papyrus swamps, and the cry of the fish eagle from its lofty perch atop a euphorbia tree is one of the unforgettable sounds of Africa. There is an abundance of fish, the rich manuring of Lake George by the hippopotamus making it one of the most productive inland fisheries of Africa.

In 1957 it was estimated that there were some 14 000 hippopotamuses in the Park, and that overgrazing and soil erosion were present as a result of overpopulation (Bere, 1959). A decision was taken by the Trustees to reduce the numbers, and this commenced in April 1958. All of the hippopotamuses were eliminated from the Mweya Peninsula, and from a number of inland wallows in the Katunguru area. With the establishment of the Nuffield Unit of Tropical Animal Ecology in 1962, park-wide reduction was started to bring the numbers to a range of set densities in selected areas, in order to determine which was the optimum number for the carrying capacity of the habitat. Thus a population at Lion Bay was totally eliminated, and the high population at Katwe Bay left undisturbed as a control. The reduction of these hippopotamuses, some 8000 in all, resulted in a number of vegetation changes, and an increase in other animal species, particularly on the Mweya Peninsula, which was kept free of hippopotamuses during the period of my study.

The Main Study Areas

The Mweya Peninsula

Kaiyura's old refuge was the main study area, in which the majority of my studies were conducted. It was chosen as an area of prime importance for study due to its high density of waterbuck, its isolated formation, and freedom from interference. The area is virtually an island of 4·4 km², surrounded by water, except for the narrow isthmus of Kanyeseswa to the north-east. West of the 33 m fault which divides the area into two, almost equal, halves, the land is low-lying, at an average height of 15 m above lake level. The fault rises steeply from these flats to the eastern plateau, where Lugard camped, and rises almost 50 m above lake level, at its highest point. Both areas are rather flat, with gentle undulations. Several deep, bushed gullies cut into the steeper slopes, generally attributed to hippopotamus activity; but which, judging from their positions, seem to represent the lines of former human footpaths, since they are matched by opposing gullies on the opposite bank of the channel. The western flats comprise grassland, with a fairly thick, but scattered, cover of *Capparis*. The "thick bush on precipitous slope" of Hale-Carpenter (1921) must refer to a rather narrow belt of dense *Azima tetracantha* along the western shore, where a steep cliff descends to the lake. The high ground in the east is almost wholly open grassland, with *Capparis* thicket along the drainage lines. Thick bush covers the steep, north-eastern bank and the eastern shore (Fig. 14).

Mweya is recorded as being almost bare of grass, until from April to August 1958 the entire hippopotamus population, of some 128 animals, was destroyed, and immigrants were prevented from re-colonizing the area. Following their removal, there was an increase in foliar cover (Thornton, 1959); exceptional rains immediately after the elimination

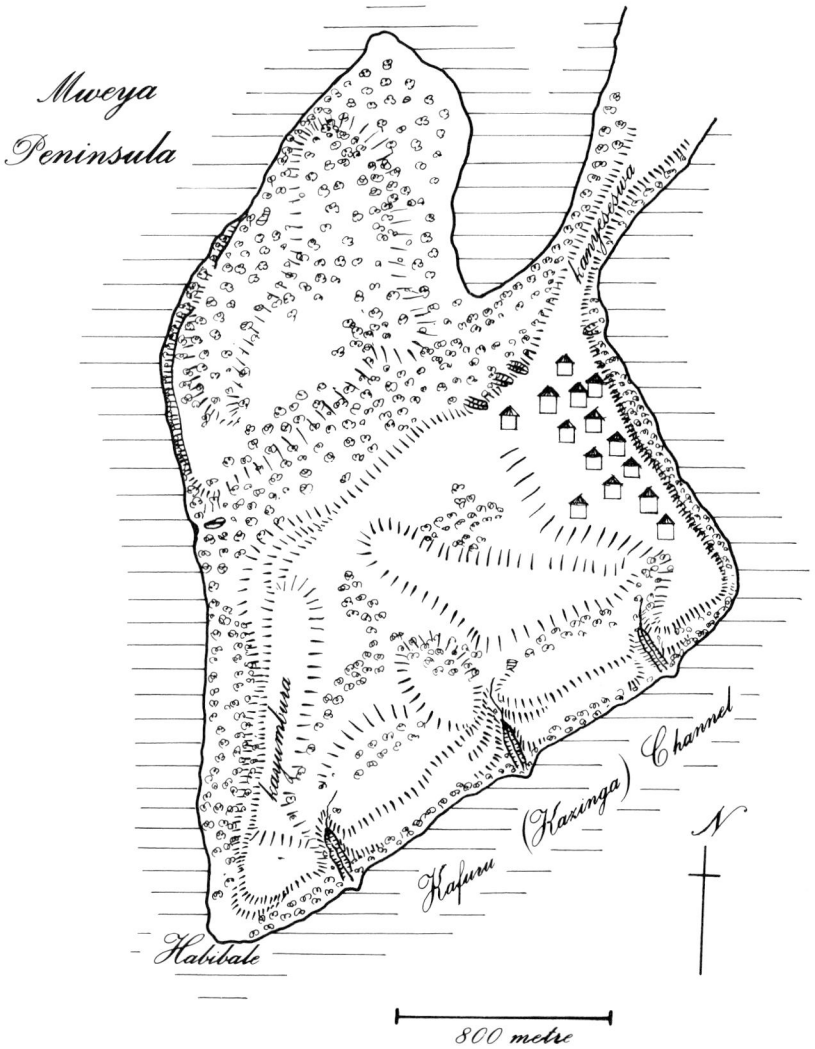

Fig. 14. Map of the Mweya Peninsula showing main topographical features.

contributed to this recovery. These rains were followed by a very dry period, during which the ground cover decreased. When the last analysis was carried out, in May 1964, basal cover had still not reverted to the percentage that it had attained in 1958. This was the result of the replacement of carpet grasses, such as *Chrysochloa orientalis*, which are encouraged by hippopotamus grazing, with tufted grasses. Notably among the latter was *Sporobolus pyramidalis*, which replaced *C. orientalis* as the dominant species in 1961 (Thornton, 1959, 1962; McKay, 1961; Lock, 1964) (Table 7).

Table 7 Changes in frequency of grass species on the Mweya Peninsula expressed as % Basal Ground Cover > 5% initial (after Thornton, 1971).

Species	Date				
	May 1958	May 1959	May 1960	May 1962	Sep. 1966[a]
Chloris orientalis	37·7	21·1	22·9	6·6	< 5·0
Cynodon dactylon	9·2	2·1	3·2	11·7	8·9
Sporobolus pyramidalis	8·0	25·2	18·9	27·6	31·0
Chloris gayana	6·4	6·8	7·2	19·4	12·2
Dactyloctenium aegyptium	5·2	0·1	2·2	0·5	< 5·0

[a] % frequency occurrence in random quadrats from Spinage, 1968.

The changes in grassland composition have been accompanied by changes in the numbers of resident grazing animals; namely buffalo and waterbuck. Buffalo increased from an average of 20·8 animals present in 1957, to 178 in 1968. Waterbuck increased from a mean of 16 to 46·5 in 1966. About 13 bull elephants are now resident throughout the greater part of the year, and their numbers have probably doubled since 1957. Herds numbering over 40 individuals move in during the wet months, spending usually two or three days foraging. The number of warthogs seems to have changed very little; the average number recorded in 1957 was 30·8, and in 1966 it was 38. But studies have shown that only about 50% of the true number is ever recorded (Clough, personal communication). We must remember, however, that there was less cover in 1957, so that a real increase may have taken place. For bushbuck, the mean number sighted in 1957 was 14·5, and in 1966 it was 14·7. Clearly this browser should have been unaffected by the grassland changes; but its true numbers are estimated to be up to 2·6 times higher than those normally recorded (Waser, 1975). Young kob occasionally make exploratory visits to the area, but never remain for very long. This is probably attributable to the absence of the grass *Themeda triandra*.

Other areas

Studies were concentrated in five additional areas; the most important of which was the Overgrazed Study Area, hereinafter referred to as the "Ogsa". This region was chosen as it adjoins the Peninsula, and there is much exchange of waterbuck between the two. It has also been subject to high intensity hippopotamus grazing, but the hippopotamuses have not been eliminated. Drs Petrides and Swank, two American Fulbright scholars who conducted preliminary studies into the hippopotamus population in 1956–1957, and made a number of waterbuck counts, named it the "Overgrazed Study Area" in contrast to their "Long Grass Study Area" of the Crater Highlands. On account of its similar vegetation, and the fact that the number of waterbuck inhabiting it approximated to the park-wide average, I considered it to be an ideal area for determining whether the Peninsula waterbuck population structure and behaviour was typical (Fig. 15).

In the north-western extremity of the Park is a region bordering Lake Edward, called Kayanja. This carried the highest density of waterbuck found in the Park. As already noted, it was characterized by the presence of a vigorous growth of *Acacia sieberiana* regeneration in parts of its area. Other than waterbuck, animal numbers were remarkably low: an aerial count in June 1966 revealed only about 226 buffalo, and 46

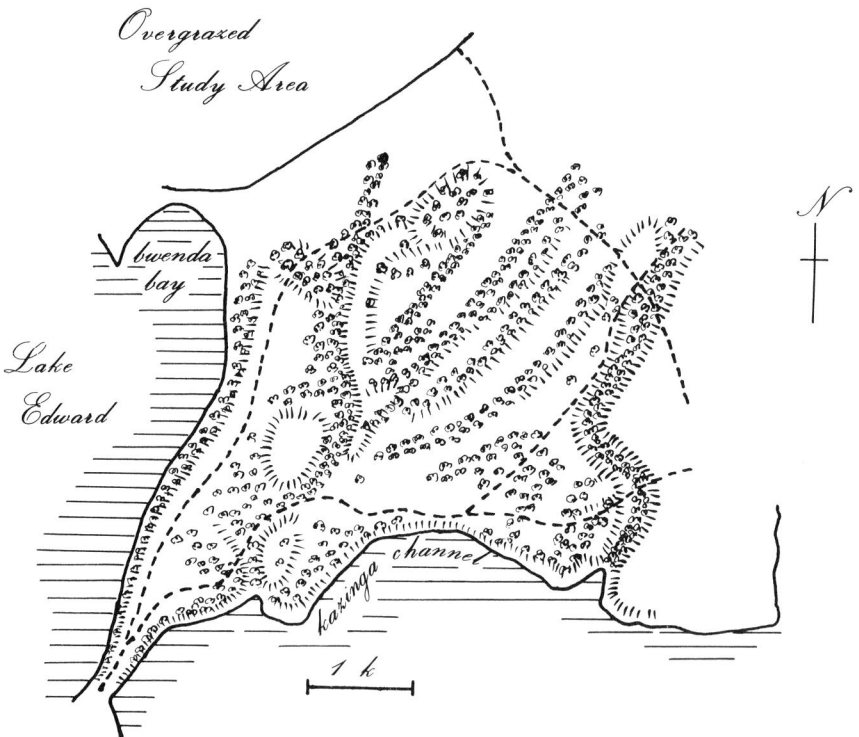

Fig. 15. Map of the Overgrazed Study Area, showing main topographical features.

elephants. Hippopotamuses were numerous along the lake shore, but 171 were removed in culling operations. Large numbers of kob were present in the south-east (Fig. 16).

As a contrast, a low density area was studied known by the name of the "Royal Circuit Area". Its main topographical differences were that a part became waterlogged during the rains, and that there was an extensive cover of *Imperata cylindrica*. Buffalo were the most numerous occupants, numbering up to 600 head, and up to 100 hippopotamuses occupied the waterlogged area during the wet seasons. Waterbuck density averaged only 1·0/km².

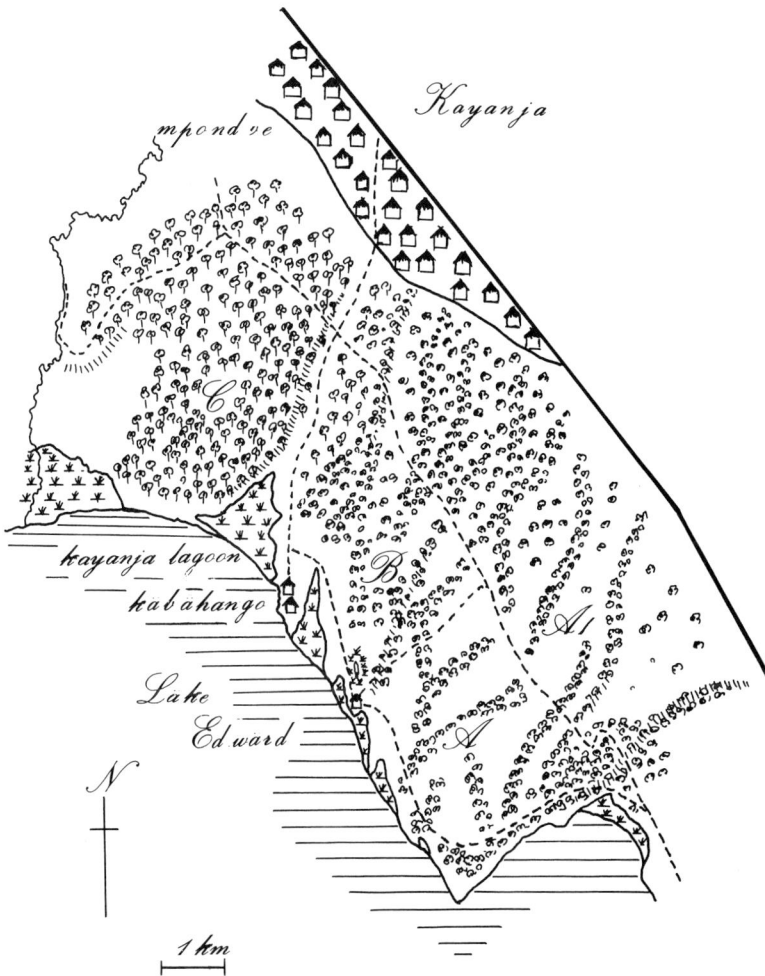

Fig. 16. Map of the Kayanja area, showing main topographical features.

A limited number of observations was made in the Nyamagasani area, adjacent to the Kayanja region, which was characterized by good numbers of buffalo and kob, and a relatively high density of waterbuck. Limited studies were also made in the Ishasha River region, the only representative of a riverine habitat. Here, the bordering plains carry high numbers of topi, kob and buffalo. Waterbuck are relatively common close to the river, but their overall density is low (Table 8).

Table 8 Summary of the main study areas.

Area	Size km²	Habitat	Vegetation	Waterbuck Density/km²
Mweya Peninsula	4·4	Lacustrine	*Sporobolus/ Capparis*	10·5
Overgrazed Study Area	14·3	Lacustrine	*Sporobolus/ Capparis*	3·7
Kayanja	26·7	Lacustrine	*Heteropogon/ Hyparrhenia/ Capparis* and *Hyparrhenia/ Themeda/Acacia*	9·9
Royal Circuit Area	20·5	Inland	*Themeda/Imperata*	1·0
Nyamagasani	14·5	Lacustrine	*Themeda*	7·0
Ishasha River	17	Riverine	Gallery forest/ *Themeda*	0·15

3. Methods of Study

Introduction

I sat quite still in my Landrover. A few metres away a waterbuck fawn stood watching me questioningly, twitching its shiny black, patent leather nose. Cautiously it advanced a step. An old doe looked up from her grazing, paused a moment, and then continued her feeding. The calf took another, uncertain, step towards me, and with painful slowness I extended my arm through the window towards it. Suddenly, boldly, it took another couple of steps forward, and abruptly plonked its wet nose against my hand. The human scent was something it had not bargained for, and, startled, it jumped back with a surprised snort. The old doe stopped her grazing again, lifted her head, and stared. The fawn regarded me unbelievingly for a few moments, and then turned and ambled away. The doe resumed her grazing, and the incident was forgotten.

This took place on the Peninsula soon after I had begun my observational studies. Once the waterbuck had become accustomed to my daily visits the fawns developed considerable curiosity, and often crowded close to the Landrover. After the bold exhibition by this particular fawn, there seemed to be no doubt, that given time and patience, I could have habituated some of the population to my presence so that I could have handled them. The seeming tameness of the waterbuck has been commented upon by the early hunters, for it is one antelope which does not flee at the sight of man, unless considerably hunted. But capture a waterbuck, and you have a most highly-strung, intractable animal on your hands. Professional trappers find it one of the most difficult of animals to keep in captivity, and its usual reaction after capture is to refuse to eat and drink. In contrast, the eland, one of the most timid of animals to approach, becomes calm as soon as it is captured.

Drug Capture

At the time of my study habituation was rather an advanced idea, although subsequent studies on red deer and big horn sheep have shown its feasibility; and that, with a little patience, it is possible to have wild animals literally eating out of one's hand. The idea of offering food to a wild grazer did not, however, seem to me to be a very feasible one at the time, and I resorted to what was another comparatively recent innovation in the study of wildlife, which held great promise. This was to capture the animals with an immobilizing drug, administered by a dart from a gun or crossbow. Anthropomorphic as it may sound, after instigating this method, the once so-trusting waterbuck of the Peninsula never trusted me again. Although they eventually got over their initial fear, they visibly jumped even at the sound of a camera shutter. Never again did a fawn try to approach the Landrover, to plant its wet nose either on it or on me!

Handling the animals was necessary to mark them for future observation, to take measurements for the study of growth, and to take tooth impressions to determine their ages. I have no doubts that the information so gained was worth the effort, although the practice of darting became a most vexatious business. It is not a method to be undertaken lightly. Confronted with a population of placid beasts, already conditioned to the extent that they barely lifted their heads when I drove up, darting looked a simple task. I did not take into account the speed with which they learn. It was not long before I was lucky if I could get close enough to dart a single animal in one day.

This was partly due to some bad luck. Dr Roger Short, a veterinary colleague from Cambridge who initiated me into the art of darting, decided to use a new, powerful drug. This was called M99, a morphine derivative said to be 1000 times more powerful than morphine itself, and exceedingly lethal to handle. As M99 possessed all the properties of an analgesic we felt it preferable to use such a drug on animals which one wished subsequently to observe. They should feel no pain under its influence, and should not be alarmed when immobilized. Furthermore, its effects in the event of an overdose, or when the work on an animal was finished, could be reversed with an antidote. Experiments conducted off the Peninsula did not have the result that we had hoped, causing great excitation of the animal. With a struggle we could handle the does, but the bucks were dangerous antagonists. Several animals died after apparently recovering from the antidote, and as a result we abandoned further attempts with it. Subsequently, workers elsewhere found that waterbuck could be successfully and satisfactorily immobilized if nearly twice the dosage that we had been using was administered (Hanks, 1967);

but this discovery was to come later, and meanwhile we turned to another drug, the muscle relaxant succinylcholine chloride. Its disadvantages are that it simply paralyses the animal, which remains fully conscious; dosage is critical, dependent upon the weight of the animal, and there is no antidote to it. But correctly applied, an animal is immobilized within 20 min and on average, 7 min after injection, and it recovers completely after 20 to 30 min with no side effects (Short and Spinage, 1967). For shooting the dart into the animal I used a 90 lb pull crossbow, a crossbow being both cheaper and making less noise than a gun, but requiring somewhat more skill in its use.

I found that the greatest cause of alarm amongst the waterbuck was not the actual darting, but the pursuit to get within range. Once an animal, or a group of animals, started to run, others, watching from vantage points, saw what was happening. In nature, running is an alarm signal, and it did not take the watchers long to identify the cause of the alarm as the Landrover following behind the running animal. I had considered the possibility of this, and so I took the precaution of always darting from a differently shaped Landrover (short-wheel base and long-wheel base, open and covered) to the one that I used for my observations. The waterbuck were still suspicious, but they could certainly recognize the difference between a long-wheel base Landrover and a short-wheel base one.

When struck by a dart an animal usually ran off; but some stood motionless as if unaware that anything had happened, and this after being struck by a sharp projectile released from a 90 lb pull crossbow! After about 7 min, on average, the animal would stagger and collapse abruptly. It was then approached cautiously, as waterbuck seemed to be very alarmed by the sound of grass rustling close to them, and would struggle to rise when they heard this. My first action was to blindfold the animal, as this had a surprisingly calming effect. Next, the tranquillizer acetylpromazine was injected, which would have a further quietening effect about 10 min after administration. The sharp horn tips of bucks were then covered with pieces of rubber hosepipe to guard against accidents (Plate 12). The head was then supported as a precaution against the possibility of the animal choking on its rumen contents; the dart was removed, and penicillin applied to the wound. In fact, the healing powers of the waterbuck's skin are such, that, by the next day, the site of the wound was indiscernible on close examination. After these preliminaries came the actual collection of data: marking, measuring and the taking of teeth impressions to determine age.

Plate 12. A drugged waterbuck.

Recognition and Marking

By the time that I had commenced darting I had already got to know many individuals by sight. The shape of the adult buck's horns varies, as we have seen in Chapter 2, and the ears often carry distinctive nicks in them. Does are much more difficult to identify as individuals, although one or two characteristic types could be defined. I recall one with limpid dark eyes, and a very shaggy neck, which always looked at me with great reproach; another had a very "oriental" look, with slanting eyes; a type that I instantly recognized later in the Akagera Park, Rwanda, 200 km away. But the method of identifying animals by natural characters breaks down when the individual is found outside its expected range. Unless the animal bears some absolutely unmistakable trait, one can never be sure whether it is the particular animal that you think it is, or not. This is why artificial marking is so necessary; added to which it usually enables individuals to be identified much more rapidly and at considerable distances.

The marking of wild animals is no new idea. Alexander the Great, some time before 300 B.C. marked a number of deer, so we are told, by placing gold collars around their necks. According to the Roman historian Pliny (A.D. 23 to 79) the deer were recovered 100 years later, their collars covered with great rolls of fat. It is unfortunate that the story became distorted in the telling, so that we shall never know the real span of time which elapsed between capture and recapture in this first experiment.

Although plastic collars have replaced Alexander's gold ones, I merely clipped a metal tag in one ear, and short, coloured plastic strips in the other; the colour combination telling me the number of the animal. These ear markers were still on many does 4 years later, but the bucks tended to lose theirs very quickly, apparently during fighting. Coloured plastic collars would have been preferable as they are much more easily seen; but their effect is disquieting to visitors in a national park, and I felt that in the case of bucks they might interfere with fighting.

Tooth Impressions

When a dentist fits a patient with false teeth he first makes an impression of the inside of the mouth with a plastic substance. This is virtually what I did with the waterbuck; for, by making impressions of an individual's teeth at yearly intervals, I could determine how much wear had taken place, and thus work out absolute ages by reference to a set of jaws covering the lifespan. This was not as simple to execute as it sounds. First one had to get the animal's jaws open. For this a modified Linton sheep-gag was constructed; but the problem was that the animal would usually clamp its jaws shut and resist every attempt to open them. Once opened, however, a tray containing a plastic substance was slipped under the left tooth row, and the animal allowed to bite, but not chew, on it. The tendency was for the animal to make a sideways movement of its jaws as soon as contact was made between the two, and this had to be avoided. Dental plastic was found to be unsatisfactory for taking impressions due to its setting time and preparation; instead, the most suitable substance was found to be slightly warmed, children's plasticene. If an impression was not suitable it could simply be rubbed out with the thumb and taken again. A plaster cast of the plasticene imprint was made while waiting for the animal's recovery from the drug, and the plasticene was immediately ready for use again.

Measurements

I marked the horns of the bucks with a small saw cut a measured distance from the tips. When the animals so marked were recaptured a year later, I could tell how much had been worn away, and whether the horns continued to grow in old animals. Horn measurements, and a number of body measurements, were taken. The latter were: chest girth around the heart, body length, shoulder height, neck girth and hindfoot length from the tuber calcis to the tip of the extended hoof. By the time that all this was accomplished the animal was often able to sit on its brisket and support its own head. Indeed, there were occasions when the animals got up and left precipitately before all was finished; for recovery can be sudden and absolute.

Reactions of Captured Animals

It was interesting to learn something of the reactions of these animals when handling them in this way. Does still on their feet would try to butt if confronted, in their natural, rather harmless method of defence. Once the animal was on its brisket the blindfold was removed, and when not blindfolded the animal was calmest if one was in its side vision. A buck would then sit perfectly still, even with one touching him; but cross in front of his field of vision and he would make every effort to impale one on his horns.

When the work was finished, and the animal could sit up unaided, I withdrew to a distance so that it would not struggle to escape; and shortly after this it was usually on its feet. Reactions then varied. Generally speaking, if it was a timid animal, or one that had been chased unduly before capture, then it was more timid after release than it was before. But animals which had been darted, followed, but did not go down (usually due to too little drug being administered), appeared to be more alarmed after the experience than those which had been captured and released. Those bucks which did not take much notice of one beforehand, showed little response afterwards. I have darted and released a buck, and then driven to within 10 m of it shortly after its recovery, the animal taking no notice of me, and this was one which had not been given a tranquilliser.

Those which were tranquillized still reacted normally to danger; a very necessary requirement of the tranquillizer. The drug I used for this was acetylpromazine, the effects of which I tested beforehand by giving one doe a very heavy dose, and then keeping her under observation. Although she walked about in a rather dreamy manner, she spotted two lionesses and avoided them accordingly at a distance at which I would not even have noticed them.

A marked animal returning to a group after release was looked at curiously for a moment or two, and then ignored. If a darted animal fell among others, their first reaction was to leap away; but then they would return, staring and sniffing inquisitively. This, of course, was interrupted by my arrival. The apparent unconcern of many after darting convinced me that, properly executed, darting was not particularly distressing to the animals.

Observations

I succeeded in darting almost all of the Peninsula bucks, and two-thirds of the does; as well as animals in other areas. When this was complete it was necessary to let them settle down again before observation could begin. Observation was always from a Landrover, for so much more ground could be covered quickly than would have been possible on foot, and one started with the advantage of the animals' unconcern for the presence of vehicles. Even if one were to habituate the study animals to your two-legged presence, there would still be others to contend with; elephant, buffalo, kob, bushbuck and warthog would all be disturbed, and communicate their alarm to the waterbuck, as well as the larger animals sometimes providing discomfiting moments for the observer.

Two daily circuits were made on the Peninsula for one month, each alternate month, between 15 October 1964 and 7 March 1967. I was also able to return in November 1968 to make a few further observations. With these visits I was able to build up a picture of the bucks' territorial arrangements, and of the activity of the does. Of course, this was not all; other areas were likewise visited to make comparative evaluations, and there were 12-h continuous observation periods to study activity rhythms, as well as many general behavioural studies. But once darting was complete, most of the work necessitated only a notebook and pencil, and a pair of binoculars.

Age Determination

In recent years teeth have attracted considerable attention with respect to determining the ages of wild animals; reviews are given by Spinage (1973) and Morris (1978). There exist, of course, those sceptics who warn against attaching too much faith to the precision which can be attained; but the use of teeth remains the only feasible manner of determining the ages of wild mammals with some certitude. The determination of age is important in wildlife studies as it is essential to the understanding of rates of growth, onset of sexual maturity, fertility peak, senescent decline and lifespan, as well as social behaviours.

It was for this reason that I took tooth impressions from the immobilized animals, but I used two main methods to determine age. The first was the relation of eruption and wear of the permanent dentition to time, and the second was the counting of incremental lines in the cementum of the tooth root.

In a seasonally breeding species a good start can be made in the establishment of age categories by reference to body size and horn growth at annual intervals from birth. With those species which breed throughout the year, no such base-line can be established, and one usually requires recourse to known-age specimens of dead or live animals. This is the case whether one is establishing eruption ages, or categories of wear of the permanent teeth. Lacking any known-age specimens in this study I had to resort to making tooth impressions from live animals at annual intervals. Firstly, I constructed a series of relative age groupings from a collection of skulls, dividing it into arbitrary age classes based upon degrees of wear of the permanent dentition; and then I related these groups to the annual estimates of wear derived from the tooth impressions. By this means I was able to determine the age of the waterbuck up to approximately 10 years old.

Cementum lines are regions which, preferably when subjected to histological staining techniques, but sometimes without, can be seen in section under the microscope as sequential dark-staining striae in the cementum of the tooth. Their physiological basis is as yet undetermined, but they appear to represent regions of interrupted or retarded growth of the ostein matrix. A correlation can be found between this interrupted, or retarded, growth and the seasons, one bold line per year being evident in temperate zones and two in equatorial zones. A fuller discussion of the phenomenon is given in Spinage (1973, 1976). In the waterbuck, in which I first identified the occurrence of bi-annual lines in equatorial animals, a regression of the number of boldly staining lines counted in the cementum of the first incisor, against age derived from the tooth impressions, showed a linear relationship of two lines per year, with a correlation coefficient (r) of $+ 0.986$. There exist, however, a number of less readily distinguishable lines, and if one counts every discernible line, whatever its intensity, a correlation coefficient of $+ 0.963$ is obtained; which is significantly different from the first ($p = < 0.001$). A discussion of this has been given in the papers referred to above; there apparently exists a rhythmic background pattern of incremental growth, which sometimes stains up well, upon which the bolder lines caused by physiological disturbances, among which what appear to be seasonal influences, are superimposed. When the mean number of lines per year class was tested against the expected number at a rate of occurrence of two lines per year, the correlation was significant for the minimum, or "bold" line count, and not significant for the maximum line count (chi sq. $= 0.672$, df 5, $p = > 9.95$, and chi sq. $= 17.37$, df 5, $p = < 0.005$ respectively). This suggested that, on average, two bold lines per year are formed in the tooth cementum; a conclusion since established for the buffalo from the same area (Grimsdell, 1973). I think that it is justified to suggest that these lines probably result from influences attributable to the bimodality of the annual rainfall.

As with other methods of age determination we can see that the interpretation of cementum lines results only in an approximate age estimate, but it can be very close to the true age, sufficient to provide a workable foundation for other studies. Line counts are especially valuable in providing us with a minimum age for old animals; minimum because in senescence the tooth root is rapidly resorbed. This does not always interfere with the line count, but caution must always be exercised when examining the teeth of old animals. In the waterbuck, line counts showed a longevity in the buck of 11 years, and in the doe of 14 years; which accords well with what we would expect from the known longevities of zoo animals (Chapter 5). However, throughout this book, where specific ages are assigned to animals, it should be borne in mind that these are only estimates, since there is no means of confirming them.

4. Growth and Senescence

Introduction

Waterbuck rank among the heaviest of the African antelopes, although not in the class of the eland; in East Africa they follow the eastern white-bearded wildebeest, roan and greater kudu, in weight. The heaviest waterbuck recorded came from the Transvaal of South Africa, with an alleged weight of 309 kg (Roberts, 1951). Size then tends to diminish as one goes north, to increase again in Uganda, where, in the region of the Rwenzori Park it may reach almost 273 kg. Here, the average buck weight was 236 kg, and height 127 cm to the withers; while the average adult doe weight was 186 kg, and height 119 cm.

In common with other African mammals, until the 1960s there were few reliable records of weight, and only with the development of a more objective interest did the appreciation of detailed weights and measurements from large samples become accepted. In 1963 Ledger and Smith collected and examined 20 buck and 20 doe waterbuck from the Kamulikwezi area of the Rwenzori Park, on the north-west shore of Lake George, as a part of a pioneer study of the meat potential of wild ungulates in comparison with East African cattle. This study was to show that wild animals, in general, had a higher killing-out percentage; that is, a higher percentage of consumable meat, the average value of 58% being comparable to that of cattle in good condition maintained on good pastures. The waterbuck, in fact, did slightly better than zebu bulls and steers, with a killing-out percentage of 59% (Ledger *et al.*, 1967). Ledger and Smith's data on waterbuck were made available to me, and in the course of further studies, 182 animals were autopsied for various purposes over a 4-year period.

Growth in Weight

Growth, expressed as weight from birth, is faster in the buck than it is in the doe. At birth the fawn weighs about 13·6 kg, and between 0 and 4 years the mean weight gain per day is 0·12 and 0·11 kg/day for the buck and doe respectively. Both sexes showed a marked drop in weight gain in years 3 to 4, which coincides with the period of most deciduous tooth replacement (3 to 3·5 years), especially the first incisors. At this time the mouth is presumably tender, interfering with grazing, as appears to be the case in the goat (S. Payne, personal communication). After this check there is some recovery in growth rate (Table 9). Waterbuck show no advantage over cattle in their rate of growth; that of steers on East African rangeland averaging 0·14 kg/day (Talbot *et al.*, 1961). The buck takes 2 years longer than the doe to approach its asymptotic weight, in view of its 21% greater average. The doe attains adult weight in her fourth year, and the buck in his sixth. Some obesity may develop with age, but although does appear to show some decline in weight between 10 and 12 years of age, there is no marked loss of weight in old age.

Table 9 Rate of growth of the waterbuck (kg/day).

Age (years)	Buck		Doe	
	Mean increase	kg/day	Mean increase	kg/day
0 — 1	63	0·17	60	0·16
1 — 2	53	0·15	48	0·13
2 — 3	50	0·14	37	0·10
3 — 4	11	0·03	16	0·04
4 — 5	27	0·07	3	0·01
5 — 6	20	0·06	10[a]	0·03

[a] Doe figures may be influenced by pregnancy in the adult.

In mammals, and in many other animals, growth follows an "S"-shaped or ogive-like curve with time. This is exemplified by a lag phase just after birth, albeit very brief, when the neonate comes to grips with its extra-uterine environment; an exponential phase of rapid growth to the approach of maturity; and, finally, an abrupt levelling-off as the asymptotic level of adult weight is approached. We have seen that there is also the possibility of an interruption to this pattern, but this is generally ignored in calculations.

The uses of fitting weight, or other growth data to such a curve, are threefold: (a) the lag phase between species or individuals can be assessed; (b) the exponential rate of growth can be compared amongst animals to determine whether one is more efficient (that is, grows faster) than another; and (c) the mean weight attained by the curve is the "real" mean weight of the animal, uninfluenced by adult fluctuations of obesity or decline. Growth, the dynamic process of anabolism and catabolism, continues after the asymptote is approached; weight varies throughout life as fat and flesh are put on in good seasons, and taken off during bad. Weight may increase in connection with reproductive activities in the buck, as well as in the doe, just as much as when food is abundant; and weight is lost in times of psychological and reproductive stress, as well as in those times of seasonal dearth.

An additional feature of semi-arid environments is the "green-grass-loss", a ruminant losing its greatest amount of weight at the onset of the rains when the grass is in fresh flush, rather than during the dry season, as is more commonly supposed (Payne, 1964). This apparent anomaly is due to the high water content of the graze, which makes its consumption rather like trying to fill oneself up on lettuce! Added to this, the development of intestinal bacteria of the *Clostridium welchii* group, which can accompany a change to a rich diet, can result in enter-otoxaemia and death in old animals. I have never observed this in waterbuck which, resorting to an increased intake of browse during the dry season, and liking a lot of water, probably do not normally experience great fluctuation in the quality of their diet in the Park, but enterotoxaemia appears to be not uncommon among buffalo and impala.

The monthly means of total body weight available to me, were insufficient to determine whether seasonal fluctuations in weight took place, and few studies on large wild herbivores have been able to do so. This is because it requires, either the destruction of large samples, or their capture and release; both methods of which can bias results. The former method may improve the lot of the survivors by removing competition for food, and the latter by subjecting the animals to stress, which repeated capture and release is likely to cause, could have an effect on their weight gain.

An ogive never goes up or down; but as the curve of increase in weight with time approximates to an ogive during the formative period of growth, it is useful to fit a logistic curve to this part of the growth period. A popular curve for animal studies is that proposed by Von Bertalanffy (1938). This curve has been discussed by Riffenburgh (1966) who considers it to be useful for comparing differential growth under varying environmental circumstances; but inappropriate when the purpose is to describe the complete growth pattern. With my data, I found that the Walford transformation, by which W_∞ was computed for the Von Bertalanffy simplification, gave excessively high values. With such variable data the best fit was obtained by using Nelder's method (Nelder, 1961), which gave values for the asymptote close to the mean empirical weight. Curves were fitted to the mean whole body weight of the buck from $\frac{1}{4}$ to 7 years; and in the doe from $\frac{1}{2}$ to 6 years. Asymptotic levels were approached at about 5 years in the doe, when $W_\infty = 186$ kg, and 6 years in the buck, when $W_\infty = 235\cdot5$ kg (Table 10, Fig. 17). This compared with a mean empirical weight of 188 kg for the doe, aged 5 years and above, and 236 kg for the buck, aged 6 years and above (Table 11).

Table 10 The logistic curve of weight increase in the waterbuck (kg).[a]

Age (years)	Buck		Doe	
	Estimated weight	Mean weight	Estimated weight	Mean weight
$\frac{1}{4}$	39·9	38·5	—	—
$\frac{1}{2}$	—	—	50·3	48·0
1	70·9	74·0	74·1	78·3
2	127·0	127·0	126·4	122·9
3	179·2	176·5	162·1	162·0
4	—	—	177·8	178·2
	225·9	215·0	183·8	181·0
6	234·2	244·5	185·1	191·0

[a] Estimated weight for the buck = $\dfrac{235\cdot5}{1 + e^{1\cdot8411 - 0\cdot9992t}}$

95% confidence limits = $\pm\, 4\%$ for the mean, and $\pm\, 11\%$ for individual values.
Estimated weight for the doe = $\dfrac{186\cdot0}{1 + e^{1\cdot5753 - 1\cdot1642t}}$

95% confidence limits = $\pm\, 4\%$ for the mean, and $\pm\, 10\%$ for individual values.

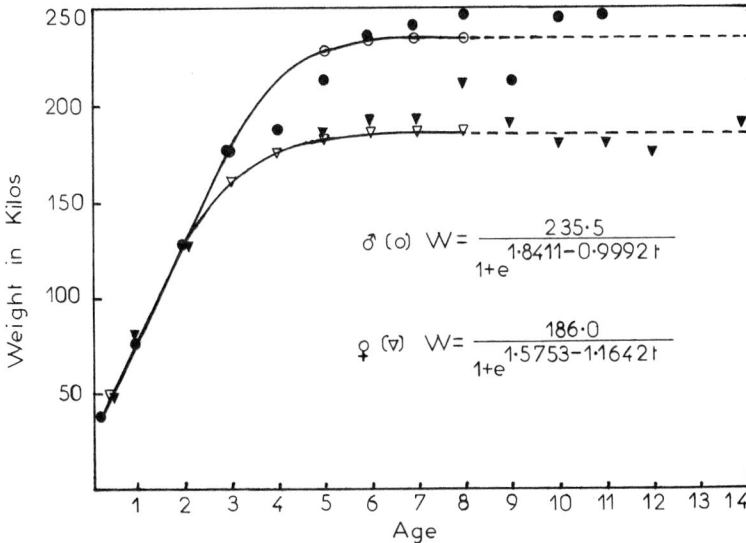

Fig. 17. The logistic curve of growth in weight for bucks — circles, and does — inverted triangles. Open symbols are calculated values, blocked symbols are observed values.

Table 11   Growth in weight of the waterbuck (kg).[a]

Age years	Buck				Doe			
	n	Mean weight	SD	Range	n	Mean weight	SD	Range
½	2	38·5	8·5	30 — 47	2	48	4·0	44 — 52
1	3	74	12·4	62 — 91	8	78	10·1	59 — 94
2	5	127	12.8	108 — 145	9	123	14·8	106 — 150
3	2	177	9·5	167 — 186	2	162	3·0	159 — 165
4	1	188	—	—	5	178	12·5	163 — 199
5	10	215	24·0	180 — 245	6	181	13·2	161 — 194
6	5	235	16·3	208 — 253	5	191	5·2	186 — 201
7	8	245	11·1	225 — 260	6	194	14·5	171 — 212
8	3	249	7·9	241 — 260	3	210	3·9	205 — 214
9	3	214	17·1	198 — 238	3	190	5·4	183 — 196
10	5	246	10·5	231 — 262	9	180	15·4	161 — 214
11	1	250	—	—	2	181	28·0	153 — 209
12					5	176	11·7	157 — 192
13					—	—		
14					5	192	11·5	177 — 211

[a] Total body weight or live weight.

Estimation Equations

Since the growth of most animal parts is allometric, it is no surprise to find that we can take almost any organ or bone from an actively growing animal, plot its change with age against some other character of growth, and find a correlation. Thus for differently aged growing individuals, an organ or bone will equal the correspondingly different weight of the entire animal. When plotted against time, any one of these characters gives us a mathematical relationship which often turns out to be a straight line, or nearly so. By using this linear regression we can predict values from the relationship of two or more different characters, when only one parameter is known. On the basis of this principle, attempts have been made to predict the weight of wild ungulates from linear measurements (Talbot and McCulloch, 1965; McCulloch and Talbot, 1965), a procedure well known in animal husbandry (Brody, 1964). This could obviate the carrying of heavy weighing equipment in the field.

In forming estimation equations for several species of African ungulates, McCulloch and Talbot (1965) found that shoulder height and body length were negligible factors in predicting body weight from linear measurements. Chest girth was the independent variable which consistently accounted for the bulk of the variance of total body weight.

Subtraction of the weight of the rumen contents gave no marked improvement in prediction. In other words, the variation between individuals was almost as great as the variation in intake of food. Furthermore, separation into sexes, age classes (that is, juvenile and adult), and stage of pregnancy, did not improve the regression equation, but merely subdivided the range of data. They therefore concluded that a simple linear regression of weight on chest girth was sufficient for weight prediction.

For the Rwenzori waterbuck a linear regression of weight (*y*) on chest girth (*x*), the latter measured around the heart just posterior to the shoulder, was calculated for 43 bucks and 64 does. The regression was calculated firstly for animals weighed entire; then for bucks minus the weight of the rumen contents, and then for does minus the weight of the total uterine contents. This should indicate what errors these variables might introduce into the relationship. It showed that, although the coefficient of determination (*r*) was high ($p = < 0.001$); in other words, the straight-line relationship was a good one, the variation was nevertheless high (Table 12). There was some, but no substantial, improvement when rumen contents were deducted from the total weight. This variation meant that the degree of accuracy, with which weight could be predicted, was poor. The 95% confidence intervals for the total buck weight showed that in predicting the weight of an individual adult, of average chest girth, there was a 95% chance of being correct to within $\pm$ 43 kg; and for a doe $\pm$ 36 kg. The remaining 5% would, of course, be further in error. This applied to predicting the weight of an individual; but if we wish to merely state the average weight of an animal from its chest girth, then the security of prediction is much better, the 95%

Table 12 Relationships of chest girth (*x*) to weight (*y*) estimated from linear regressions where $W = a + bC$, when W = weight in kilos, and C = chest girths in cms.

	Weight range	*n*	Relationship	*r*	Conf.1[a]	Conf.2[b]
Buck, total weight:	30—267	43	$2.745C - 176.94$	0.90	$\pm$ 42.5	$\pm$ 6.5
Total weight minus rumen contents:	26—241	43	$2.476C - 162.21$	0.92	$\pm$ 33.9	$\pm$ 5.2
Doe, total weight:	73—214	64	$1.964C - 85.21$	0.68	$\pm$ 35.9	$\pm$ 4.5
Total weight minus uterine contents:	73—200	64	$1.794C - 70.0$	0.64	$\pm$ 35.5	$\pm$ 4.4

[a] Conf.1 = 95% confidence limits for an individually predicted weight at mean chest girth.
[b] Conf.2 = 95% confidence limits for a mean predicted weight at mean chest girth.

confidence limits being 6·5 kg and 4·5 kg for the buck and doe respectively.

Talbot and McCulloch (1965) achieved much better security of prediction than this in species which they examined; but Smith and Ledger (1965) found that the hind leg weight of the waterbuck, expressed as a percentage of total body weight, possessed a standard deviation two times greater than that of the highest recorded among ten other species of ungulate. This seemed to be attributable to the weight of the hind leg decreasing in relation to the total body weight, with increasing age, and suggested that the waterbuck differed from the other species examined, in that weight was either being lost from the hindquarters, or being added elsewhere.

The continuous growth of the buck's horns, although it may not result in an increase in their length due to the tips being abraded, nevertheless is expressed in an increase in basal circumference. It seemed likely to me that the resultant weight from continued horn growth might be reflected in an increased neck girth, as well as the fact that it is the neck which is so important to the mature buck in fighting; thus one might expect greater development of the neck with increasing age. The regression of body weight on neck girth was therefore examined in 18 bucks and 36 does, and was found to reduce the confidence interval by almost 18% for the buck, and 13% for the doe (Table 13). A multiple regression was thus calculated for the 18 bucks and 36 does of weight on chest plus neck girth. This had the result of reducing the confidence interval by 44% for the buck, and 24% for the doe, for an individually predicted weight at mean chest and neck girth. This meant that the weight of a given buck of mean chest and neck girth could be predicted, at the 95% confidence level, to within ± 25·5 kg, and for the doe ± 25·3 kg. In other words, an average weight could be predicted to within ± 5·8 kg and ± 4·2 kg for bucks and does respectively. Although the confidence interval was similar for both sexes, the improvement obtained for the doe was 20% less than that for the buck.

For predicting weight from linear measurements in the waterbuck, neck girth is thus preferable to chest girth, but a combination of the two is better than either one used alone.

The regression of weight on neck girth in the buck showed an improvement in prediction for an individual weight at mean neck girth of 17·6 kg, compared with chest girth; against an improvement of 12·6 kg in the doe. If buck and doe measurements are analysed together a loss of precision in prediction results for both sexes. For the regression of weight on chest girth, however, a small loss in precision for the doe is compensated by a gain in the buck. This suggests that the relationship of neck girth to weight in the buck is a secondary sexual character which is reflected to a limited extent in the doe also (Spinage, 1969). In the field the thick necks and heavy horns of the bucks are obvious enough; but that the same pattern of growth, albeit to a reduced extent, also exists in the doe, is less obvious.

Table 13 Relationships of chest girth (x_1) and neck girth (x_2) to weight (y) estimated from linear regressions where $W = a + bC$; $W = a + bN$; and $W = a + b_1C + b_2N$, when N = neck girth in cm.

	Weight range	n	Relationship	r	Conf.1	Conf.2
Buck, total weight:						
Weight × Chest	91 − 260	18	$3.341C - 262.24$	0.85	± 44.8	± 10.3
Weight × Neck			$3.033N - 35.23$	0.89	± 36.9	± 8.5
Weight × Chest + Neck			$-167.14 - 1.608C + 1.85N$	0.95	± 25.5	± 5.8
Total weight minus rumen contents						
Weight × Chest × Neck				0.95	± 23.9	± 5.5
Doe, total weight:						
Weight × Chest	150 − 214	36	$1.700C - 44.46$	0.28	± 33.2	± 5.5
Weight × Neck			$3.614N - 24.87$	0.48	± 29.0	± 4.8
Weight × Chest × Neck			$-152.02 + 1.228C + 0.62 + 3.138N$		± 25.3	± 4.2
Total weight minus rumen contents						
Weight × Chest × Neck				0.60	± 24.6	± 4.1

Notation as for Table 12.

Skeletal Growth

An adult buck stands 6.7% higher at the shoulder than the doe, with mean heights of 126 cm (SD 2.8) and 117 cm (SD 4.0) respectively. The buck is also 8.2% longer in the body, the mean buck and doe length measurements being 243 cm and 223 cm respectively.

Lower leg length, measured from the point of the tuber calcis to the tip of the hoof, represents approximately 43% of adult height, declining from 47% in the first year of life. Adult lower leg length is 3% longer in the buck, with means of 54.4 cm (SD 2.9) and 51.7 cm (SD 1.9) in the buck and doe respectively. Adult length of the lower leg is attained in the third year; thus although it is a measurement which is often taken in mammals, it is not, in this species, a useful indicator of age beyond the early stages of growth. This rapid growth of the lower leg is a reflection of the extremely fast elongation of the metacarpal and metatarsal bones, to provide the young animal with a stilt-like locomotion, giving a long pace with little muscular effort. Such a pace enables the young animal to run as fast as, or at least to keep up with, the adult when fleeing from predators; although, of course, it lacks the energy reserves to sustain it over long periods of such activity. The relative decline in "stilt-length" in relation to height is presumably compensated for by increased muscular volume. Full growth in height is attained between the fourth and fifth years of life, while growth in body length ceases in about the third year. As we have seen, adult weight is not reached until the fifth and sixth year in does and bucks respectively.

Early arrest of skeletal growth is confirmed by measurements of both skull and mandible lengths, which approach their asymptotes in the third to fourth year. No apparent reduction in mandible length takes place with age; neither apparently is there continued growth, contrary to that reported for red deer (Staines, 1978). Mean mandible length per age class was, however, very variable in the sample, possibly obscuring any trends (Table 14); although I would not expect it to increase with age (see Chapter 12).

Table 14   Growth in length of the mandible (mm).

Age (years)		Buck				Doe		
	n	Mean length	Range	SD	n	Mean length	Range	SD
$\frac{1}{2}$	2	174	214—236	98	2	186	180—192	72
1	3	231	225—238	43	10	220·5	205—242	138·6
2	5	266·6	251—279	117·3	7	268·1	247—278	103·8
3	2	290·5	290—291	0·5	2	285	281—289	32
4	1	305	—	—	5	298·2	295—301	5·2
5	8	301·1	290—308	50·4	6	293·8	283—312	95·8
6	5	308·6	296—315	63·3	5	298·2	285—305	69·7
7	8	308·4	294—319	57·7	6	299·8	284—311	109
8	3	315	310—320	25	3	305·7	300—311	30·3
9	3	301·7	295—324	217	3	300	293—311	93
10	5	308·8	306—313	8·7	7	301·1	285—313	101·1
11	1	306	—		2	294	288—300	72
12	—	—	—		4	297·3	280—321	296·9
13	—	—	—		—	—		
14	—	—	—		5	306	298—312	26·5

The Viscera

The first anatomist to undertake a comparative study of wild animal organ weights was an American anatomist, Dr George Crile. In 1935 he autopsied several species of African mammal in, and near, present-day Lake Manyara National Park in Tanzania, although these did not include waterbuck. Other than his meagre samples (Crile and Quiring, 1940), visceral weights of most species of African ungulate were non-existent until the studies of Ledger at the end of the 1950s (Ledger *et al.*, 1961). Ledger's studies, as we have seen, were directed towards the determination of the relative efficiencies of wild animals and domestic stock as meat producers; but a basic inventory of gland and organ weights for a species could provide a useful baseline for the identification of pathologic abnormalities at autopsy, which may have value in management considerations. Of course not all organs are important in this respect.

A comparison of the weights of waterbuck vital organs expressed as a percentage of total body weight, namely the heart, lungs, liver and spleen, was made with those of the Uganda kob and the domestic ox. This suggested that, in general, the wild animals tended to have proportionately larger organs in relation to their body weight, except for the liver which was similar in size in all three species, with a constant ratio of 1·2% (Table 15).

Table 15 A comparison of some visceral weights as % liveweight.

| | Species | | |
Viscera	*Bos taurus*[a]	*Kobus kob*[b]	*Kobus ellipsiprymnus*
n	198	40	102
Heart	0·4	0·6	0·8
Lungs	0·7	1·1	1·2
Liver	1·2	1·2	1·2
Spleen	0·2	0·2	0·4

[a] From Spector (1956), all cows.
[b] From Ledger and Smith (1964), both sexes.

The liver

The liver is an important organ of many functions; as expressed by Young it constitutes a chemical workshop for the body (Young, 1957). It has several functions, one of which is that excess carbohydrate may be converted into glycogen and stored there. Thus we might anticipate that animals on a high nutritional plane would tend to have livers of above normal weight. When gross liver weight of the doe was examined, I found that the mean weight for the months of May to July, and October to December, was significantly greater than that for the other months ($t = 2.78$, df 46, $p = > 0.99 < 0.999$); the difference amounting to 20.7%. However the amount of rainfall and the number of births and conceptions is not significantly different between these months. But if we correct for a two months' lag between months of high liver weight and high rainfall, then there is a highly significant difference in the amount of rainfall in those months preceding months of high liver weight ($t = 4.072$, df 10, $p = > 0.99 < 0.999$); namely the months of March to May and August to October, compared with those following. This suggests that the increased availability of carbohydrate in green grass following rainfall does result in higher liver weights; but that this weight increase lags behind the birth peaks suggests that it is not important in this respect. In other words does are able to maintain sufficiently high levels of liver carbohydrate throughout the year in this area. The buck showed only a peak in December (two months after rainfall peaks), but although doe conceptions are high in this month, and thus it may be supposed to be a time of increased activity in the buck, there is no rise corresponding with the June conception peak.

The heart

The mean weight of the heart was 1.64 kg and 1.38 kg, in the buck and doe respectively, forming 0.7% of the total body weight, with no significant difference between the sexes ($t = -0.134$, df 82, $p = > 0.9 < 0.8$).

The spleen

The functions of the spleen are those of forming a reservoir of red blood corpuscles, destroying wornout ones, and acting as lymphoid tissue. Its latter role signifies its importance in disease, and an enlarged spleen indicates that the animal has been subject to infection. A very compact, "leathery" organ, the spleen is easily dissected out and weighed accurately. In the post-pubertal buck and doe (that is, animals greater than 2 years of age), the average weight of the spleen is 0.45% and 0.43% of the total body weight respectively; with the doe showing the greater variance (SD 0.002 and 0.017 respectively). Both sexes showed less variance at this age than did younger animals in the 0 to 2 years age group ($n = 7$ for the buck and 10 for the doe). The youngest animals, a doe of one week and a buck of one month, had spleens weighing only 0.08% and 0.09% of their body weight. But this age grouping also showed the highest weights, up to 2·3% and 0·95% in the buck and doe respectively. This could suggest that the young animal is at first dependent upon the transfer of maternal gamma globulins for its immunity to disease; but with their disappearance the spleen rapidly enlarges, infections perhaps accounting for the largest departures from normality. After puberty there is no significant change in size with age ($p = \gg 0·10$ for the buck, and $p = \gg 0·10$ for the doe); mean weights being 1·02 kg and 0·84 kg for the buck and doe respectively.

The thyroid gland

The thyroid gland is sometimes single, but is more commonly a bi-lobed body lying on either side of the larynx. This important endocrine organ has the property, among other things, of controlling the basal metabolic rate. This faculty seems mainly to be related to the attraction and mobilization of iodine as a hormone, probably as the tyrosine derivative tri-iodothyronine. The amount of hormone secreted is related to the body's demand for it, and may also be controlled by the nervous system through its control of the blood supply. The anterior lobe of the pituitary also produces a thyrotropic hormone resulting in symptoms of increased thyroid functioning. Being influenced by so many factors the gland fluctuates considerably in size, and changes may be observed during puberty or pregnancy, for example.

In the waterbuck, thyroid gland weight in the doe exceeds that in the buck from 2 years of age, the mean adult weights being 7·8 g and 6·1 g respectively, or 0·004% and 0·003% of total body weight. There is a tendency for some animals to produce large thyroids in old age, but it is not consistent. One 10-year-old doe was found to have a cystic thyroid weighing 19 g, of which 17 g was fluid, and a 12-year-old doe possessed an apparently healthy gland weighing 50·8 g (fixed weight). The latter doe was above average weight (192 kg), which one might expect in view of her age, but unable to lactate (see Chapter 5) otherwise both of these animals seemed to be healthy.

The mean weight of the gland in the doe during the final 2 months of pregnancy appears to be slightly greater than that of the previous 7·5 months, 8·4 g (SD 17·0) compared with 7·5 g (SD 10·7), fixed weights, but this was not significant at the 95% level of probability ($t = 1·66$, df 40, $p = > 0·8 < 0·9$).

The adrenal glands

Adrenal glands are paired bodies lying close to the kidneys, whose function in relation to stress is well known; they produce corticosteroids which elicit suitable reactions in the body concerned with "fight or flight". The medulla produces epinephrine, which effects changes in blood distribution, level of blood sugar, and other reactions; while the cortex responds to slower regulatory adjustments to maintain homeostasis, thus assisting the body to resist adverse conditions and infections. The adrenals are also involved in sexual functions; abnormal activity of the glands in man usually being associated with abnormal sexual development and functioning. They are also involved in the maintenance of the oestrus cycle.

It has been found in some species that, when an animal is under stress from some factor, such as overpopulation, the increased production of corticosteroids which ensues, causes the adrenals to enlarge. Significant changes in the size of the glands might therefore provide clues to population conditions.

In the waterbuck the adrenal gland weights for both sexes are comparable until the fourth year of life, with mean weights of 6·5 g and 6·6 g in the buck and doe respectively (fixed weight). After this, the doe's gland weight exceeds that of the buck by an average of 64%, with mean weights of 13·5 g and 8·7 g, a significant difference at the 95% level of probability ($t = 5·33, p = \ll 0·001$), and an indication of the important role of this gland in the doe's reproductive activity. In the doe the glands increase in size up to five years of age, and thereafter remain relatively constant in size, with no significant decline in old age (years 12 to 14), at the 95% level of probability ($t = 0·44, p = < 0·7 > 0·6$). The glands are, however, very variable in size with a standard deviation of 22·2% of the mean, over the age span of 6 to 11 years. Those of the buck are even more variable, with a standard deviation of 29·8% of the mean. In his fifth year, when a buck prepares to leave the bachelor group and enters into his first serious engagements with territory owners, when an increase in the size of these glands might have been anticipated, on average there is no significant difference at the 95% level of probability between this age group and those adults of 6 to 9 years ($t = 1·44, p = < 0·2 > 0·1$). Several individuals studied had glands of above average size; one for example had glands totalling 16·2 g, compared with the mean adult weight of 8·7 g (SD 2·59). Ten-year-olds show a significant increase in size at the 95% level of probability when compared with the 6- to 9-year-old group ($t = 3·12, p = < 0·01 > 0·001$). This could be attributable to stress conditions in the old animal, resulting from his attempts to maintain territory against the younger usurpers (Table 16).

In the doe there is no apparent enlargement during pregnancy when the last 2 months are compared with the previous 7·5 months, and with non-pregnant animals ($t = 0·972, df 44, p = > 0·6 < 0·7$).

Table 16 Mean adrenal weight (g) for age.

Age (years)		Buck				Doe		
	n	Mean weight	SD	Range	n	Mean weight	SD	Range
$\frac{1}{2}$	2	2·7	0·02	2·6 — 2·8	2	2·9	0·61	2·3 — 3·4
1	3	4·8	0·85	3·8 — 5·6	8	4·1	1·85	3·0 — 7·0
2	5	5·9	1·72	4·3 — 7·9	8	6·0	1·30	5·0 — 7·8
3	2	8·7	1·80	7·7 — 9·6	2	9·7	0·18	9·4 — 10·0
4	1	6·5	—	—	5	9·0	0·44	7·3 — 12·0
5	9	9·9	8·76	7·0 — 16·0	6	9·8	5·34	7·0 — 12·0
6	4	6·9	2·25	5·5 — 9·0	5	11·3	2·92	8·5 — 12·6
7	4	7·7	10·7	4·5 — 10·5	6	13·3	20·7	8·8 — 21·5
8	3	10·5	24·8	7·2 — 16·2	3	16·3	24·1	12·9 — 21·9
9	3	8·0	1·44	6·6 — 8·8	3	13·1	18·2	8·7 — 17·2
10	5	12·5	2·62	11·0 — 14·7	9	15·1	15·5	10·0 — 21·0
11	1	5·5	—	—	2	15·7	0·24	15·3 — 16·0
12					5	13·1	8·12	10·0 — 17·0
13					—	—	—	—
14					4	13·7	3·69	12·2 — 16·3

The kidney

Examination of the kidney is of interest because of the waterbuck's close dependence upon water, although its scrutiny portrays no evidence of this habit. When I plotted mean total kidney weights of 10 species of African ungulate, from the hydrophobic gerenuk to the hydrophilic hippo, against mean metabolic body weight ($W^{0.73}$), using a number of figures kindly supplied to me by H. Ledger, the points were found to approximate to an exponential relationship of the form:

$$Y_t = Y_o e^{\kappa g t}.$$

Substituting, this gave a value for Y of $2·548e^{3·067w}$, where Y was equal to kidney weight in g, and W was equal to metabolic body weight in kg. Thus animals of below about 30 kg adult body weight appear to have larger kidneys relative to their body size, and habit; whether the animal is dependent upon water or not, has no apparent part in it. The weight of the waterbuck's kidneys is close to the expected value, and probably an even closer correspondence with the function would be shown by all if adjustments were made to the exponential of 0·73 for metabolic body weight (Fig. 18). Rogerson (1966), for example, has drawn attention to the unlikelihood of this value representing the exponential for the wildebeest.

The waterbuck's kidneys have a mean weight in the buck of 404·4 g (SD 78·2) and in the doe 390·1 g (SD 76·5), equivalent to 0·2% of the total bodyweight; against an expected mean value for both sexes combined of 446 g; but we can see that there is considerable variation in weight.

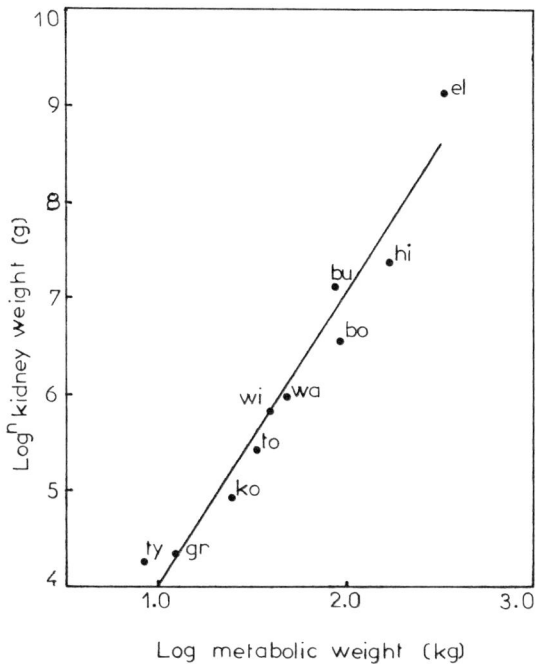

Fig. 18. Graph of $\log^n$ kidney weight against log metabolic weight. el — elephant, hi — hippopotamus, bu — buffalo, bo — boran steer, wa — waterbuck, wi — wildebeest, to — topi, ko — kob, gr — Grant's gazelle, ty—Thompson's gazelle.

Fat deposits

Fat deposits are an indication of animal health, implying when present, an adequate nutrition and lack of demand upon reserves, during certain periods of the yearly cycle; but in tropical ungulates this role is often misinterpreted. My observations on the waterbuck showed that fat deposits may be laid down in the dry season and mobilized in the wet, or vice versa. It is not simply laid down in the wet season for use in the dry season. The fat could be seen as an insurance against the failure of the rains at the end of the long dry season, only being used if the dry season was prolonged. In fact, when it rains after the long dry season, the fat deposits disappear within a matter of days, and upon visceral examination the animals look in poor health at this time, although the rains have brought on new food. This has often puzzled workers autopsying animals at this time; but fat mobilization may be due to the "green-grass-loss", to which reference has already been made.

Ledger (1968) was the first to point out that tropical ruminants do not lay down their fat within the muscle, as cattle do. This is what gives game meat its well known "dry" nature; but Ledger further considered that there was no evidence to indicate that tropical ruminants compensated for low carcase fat by increased deposits of internal fat; such as the pericardial, omental and mesenteric deposits. But this conclusion seems to stem from the fact that Ledger's sampling of game animals was restricted to one period, such that the changing nature of fat deposits through the seasons could not be properly appreciated. The want of intramuscular and subcutaneous fat in tropical ungulates explains the farmer's traditional complaint that wild animals always look fat and healthy in a drought, when cattle, on the other hand, may be in poor condition. Although the wild animals are better adapted to make the best use of their environmental resources, there often may not in fact, be much to choose between the cattle and game in regard to their fat deposits at such times. The difference is that the cattle mobilize it from where it can be seen, under the skin and in the muscle, thus leaving the animal with a shrunken, flabby look. To attain the same appearance the wild animal would needs be using up its muscle reserves, a strategy of a relatively advanced state of starvation. Its fat is instead mobilized from sites which do not alter its outward appearance, around the heart, kidneys, and mesenteries. As to why this difference in fat location exists, Ledger (1959, 1968) has implied that it is related to the physiology of heat dissipation, temperate cattle, for example, by laying down fat under the skin have an insulator against low temperature. The tropical animal stores it where it is not going to form an insulator, which would interfere with heat loss.

These fat deposits can be measured in two simple ways: (1) by weighing the omentum fat which is found adhering to the mesenteries covering the rumen and the intestines, and (2) by weighing the fat deposited around the kidneys. Fat seems to be deposited around the heart last of all. In the waterbuck the omentum appears to be the most important site, carrying nearly three times as much fat as the kidneys. Does have consistently more fat than bucks, a pattern probably related to pregnancy. The mean weight of the omentum fat in the pubertal buck was

159 g, and in the doe 363 g; which compared with mean kidney fat weights of 58·5 g and 133·1 g respectively.

Some workers have made use of a "condition index" (a review is given by Smith, 1970), expressing fat content as a percentage of the total body weight, or as a percentage of the fat weight divided by the weight of the encinctured organ; for example, perinephric fat divided by kidney weight.

A condition index of omentum fat expressed as a percentage of total body weight shows no significant change with age in the adult buck ($t = 1·495, p = < 0·8 > 0·9$); and no significant change in quantity in wet and dry seasons ($t = 0·023, p = \gg 0·9$), although the sample size of the latter was small. The doe also shows no evidence of change with season ($t = 0·649, p = < 0·6 > 0·5$); neither does she show any significant change during pregnancy, and there is no difference between pregnant and non-pregnant animals (Tables 17 and 18). It comes as a surprise that this important fat deposit shows no apparent relation to season or to pregnancy; but this could be due to its diffuse nature, which makes it difficult to collect accurately.

A more positive relationship with season can be shown by the kidney fat index. Unfortunately there were no buck samples from the long dry season of December to February, except for the first month; but the April fat deposits, coincident with the time of "green-grass-loss", were significantly lower than those of September, during the height of the long rains. April's deposits were almost significantly different, at the 95% level of probability, from those of June and December, these months signifying the beginning of the short and long dry seasons respectively. These months also coincide with the months of peak conceptions in the

Table 17 Seasonal omentum fat index $\left(\dfrac{\text{weight of omentum fat} \times 1000}{\text{total body weight}}\right)$.

Season	n	Mean	Range	SD	t	p
Buck						
Aug–Nov; Mar–May (Wet)	17	0·844	5·291 — 0·275	1·356		
Dec–Feb (Dry)	8	0·683	1·744 — 0·039	0·256	+ 0·372	< 0·8 > 0·7
Jun–Jly (Dry)	10	0·990	2·022 — 0·311	0·288	+ 0·372	< 0·8 > 0·7
Total Dry	18	0·844	2·022 — 0·039	0·257	+ 0·023	≫ 0·9
Doe						
Aug–Nov; Mar–May (Wet)	38	2·130	7·489 — 0·472	6·245		
Dec–Feb (Dry)	13	1·572	3·539 — 0·338	0·771	+ 1·246	< 0·3 > 0·2
Jun–Jly (Dry)	4	2·146	4·034 — 0·912	1·907	− 0·020	≫ 0·9
Total Dry	17	1·859	4·034 — 0·338	2·323	+ 0·649	< 0·6 > 0·5

Table 18 Omentum fat index related to stage of pregnancy in the doe.

Fetal weight (kg)	n	Mean	Range	SD	t	p
(i)						
0[a]	10	1·425	3·784 — 0·338	1·043		
0·001 — 0·453	13	1·615	4·043 — 0·489	1·437	+ 0·401	< 0·7 > 0·6
0·454 — 3·59	11	2·343	4·034 — 0·753	1·418	− 1·887	< 0·1 > 0·05
3·60 — 7·29	8	2·463	7·489 — 0·472	5·027	− 0·785	< 0·5 > 0·4
7·30 — 11·0	8	2·466	4·173 — 0·932	1·418	− 1·818	< 0·1 > 0·05
11·1 — 14·5	4	2·124	3·432 — 0·379	2·369	− 1·008	< 0·4 > 0·3
Total pregnant does	54	2·202	7·489 — 0·379	2·334	− 0·777	< 0·2 > 0·1
(ii)						
11·1 — 14·5	4	2·124	3·432 — 0·379	2·369		
7·30 — 11·0	8	2·466	4·173 — 0·932	1·418	− 0·428	< 0·7 > 0·6
3·60 — 7·29	8	2·463	7·489 — 0·472	5·027	− 0·269	< 0·8 > 0·7
0·454 — 3·59	11	2·343	4·034 — 0·753	1·418	− 0·293	< 0·8 > 0·7
0·001 — 0·453	13	1·615	4·043 — 0·490	1·043	+ 0·699	< 0·5 > 0·4

(i) Non-pregnant does compared with stages of pregnancy.
(ii) Stages of pregnancy compared.
[a] Does greater than 2 years of age.

doe (see Chapter 6), and thus imply months of greater sexual activity on the part of the bucks (Table 19, Fig. 19).

Table 19 Seasonal kidney fat index $\left(\dfrac{\text{weight of epinephric fat} \times 100}{\text{total kidney weight}}\right)$.

Season		n	Index	SD	t	p
Buck						
Mar–May	(Wet)	8	12·67	17.27		
Jun–Jly	(Dry)	13	18·34	17·95	− 0·713	> 0·5 < 0·6
Aug–Nov	(Wet)	14	14·89	9·64	− 0·390	> 0·2 < 0·3
Jun–Jly	(Dry)	13	18·34	17·95		
Aug–Nov	(Wet)	14	14·89	9·64	0·629	> 0·4 < 0·5
Doe						
Dec–Feb	(Dry)	11	29·82	25·27		
Mar–May	(Wet)	24	19·52	12·21	1·640	> 0·8 < 0·9
Jun–Jly	(Dry)	11	20·47	14·83	1·058	> 0·6 < 0·7
Aug–Nov	(Wet)	24	33·56	27·21	− 0·386	> 0·2 < 0·3
Mar–May	(Wet)	24	19·52	12·21		
Jun–Jly	(Dry)	11	20·47	14·83	− 0·120	> 0·1 < 0·2
Aug–Nov	(Wet)	24	33·56	27·21	− 2·306	> 0·95 < 0·98
Jun–Jly	(Dry)	11	20·47	14·83		
Aug-Nov	(Wet)	24	33·56	27·21	1·489	> 0·8 < 0·9

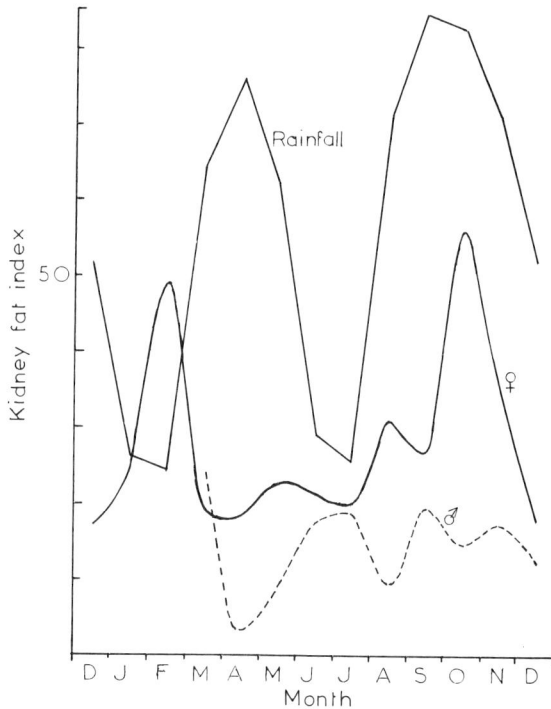

Fig. 19. Mean monthly kidney fat index for bucks and does, with mean monthly rainfall.

In the doe, climate appears to be more important than pregnancy in determining kidney fat levels. When non-pregnant does were compared with does at different stages of pregnancy, and with total pregnant does. there was no significant difference at the 95% level of probability. But there was a significant difference between the first stage of pregnancy and the final one ($t = 2.461$, df 16, $p = > 0.95 < 0.98$) (Table 20). Thus, although there is some relation between pregnancy and kidney fat levels, climate appears to be the over-riding factor; as was demonstrated by the fact that the April deposits were significantly lower than those of February, October and November, at the 95% level of probability. Those of the short rains as a whole, March to May, were significantly lower than those of the long rains of August to November, but not from those of the short and long dry seasons.

At first these results appear to be anomalous, but it would seem that fat deposits in the doe build up steadily during the long rains to a peak in October, then decline as the rains diminish; but respond rapidly to the first fresh flushes at the onset of the first rains (or perhaps to fresh flushes

Table 20 Kidney fat index related to stage of pregnancy in the doe.

Fetal weight (kg)	n	Mean	Range	SD	t	p
(i)						
0[a]	9	28·84	9·38 — 94·86	27·04		
0·001 — 0·453	14	21·56	1·33 — 37·78	11.40	0·899	> 0·6 < 0·7
0·454 — 3·59	8	33·96	11·91 — 50·0	14·0	− 0·480	> 0·3 < 0·4
3·60 — 7·29	8	32·35	9·68 — 56·25	17·38	− 0·314	> 0·2 < 0·3
7·30 — 11·0	8	39·87	10·81 — 100·0	30·36	− 0·793	> 0·5 < 0·6
11·1 — 14·5	4	49·12	14·58 — 100·0	38·96	− 1·098	> 0·7 < 0·8
Total pregnant does	42	32·09	1·33 — 100·0	21·76	− 0·390	> 0·3 < 0·4
(ii)						
11·1 — 14·5	4	49·12	14·58 — 100·0	38·96		
7·30 — 11·0	8	39·87	10·81 — 100·0	30·36	0·455	> 0·3 < 0·4
3·60 — 7·29	8	32·35	9·68 — 56·25	17·38	1·061	> 0·6 < 0·7
0·454 — 3·59	8	33·96	11·91 — 50·0	14·0	1·017	> 0·6 < 0·7
0·001 — 0·453	14	21·56	1·33 — 37·78	11·40	2·461	> 0·95 < 0·98

(i) Non-pregnant does compared with stages of pregnancy.
(ii) Stages of pregnancy compared.
[a] Does greater than 2 years of age.

induced by burning at the end of the dry season). However, as a result of the long dry season, when the rains break, the animals suffer the "green-grass-loss", and deposits decline to a minimum. There is then little increase in the short dry season, deposits increasing again with the first long rains in August. A check in September may reflect a secondary "green-grass-loss" of less magnitude (Table 19, Fig. 19). On the other hand, these declines with the rains may be related to the peaks in birth at these times.

We must remember, when considering these results, that the waterbuck in the study area inhabit a relatively benign environment, and no excessively harsh seasonal conditions were experienced that might put the animals to test. We shall see also that the waterbuck is able to maintain the quality of its food between the seasons, by using higher proportions of browse to graze when the latter is low in protein; and water, of course, is never a limiting factor.

The Horns

Compared with some species of ungulate in which the male is horned, for example the impala, whose horns first appear at 4 months of age, the horns of the waterbuck make their appearance relatively late. The points of keratinization first become evident at about 8 months of age, confirming the observation of Jackson (1926). Thereafter, they grow rapidly, at an average rate of 16 cm per year until the animal is 5 years old; representing an annual specific growth rate of 44·5%. The rate of growth then slows abruptly to approach an asymptotic level of 76·3 cm at 6 years of age, a length which would be considered as of "record" size in some races (Table 21). This cessation of continued increase in length is not due to a halt in growth; growth continues, albeit at a greatly retarded rate, but tends to be matched by wear at the tips. In the four specimens which I examined, wear at the tips, matched by growth in length, averaged 0·59 cm per year. In old age, estimated at 11 years and over, growth does tend to cease altogether, or at least is no longer measurable, but wear continues so that very old animals have increasingly shortened horns. This wear results from the buck's habit of horning the earth to keep the points sharp. Although this was an activity which was not often seen, the rate of wear on the tips testifies to its frequency of occurrence, as also does the frequent presence of mud on the horn tips during the wet seasons. Naturally, a non-aggressive buck which does not indulge regularly in this activity, is likely to produce a more sought-after trophy from the sportsman's point of view.

Table 21 Horn growth (cm).

Age (years)	n	Mean length	Increase per year	SD	Ridges	Increase per year	SD
1	1	8·25	8·25	—	0	0	—
2	3	25·60	17·35	1·3	1·3	1·3	—
3	5	49·60	24·0	12·2	12·2	10·9	2·3
4	5	61·75	11·15	18·6	18·6	6·4	3·8
5	8	74·96	13·21	23·3	23·3	4·7	2·9
6	12	77·10	2·14	25·6	25·6	2·3	1·3
7	11	76·90	− 0·20	25·4	25·4	− 0·2	2·3
8½	13	75·30	− 1·60	26·4	26·4	1·0	2·5
10	14	76·20	0·90	26·9	26·9	0·5	2·6
11 +	5	76·20	0	24·4	24·4	− 2·5	3·0

Trophies, as the sportsman terms his abnormally long horns, are derived in two ways: either from lack of wear as described above, or from a genetically determined greater-than-average length, in which case, the younger the adult the greater the chance of finding long horns. In the former case it is the older buck which has the longer horn. The third longest pair of horns, which were found in the Park in 1967, were measured in the dry state (as shrinkage occurs when fresh horn dries out) as 97·1 and 99·5 cm on the curve, left and right respectively; although this is now given in Best and Best (1977) as 99·06 cm, greatest length. This animal was in the "weak buck" category, having an estimated age of greater than 11 years. The second "world record" also comes from a skull found in the Park, in 1958, measuring 99·38 cm greatest length (Best and Best, 1977) but the jaws were unfortunately not available for examination. Pride of place as the "world record" holder goes to a length of 99·7 cm, for a waterbuck reported as coming from Toru. These animals, however, probably represent a combination of inherited greater-than-average length as well as a lack of aggressive disposition, since their horns are far greater than the theoretical average possible length in the absence of wear, which would be only 79·3 cm in a 12-year-old buck. Generally one would expect the "weak buck" category to be much rarer in occurrence than the genetically determined long horned bucks, as natural selection operates against them. Table 22 shows indeed how the latter category produces more long horned individuals, and how these horns are found in the youngest adult age group (5 to 6 years), before they become worn with use.

Table 22 Percentage frequency distribution of adult horn length with age.

Horn length (cm)	Age (years)				
	5–6	7–8	9–10	11 +	Mean
71 — 75	26·0	17·5	55·0	16·5	29·5
76 — 80	35·0	53·0	25·0	44·5	38·5
81 — 85	26·0	29·5	20·0	33·5	26·9
86 — 90	13·0	0	0	0	3·9
91 — 95	0	0	0	0	0
96 — 100	0	0	0	5·5	1·2

The horns of the waterbuck are delicately poised rapiers, able to inflict a mortal wound with their sharpened points and 250 kg of weight behind them; but they can only withstand such a strain if they are properly matched. The bow of the horns may vary, but it must be correctly balanced. Animals with irregularly matched horns, a not uncommon defect being for one horn to be twisted slightly outwards, almost invariably have the tip of the asymmetric horn broken off.

On the anterior face of the horn are prominent ridges; strictly speaking they are not annulations, as they are sometimes called, for they do not encircle the horn. These ridges are not, at least at this latitude, related to seasonal variation in growth rate, but are related directly to the length of the horn, forming at an average rate of one for every 13·6 cm of length, and increasing rapidly to one for every 2·1 cm of horn as it becomes more curved. The mean adult number is 25·7 ridges. Except in the earliest stages of growth they are of no help in determining the age of the animal, although at this stage length itself is a suitable indicator (Fig. 20, Plates 13 to 18). The function of these ridges is not decorative, despite their elegant

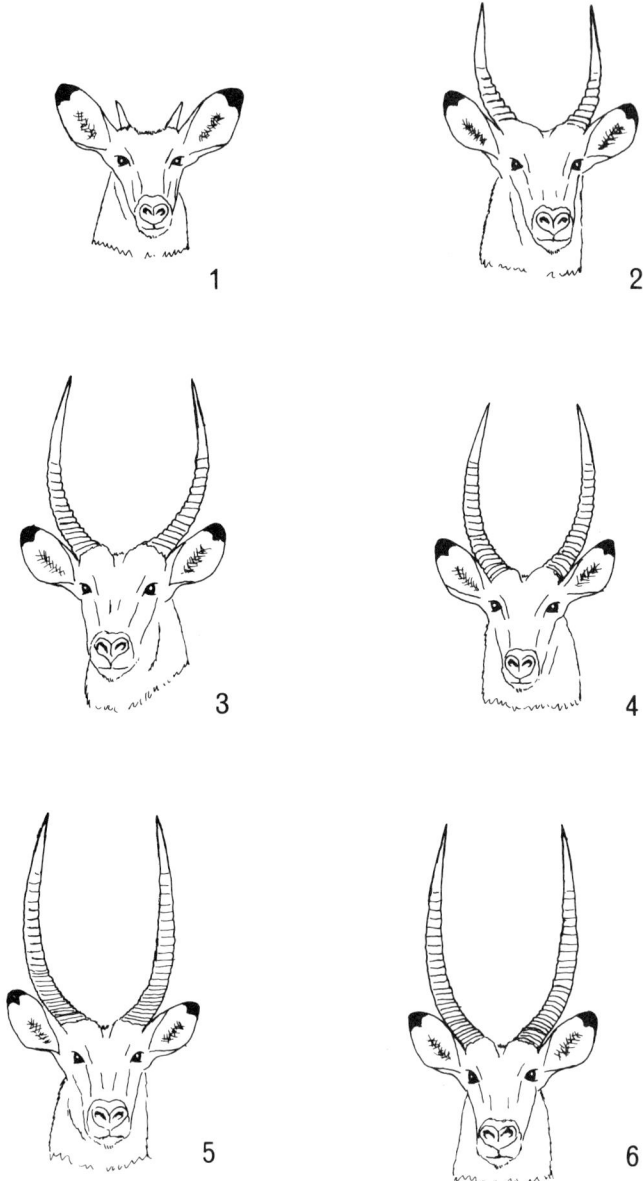

Fig. 20. Growth in horn length of waterbuck from 1 to 6 years.

Plate 13. A 9-month-old buck with the horns just appearing.

Plate 14. 10- and 11-month-old bucks.

Plate 15. A 2-year-old buck.

Plate 16. A 5-year-old buck, still in the bachelor group. Compare the body conformation with the adult in Plate 1.

Plate 17. A horned doe.

Plate 18. A doe with completely male-like horns, one of which had been broken off. Picture by F.E. Guinness.

appearance, but appears to be mechanical, containing compression stresses when the horn is thrust against a resistant object, as it normally is during fighting. At the same time, they may aid in deflecting the points of an opponent's horns from a target point. If the horns were smooth, they would slide against each other in combat; but because they are ridged, when one is rubbed along the length of the other, vibration is set up, which is greatest at the tip of the horn. In contrast to those of the waterbuck, the horns of the sable antelope curve backwards, but still possess ridges on their anterior aspect. In order to use the points of its horns in fighting, the sable antelope must hook with them; in so doing it will set up stresses on the outer curve of the horn which will be resisted by the ridges in the same manner as they are in the waterbuck, in both cases preventing stretching and fracture of the opposite surface (Fig. 21).

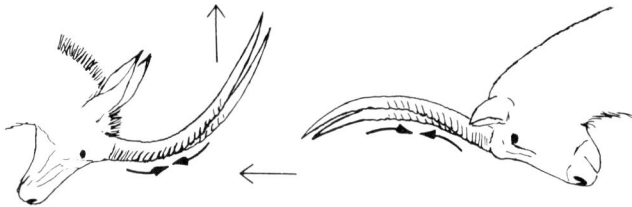

Fig. 21. Fighting attitudes of the waterbuck and sable, showing how, although the horns are inversely curved in the two species, the ridges oppose stresses in the same way.

Freak horns, usually deformed in structure, occasionally occur in does. Jenkins reported such a case in 1929 for the common waterbuck (Jenkins, 1929), and I observed a similar one in the Kayanja area. The horns appeared to lack a bony core, wobbling as the animal ran. After my study had finished, I happened to return to the Park more than a year later, when a horned doe was reported to me. I found the animal to have one full-length horn, the other having been broken off. To all intents and purposes a buck, its sex was determined by its manner of urination. This animal, which appeared to be fully adult, must have led an extraordinarily unhappy life, if one may be permitted the use of such an anthropomorphic term, being attacked by adult bucks wherever it went. The broken horn was no doubt the result of such an encounter, and suggests that the bucks react more to what they see, rather than to the smells and behaviour of the does (Plates 17 and 18).

Almost all doe skulls bear small bony rudiments at the site of the buck horns. These are already present by 2 years of age, and may sometimes be pronounced, showing no apparent increase in size with age (Table 23).

Table 23 The occurrence of horn rudiments in does.

Age	Sample size	% occurrence	Size frequency		
			*	**	***
1	14	0	0	0	0
2	14	14	2	0	0
3	5	80	4	0	0
4	17	77	12	1	0
5	11	82	7	1	1
6	16	94	10	5	0
7	7	86	6	0	0
8–9	13	100	10	3	0
10	19	84	10	6	0
11	4	100	2	2	0

* = small.
** = medium.
*** = large.

The Teeth

The dentition of the waterbuck is of the typical pecoran hypso-selenodont form, having a complement of 32 permanent teeth; incisors and canines being always lacking in the upper jaw. The deciduous dental formula is:

$$\mathrm{Di}\frac{0}{3} \quad \mathrm{Dc}\frac{0}{1} \quad \mathrm{Dp}\frac{3}{3}$$

and the permanent is:

$$\mathrm{I}\frac{0}{3} \quad \mathrm{C}\frac{0}{1} \quad \mathrm{Pm}\frac{3}{3} \quad \mathrm{M}\frac{3}{3}.$$

Departures from the dental formula are rare.

In the waterbuck the deciduous first incisor cuts the gum about 2 days after birth, and is followed by the appearance of the remaining teeth of the deciduous dentition within one week. Of the permanent dentition, M_1 is the first to erupt, appearing after one year, followed by M_2 after 2 years. In the third year, I_1, Pm_2 and M_3 all erupt, followed within 6 months by I_2, I_3, C, Pm_3 and Pm_4. Thus almost all replacement takes place between the ages of 3 and 3·5 years, which, as we have seen, probably has a significant effect upon the animal's rate of growth.

In the bovids the time of eruption of the permanent dentition probably has some relation to the size of the animal. It is, for example, much earlier in the small impala (Child, 1964), and later in the larger buffalo (Grimsdell, 1973). Sinclair (1977) found some difference in replacement times between the upper and lower jaws of the buffalo. No such difference was found in this study, but no captive animals were available to follow eruption stages in detail. Some workers, for example Caughley (1977), consider eruption timing unsuitable as a means of age determination, due to the variation which some species show. The nutritional plane does seem to affect eruption timing, but the value of the timing as a means of age determination depends upon whether one is calculating age to the nearest month, or to the nearest year, the variation usually being expressible in months.

Age changes in teeth

Unlike the premolars, which erupt under the deciduous molars, pushing the latter out, the permanent molars erupt from behind and move forward in the jaw. This mesial drift continues throughout life in herbivorous ungulates, causing pressure upon the juxtaposed tooth faces which results in abrasion and shortening of the tooth-row. This has the important effect of maintaining a united grinding surface. In a single individual it may be possible to follow this shortening throughout an animal's lifespan, but in the waterbuck the overall variation in mandibular and maxillary tooth-row lengths is too great to identify any significant changes in length of tooth-row with age. But significant changes in weight take place, the combined weights of buck and doe tooth-rows from one side of the jaw decreasing from a mean weight of 125 g at the age of 5 years, to 62·2 g at the age of 12 + years. This decrease is due primarily to the attrition of the crowns, but in senescence there is also a rather rapid resorption of the roots. The incisors always decrease in proportion with the cheek teeth, maintaining a ratio of 4·8% of the weight of the latter. When the cusps of the incisors become flattened, the animal tends to pull the grass against the sides of the first incisors, producing grooves in the unprotected dentine, which may eventually undercut the crown (Fig. 22).

A thick scale may form on the teeth, and this often becomes burnished, with an appearance like gold. I have also noted this in Grant's gazelle in Tanzania. It seems not uncommon among grazing animals, for in 1609 the mediaeval traveller William Lithgow reported that sheep on Mount Ida in Crete, had their teeth "gilded like to the colour of gold" (Phelps, 1974). I did not examine this phenomenon in detail, but the golden appearance is probably due to reflection from polished silica bodies derived from the grass.

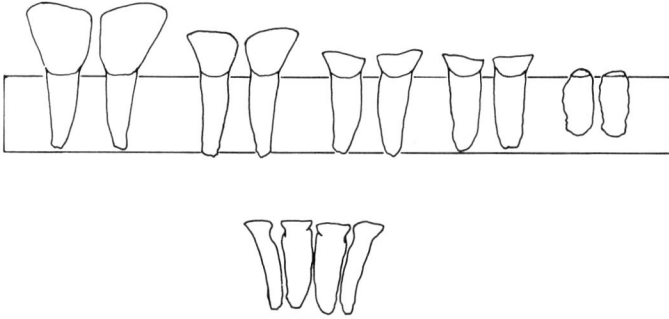

Fig. 22. Wear of the first incisors, showing root absorption in old age, and the undercutting of the crowns.

Morphological Changes with Age

The coat of the newborn fawn is fluffy in appearance, and lighter in colour than that of the adult. Its hair changes between the sixth and seventh month to the coarser textured hair of the adult, with long hair developing on the neck. Flecks of grey hair, differing from the silver-grey haired variants, develop at about 10 years of age. Old does may also exhibit large, bare callosities on the belly. None were found in does under 9 years of age, but they occurred on 94% in those animals examined of 14 years and older. One 12-year-old also bore callosities on her foreknees. No such callosities were found on bucks.

Longevity

Recorded longevity in the New York Zoo of a defassa buck was 14 years, and for a common buck 16·5 years (Crandall, 1965). Flower (1931) stated that of 40 animals selected from the records of the longest-lived kobs, the average longevity was 9 years and 2 months; but 16 animals were still alive. Unfortunately he did not distinguish between lechwes, kobs and waterbucks, but concluded that 10 to 12 years was not uncommon, and that records of animals over 12 years were rare. Four records representing the longest-lived animals were:

K. defassa: Giza Zoological Gardens, maximum for a buck 13 years, 1 month, 22 days; maximum for a doe, 16 years, 8 months.

K. defassa unctuosus: London Zoo, longest lived doe, 15 years, 2 months, 25 days.

K. d. unctuosus × *K. ellipsiprymnus*: London Zoo, longest lived buck, killed at 14 years, 6 months, |12| days.

These records all accord well with the expected longevity deduced from the lifespans observed in the wild. Exceptions naturally occur, and one Peninsula doe which I observed for 11 months before she died from natural causes, had an estimated age from her tooth wear of 9 to 10 years. A cementum line count gave a suggested age of 18·5 years, which I believed to be correct. This animal had every appearance of advanced senescence; she had many grey hairs, was gaunt, and suffered acute stiffness at the onset of the rains in March 1965. A veterinary colleague who saw her at that time commented that she was a very sick animal, and would not survive the wet season. But she did, recovering from her stiffness as the rains progressed, to live out another season, calving into the bargain. I saw no comparable animal in the Park, and believe her to have been an exceptionally long-lived specimen, probably surviving to beyond normal age because her teeth were harder than average.

5. Reproduction and Reproductive Behaviour

Reproductive Anatomy and Physiology

The buck

From each autopsied buck, the testes, epididymides and seminal vesicles were weighed at collection; portions were then fixed in formalin or Bouin's fluid, and later sectioned and stained for histological examination, using standard procedures.

The penis of the buck was described for the first time by Gerhardt (1933), and is of the fibro-elastic type similar to that of the ram. Its glans is pigmented and bears the urethral orifice at the extremity of a *processus urethrae,* which is erected by vascular pressure. The adult's paired testes have a mean weight of 147·9 g (SD 20·25), and the epididymides have a mean combined weight of 37·3 g (SD 36·0). Adult testis weight is reached in the fifth to sixth years, and there is no apparent decline in weight with age. The seminal vesicles are paired, lobulated bodies, with a mean weight of 21·3 g (SD 14·7), in animals of 6 years and older. The ampullae of the *vasa deferentia* (which enlarge in some species during phases of sexual hyperactivity), average 6·6 g (SD 4·6) in the adult.

The fetal testes descend to the scrotal entrance by mid-gestation, where they remain until after birth. In a 217-day fetus the germinal cells are similar in size to those found in a young buck in which spermatogenesis has not yet commenced, but the fetal gonad contains more interstitial tissue. Six months after birth the primary spermatocytes are in division, and by one year of age some spermatozoa can be found, while the seminiferous tubules have developed lumina, but there are, as yet, no contents in the epididymides. It is not until 6 months later, when the animal is 18 months old that a few spermatozoa may be found.

Spermatozoa continue to be rare until 3 years of age, when most lumina then have them, and the epididymides become packed. At 4 years of age all tubules show active spermatogenesis, and this showed no decline in the oldest animals examined (11 years), the epididymides always being tightly packed with spermatozoa.

When examined on a seasonal basis, testis tubule diameter showed no significant difference between any month ($p = > 0.05$); but testis weight did show significantly low weights in July, the driest month of the year ($p = < 0.05 > 0.025$) (Table 24).

Table 24   Mean paired testes weights (g) per month of collection.

Month	Wet/dry	Weight $+$[a]	SD	n	Weight $-$[b]	SD
Jan	Dry	215	450	2	175	450
Feb	Dry	170	1800	2	135	1250
Mar	Wet	220	2600	4	158	1292
Apr	Wet	190	0	3	163	133
May	Wet	—	—	—	—	—
Jun	Dry	180	0	2	160	0
Jly	Dry	153	533	3	125	945
Aug	Wet	183	1233	3	155	1075
Sep	Wet	150	0	2	125	50
Oct	Wet	200	0	1	150	0
Nov	Wet	210	0	1	190	0
Dec	Dry	215	1250	2	155	50

[a] Weight $+$ = testes weight plus epididymides.
[b] Weight $-$ = testes weight minus epididymides.

The doe

The reproductive tract of the doe is hitherto undescribed, but is similar to that of the ewe. It had a mean weight of 230 g in non-pregnant animals. The surfaces of the paired ovaries are smooth, interrupted by the pink swellings of the corpus luteum when present, and the smaller, vesicle-like swellings of the follicles.

In common with many other mammals, the waterbuck exhibits hypertrophy of the fetal feminine gonad, accompanied by the development of follicles with antrum formation; a phenomenon apparently resulting from the passage of maternal oestrogens and gonadotrophins across the placental barrier. The ovaries of a 51-day *ante-partum* fetus were found to be small in size, consisting of germinal cells and connective tissue. In contrast, those of a fetus 33 days older, showed a vigorous development of Graafian follicles, and the ovaries weighed 0.5 and 0.49 g.

Regression of the ovary apparently occurs in the neonate, and a 5-month-old fawn had ovaries of only 0·26 and 0·21 g, left and right respectively (Plate 19). Although Graafian follicles were present, these ovaries had a greater amount of connective tissue than did the fetal ovaries. At 6 months of age a young doe had ovaries of 0·6 and 0·35 g, left and right, with much larger follicles.

No animals less than one year old ($n = 9$) showed any signs of *corpora albicantia*, which would have indicated that ovulation had taken place; and they were found in only two out of 12 animals aged from 1 to 2 years.

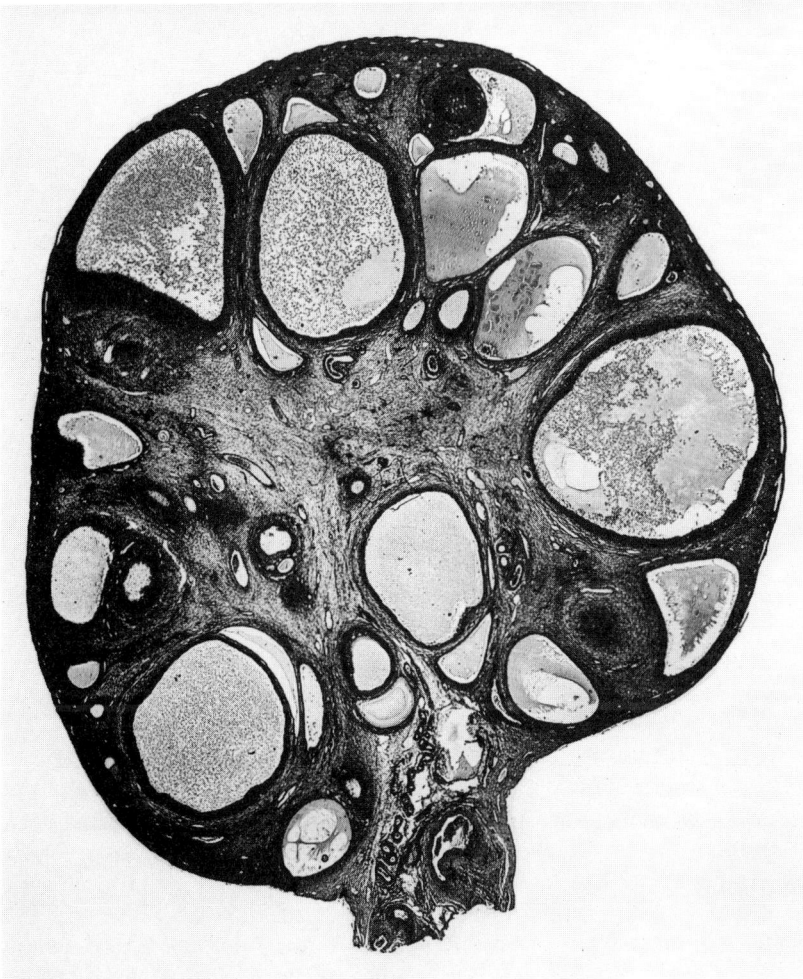

Plate 19. Hypertrophy of the female fetal gonad; Graafian follicles in the ovary of a 194-day-old fetus.

In the first year the mean paired ovary weight was 1·02 g, rising to 2·3 g in the second year. Pregnant does had ovaries exceeding 3 g in weight. By the third year, the mean adult ovary weight of 2·81 and 2·36 g left and right respectively is reached; the left being significantly larger than the right ($n = 80$, $p = < 0·05 > 0·025$), a disparity which seems to have commenced at the time of the regression of the hypertrophied fetal gonad.

Follicular development appears to remain constant throughout life, no decline being detected in old animals; it is also constant throughout most of the period of gestation. But the number of enlarging follicles, exceeding an arbitrary 5 mm in diameter, which average 1·4 per pair of ovaries, changes during gestation. The number rises at the beginning of pregnancy to a mean of 1·8, then declines steadily to almost zero at full term, developing again in the *post partum* period (Table 25). Unlike in some mammals, therefore, lactation does not inhibit follicular development, for some 50% of the does examined were lactating, and almost all multiparous does would be if they had not lost their fawns.

Table 25 Development of follicles and the corpus luteum of pregnancy.

Days of pregnancy	n	Number of follicles > 5 mm diameter	Range	SD	Mean diameter of corpus luteum (mm)	Range	SD
0	10	0·7	0–3	1·0	11·5	6a–18	12·1
0–56	6	1·7	1–2	0·3	15·6	14–19	3·3
57–113	18	1·8	1–3	0·4	14·9	13–17	1·6
114–170	8	1·3	0–2	0·6	14·6	13–16	1·7
171–227	17	0·7	0–2	0·7	15·4	13–18·5	2·1
228–284	19	0·05	0–1	0·05	15·4	12–22	6·1
Post partum	3	0·7	0–1	0·3	12·0	11–14	3·0

[a] Recent ovulation, egg in 2-cell stage.

The Graafian follicle is about 6 mm in diameter immediately after extrusion of the ovum, and the corpus luteum, the endocrine organ which forms in the Graafian follicle after ovulation, is 6 mm in diameter when the fertilized ovum is in the 2-cell stage. By the time that the 4- to 8-cell stage is reached, it attains a diameter of 10 mm (Plate 20). It is maximal in size by the 56th day after ovulation, and then remains constant in size to full term; regressing rapidly after parturition (Table 25); as Morrison (1971) found in the Uganda kob. In this respect these species differ from cattle, which retain the full size of the corpus luteum

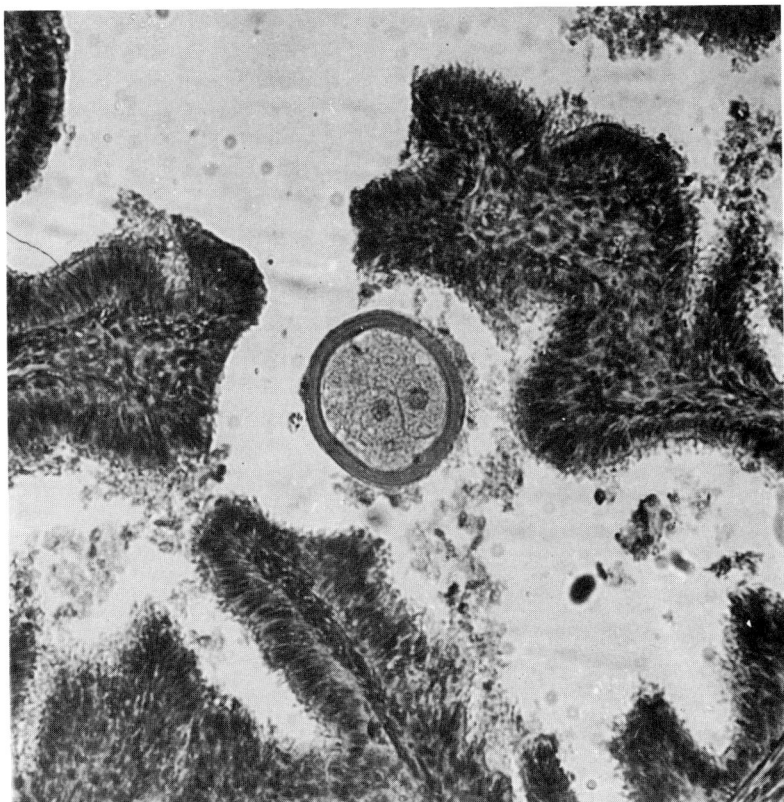

Plate 20. A waterbuck egg in the 8-cell stage.

for 30 to 60 days after parturition (Hammond, 1927). The corpus luteum produces the steroid hormone progesterone, which prepares, and maintains, the body for pregnancy. The gland's consistent size throughout pregnancy indicates that it has an active role throughout this period, unlike in some antelopes, such as the Grant's gazelle, where, once pregnancy is established, the corpus luteum degenerates (Spinage, unpublished), its role being taken over by another site; probably the placenta. In some species corpora lutea form in the absence of fertilization, but no such accessory corpora lutea were detected in the waterbuck.

Scars of corpora albicantia have been used by some workers to indicate the number of pregnancies that a female has undergone, the scar representing the site of a regressed corpus luteum. My impression was, that whether the corpus albicans scar persisted or not in the ovary, was due to chance; dependent upon whether or not other corpora lutea formed close to it, in which case they eventually obscured it. Counts of corpora albicantia scars ranged from zero in a 9-year-old animal, to 10 in a 10-year-old. There was no correlation between the number of observed corpora albicantia and the expected number of pregnancies (Table 26). Animals of 10 years and older have tended to have more corpora albicantia scars as these animals apparently ovulate more frequently without fertilization taking place. Nevertheless does are capable of breeding throughout life, and even the 18·5-year-old doe we saw gave birth 2 months before her death, although she failed to rear the fawn.

Table 26   The mean number of corpora albicantia scars each year of life against the expected number of pregnancies.

Age (years)	Sample size	Mean number of scars	Range	Expected number of scars[a]
1	9	0	0	0
2	12	0·4	2 — 3	0
3	3	0·7	0 — 2	1
4	8	1·5	0 — 3	2
5	9	2·0	0 — 6	3
6	7	1·5	0 — 3	4
7	11	2·0	0 — 4	5
8	3	2·0	1 — 3	7
9	6	3·0	0 — 5	8
10	9	4·0	2 — 10	9
11	4	2·5	1 — 4	10
12	5	4·5	3 — 8	11
13	1	4·0	—	13
14	7	4·0	1 — 6	14

[a] Assumes a 10-month interval between pregnancies.

In domestic species a "silent oestrus" is normally experienced, in which the animal undergoes ovulation without overt sign, so that the male is not attracted for mating. Robinson (1951) suggests, that in the ewe, a waning corpus luteum is necessary for the full manifestation of oestrus. To confirm or deny the occurrence of silent oestrus in the waterbuck, two marked does, both of which had lost their fawns, were killed at 61 and 62 days *post partum* respectively. The first, of an estimated age of 10 years, was found to have had a recent ovulation, signified by the presence of a corpus luteum of 11·5 mm in diameter. The corpus albicans of the previous ovulation, almost certainly the first *post partum* ovulation, measured 6·5 mm in diameter; thus this animal could have undergone a silent oestrus, or may simply have failed to conceive at normal oestrus. The second doe was estimated to be 12 years old, and was bearing a fetus calculated to be 43 days old. Her ovaries had a corpus luteum of 13 mm diameter, and eight corpora albicantia scars. The absence of a corpus albicans suggested that she must therefore have been fertilized in the first oestrus following parturition. Two other does examined, both in their first pregnancies, with fetuses of 72 and 176 days, bore no corpora albicantia scars. The evidence from the 12-year-old animal suggests that conception could take place at the first *post partum* oestrus, and that from the two primiparous does suggests that conception could be effected at the first oestrus. There is no evidence of the regular occurrence of a silent oestrus, a phenomenon which may only be necessary in seasonal breeders required to come into condition again after interruption of the breeding cycle by a long anoestrus phase. In the waterbuck, at the Equator, there may be only a comparatively brief interval between the decline of the corpus luteum of one pregnancy, and the growth of the next; as little as 45 days.

Counts of corpora lutea, plus corpora albicantia, per ovary, show that the left ovary is 1·2 times more active than the right; which is also confirmed by its larger size. But of 47 animals examined, implantation had taken place in the left uterine horn in four cases, and in the right in 43; a preference for implantation in the right of 92%. Unilateral implantation is not unusual in wild ungulates, some advantage apparently being gained by the fetus developing on the opposite side to the rumen. This has also been reported in such species as the dik-dik (Kellas, 1954), the Uganda kob (Buechner, 1961a), and the impala (Mossman and Mossman, 1962). The situation in the African buffalo differs markedly from that in the waterbuck, where the preference for asymmetrical implantation is not so evident. Sinclair (1977) provides an average figure of 64% preference for implantation in the right horn; but the right ovary is the most active.

Placentation, first described by Amoroso *et al.* (1954), is of the cotyledonary, syndesmochorial type. The cotyledons were well-formed by the 43rd day of gestation, their number ranging from 20 to 48 in the animals examined. Before the cotyledonary villi have attached themselves to the uterine wall, the blastocyst has developed into a trophoblast of about 20 cm in length.

No cases of intrauterine twinning were recorded. Young fawns often associate together, and thus two may not infrequently be seen following one doe. This probably accounts for statements in the literature such as: "Twins perhaps not infrequent" (Ansell, 1960). Riney and Kettlitz (1964) record that one of four doe common waterbuck introduced into the Loskop Dam Nature Reserve of the Transvaal, produced twins. An oft-quoted reference is that of Frechkop (1954), who found three embryos in one doe, a rare departure from normal, as the waterbuck is undoubtedly monotocous.

Of 60 fetuses older than 56 days, which could be sexed, 30 were found to be male and 30 female.

I was unable to confirm the length of gestation. This has been given as 240 days by several authors for the common waterbuck, probably a repeat of Stevenson-Hamilton's "about eight months" (Stevenson-Hamilton, 1912), although Elliott (1976) claims to have observed an eight-month period, while Dekeyser (1955) gives 243 days for the defassa waterbuck. However, Heinroth (1908), from the observation of six births in captivity, gave a range of from 272 to 287 days for the defassa, suggesting a mean of about 280 days. Dittrich's (1972) vindication of figures for other species presented by Heinroth, provides credence for this figure of 280 days, in preference to the estimations of subsequent writers. Pocock (1904) gives 283 days for a defassa × common waterbuck hybrid. There may be no argument in this, it may well be that a real difference exists between the common and defassa, and that these animals are further apart than the taxonomists would have us believe. If such a difference exists, it fits well into Geist's (1974) theory that speciation may result from neotenization, suggesting that the defassa is derived from a neotenized common waterbuck; an extended gestation length according with the theory of slowing of growth that neotenization implies.

Fetal Development and Birth Prediction

In the fetal buck, dark-coloured papillae of the horn primordia have appeared by the 94th day. By the 143rd day vibrissae are present on the snout and eyelids, and the testes are positioned above the scrotum. Both sexes are fully hirsute by the 219th day. The eyes have opened by about the 230th day, and in the final week preceding parturition the first incisors may have cut the gums.

Huggett and Widdas (1951) described how the regression of the cube root of fetal weight, plotted against days of gestation, gives an almost straight line for a wide range of mammals. Knowing gestation length, and weight at birth of the fetus, one only requires to know the x intercept, or t_o: the period of time during which the weight of the fetus is too small to contribute to the regression. From knowledge of the specific fetal growth velocity for a range of mammals, Huggett and Widdas proposed, that for gestation times of 100 to 400 days, t_o was approximately equal to 0·2 times gestation length. In Spinage (1969a) I found that a factor of 0·13 fitted the observed data more closely, assuming the gestation length of

240 days as given by other authors. But, adopting Heinroth's estimate of 280 days gestation, then Huggett and Widdas' factor of 0·2 times this gestation length, or 56 days, gives a satisfactory fit to the known data.

This data was as follows: doe Y120 gave birth on 9 December 1966, she was killed, after she had lost her fawn, on 9 February 1967, 62 days *post partum*. She was found to be bearing a fetus weighing 1·87 g ($3 \sqrt{W} = 1\cdot2$). Another doe, Y139, who had also lost her fawn, was killed 104 days *post partum*, and found to be bearing a fetus weighing 88·5 g ($3 \sqrt{W} = 4\cdot4$). Using the transformation of Huggett and Widdas' formula:

$$\text{Fetal age} = \frac{W^1/3}{0\cdot1094} + 56 \text{ days},$$

where the divisor is equal to: $\frac{24\cdot5}{224}$; the expected ages of the fetuses of Y120 and Y139 were then calculated as 67 and 96 days respectively. These compared with observed *post partum* intervals of 62 and 104 days, which, allowing for the error inherent in the method, suggests that both does conceived shortly after parturition.

In their detailed studies on the red deer in Scotland, Guinness *et al.* (1978a) have demonstrated that neonatal weight may vary by as much as 300%. This could obviously lead to errors in determining the slope of the regression of fetal weight on fetal age, when the heaviest recorded fetal weight is taken as weight at birth. But temperate zone species, existing in habitats greatly altered by man, and exposed to greater seasonal extremes of food supply, may, possibly, show greater variability in reproductive processes than might tropical species.

Heape (1901) gave the length of the anoestrus cycle for the common waterbuck as 21 days; Y120 had previously given birth on 16 April 1965, 11 February 1966 and 9 December 1966, suggesting two *post partum* intervals of 21 days. Another doe, Y112, calved on 7 March 1966, and was seen to have calved again by 15 January 1967. She was killed on 6 March 1967, when she no longer had her fawn, a date which should have been 84 days *post partum* from 13 December 1966. She was found not to be pregnant, had recently ovulated, and had a regressing corpus albicans. It appears that this doe may have undergone four anoestrus cycles each of approximately 21 days. In both cases, a gestation length of 280 days fits the data better than does one of 240 days, which would imply that the first doe underwent three 21-day cycles before conceiving, and the second six anoestrus cycles.

Using an *x* intercept of 56 days, I predicted the dates of birth of 45 fetuses. This showed that births took place throughout the year, but with peaks in August, and November–December. Of the total number of births 62·2% fell in the wet seasons. Of 50 observed births, 72% took place in the wet seasons, thus observation was in agreement with the predicted results. When lumped together, these results showed that 67·4% of births took place in the wet seasons, approximately the same number taking place each month in both short and long rains, namely, 9% and 9·2% respectively (Fig. 23).

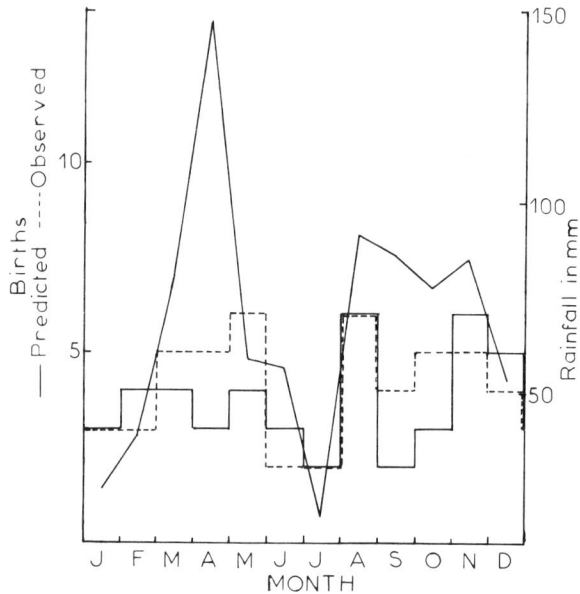

Fig. 23. Predicted and observed births each month, with mean monthly rainfall.

When fetal data were predicted back to time of conception (observed birth dates were not precise enough to do this with confidence), it was found that conceptions occurred in all months of the year, with peaks in February–March, June and November. Of the total number of conceptions 42·2% took place in the dry seasons, resulting in 62·2% of births in the wet seasons. Most conceptions took place at the beginning of the short rains and the end of the long rains (Fig. 23). As there are, on average, 5 dry months and 7 wet months in the year, it is necessary to determine, whether or not, assuming an equal chance of conception or birth occurring in any month of the year, that this distribution is due to chance. Analysis shows that it is ($t = 0·05$, df 8, $p = \ll 0·1$, and $t = 0·06$, df 12, $p = \ll 0·1$ for dry and wet months respectively), thus the fact that there are more births in the wet seasons is simply due to the fact that there are more wet months in the year than there are dry ones.

Ansell (1960) commented that, in the Luangwa Valley of northern Zambia, matings of the common waterbuck were generally seen in June and August, with most births taking place during the dry season in May to August; a period which fits a gestation length of 280 days better than it does one of 240 days. However, in the Kafue National Park of southern Zambia, 3° latitude further south, Hanks *et al.* (1969) found the defassa waterbuck to have a calving peak at the height of the rains, in January; with a spread from mid-November to April or June. Copulations were mainly observed from late February to mid-August, suggesting to these authors that 240 days was the correct gestation length, but their data, as given here, suggests to me the period of 280 days as more likely to be correct in this instance also.

A number of studies on reproduction in other species has been conducted in the Rwenzori Park, namely on the hippopotamus (Laws and Clough, 1966), the buffalo (Grimsdell, 1969, 1973a), and the warthog (Clough, 1969, and personal communication). In the buffalo Grimsdell found that peak conceptions occurred at the end of the short rains, and with a much greater spread. He thus postulated that the cows would be in much better condition at this time, resulting in more ovulations. This gives rise to calving peaks at the height of each of the two rainy seasons. Laws and Clough found in the hippopotamus that conception peaks occurred in December and July, with the least number of conceptions taking place in October to November. This pattern also produces most births during the wet seasons. Finally, the warthog shows a quite different pattern, with one sharp peak in conceptions at the beginning of the rains in August, which produces a peak in births at the end of the long dry season in February. But in this species the young do not come above ground for three weeks, which would be when the short rains are well under way (Fig. 24a, b). Thus the evidence indicates the significance of the wet seasons for the rearing of offspring in most species; but since each of these species has a different gestation length, ranging from 5·8 months in the warthog, to 11 months in the buffalo, and yet each gears its rhythm to produce offspring at much the same time of the year, with the possible exception of the waterbuck in which there are elements suggestive of chance, the proximal stimulus to ovulation must be different in each, or, the same stimulus must be used in different ways.

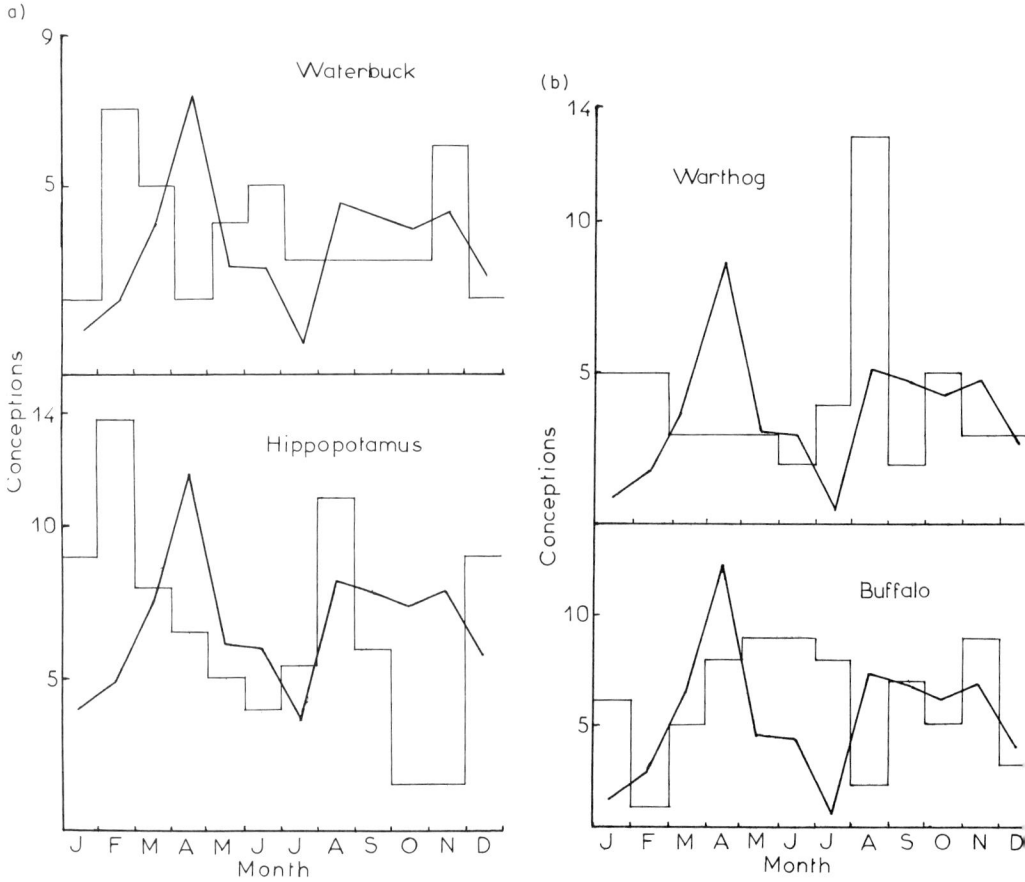

Fig. 24. Monthly conceptions in (a) waterbuck, hippopotamus, (b) warthog and buffalo, with mean monthly rainfall.

Time of Birth

If we assume that the waterbucks' peaks of births in the wet seasons are as contrived as they appear to be in some other species from the same area, this still does not tell us what factor of the environment may provide the cue for ovulation, anticipating the time of birth more than 9 months beforehand. In Spinage (1969a) I argued that the amount and quality of food as a factor influencing conception was debatable in the waterbuck, although indices of kidney fat do seem to show some correlation with conceptions' being at their highest in February and October (Fig. 25). The phenomenon of "flushing", whereby cattle tend to ovulate if changed to a higher plane of nutrition, is well known, and may well operate if the natural timing of ovulation has been suppressed by poor nutrition, but favourable conditions producing a plethora of ovulations does not imply favourable conditions 9 months in the future, when gestation is complete.

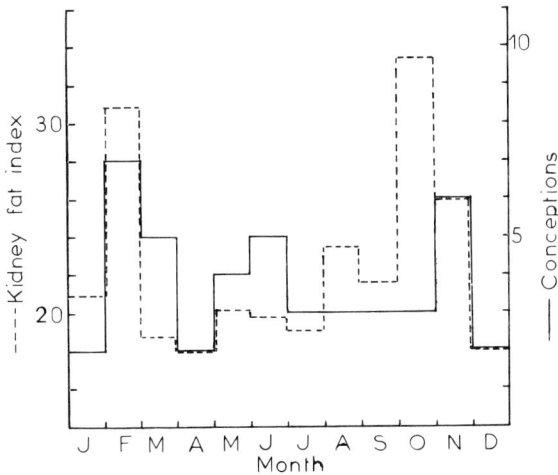

Fig. 25. Monthly conceptions compared with the mean monthly kidney fat index.

Sexual periodicity must assume either, that natural selection has operated to favour births at the most propitious time of the year; this being a teleological assumption in the absence of any knowledge of comparative survival rates, or, that some environmental factor has a direct effect upon mating or conceptions. The former alternative is obviously inflexible, for, if the seasons change, for example the rains fail, then the young will be born at the wrong time. Perhaps this is not really significant as a pair of waterbuck could lose their offspring seven times in succession, and still have time enough to replace themselves. In the waterbuck, at the Equator, there is no evidence to suggest, however, that normal dry seasons are unfavourable for the survival of fawns. There are many statements in the literature to the effect that it seems logical to produce young when environmental conditions are optimal for the survival both of mother and offspring (Amoroso and Marshall, 1960; Sadleir, 1969; Lofts, 1970), and thus natural selection will favour those individuals producing young at the most propitious time. Perhaps survival of the parent is the important factor, for in the favourable conditions for survival which the tropics can offer, over-production could threaten the continued existence of the whole species. Hence the random survival of young might be equally selected for.

In Loft's (1970) view, progeny reared at less favourable times will suffer a high and wasteful mortality, but in most natural populations of wild animals progeny losses are at least 50%, rising to 80% in healthy populations. A high survival rate of young means that the adults are not surviving, except in expanding populations. Phases of increase in recruitment to the adult stock, followed by decline, are more wasteful of resources than if the progeny die at birth. One might thus ask, why continuously produce the young at all in that case, in view of the expense and risk to the mother? But by producing the young at regular intervals the population is always ready to respond instantly to any change leading to more favourable circumstances for survival. This is why tropical ungulates have such tremendous powers of response. The waterbuck in the Rwenzori Park, like other ungulates there, such as the buffalo, kob and hippopotamus, has maximum insurance for its population, producing young throughout the year, but with an increased production during the rains, when the mother's high nutritional requirements for lactation can be more easily met.

In an earlier paper (Spinage, 1973a) I suggested that for several species of African ungulates, elements existed suggesting that a proximate photoperiodic stimulus is operating; this response being obscured at the Equator due to the lack of marked astronomical changes. Daylength on the Equator is constant to within ± 2 min throughout the year, although times of sunrise and sunset vary 30 min between February and the end of October. When monthly conception totals in the waterbuck are plotted against this change in time of sunset and sunrise throughout the year, we find some degree of correlation, although this is not wholly surprising as the seasons relate to these astronomical changes. Data on other species show, in some cases, a much more convincing relationship than does that of the waterbuck. Grimsdell's data on buffalo conceptions show markedly coincident wavelengths, while hippopotamus conceptions show a clear negative correlation. Warthog conceptions show no clear relationship at all (Fig. 26). Supposing there to be a real relationship between these astronomical changes and conceptions, we must still postulate a cue which would enable the animal to identify these changes, since it has no watch to guide it. This cue could perhaps be provided by the changing position of the overhead sun, which alters by $23 \cdot 5°$ with respect to the Equator each 6 months; being furthest north on 21 June, and furthest south on 21 December.

Temperature as a stimulus should not be ignored, for the fluctuation in daily temperature at the Equator is much greater than mean minimum and maximum temperature records indicate. Added to which, a

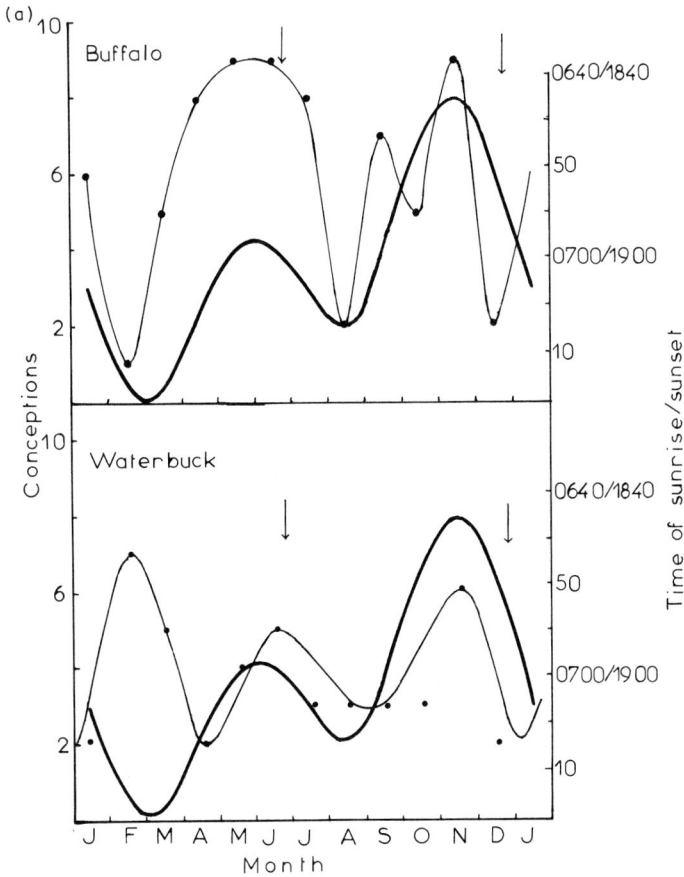

Fig. 26. Smoothed curves of the monthly conceptions in (a) buffalo, waterbuck, (b) hippopotamus and warthog, compared with time of sunrise and sunset at the equator. Arrows indicate the summer and winter solstices.

homoiotherm accustomed to a narrow temperature range is likely to be more sensitive to small changes than are those accustomed to more rigorous climatic changes. But its role as a proximate factor initiating conception shows no obvious correlation. Wodzicka-Tomaszewska *et al.* (1967) have shown that temperature changes did not affect oestrus ewes in Australia in the absence of changes in daylength, and Ducker *et al.* (1973) go so far as to say "and there is no evidence of other environmental factors doing so".

Species on the Equator tend to breed continuously through the year, but with two, sometimes three, peaks of increased activity. This is typified by the domestic sheep, which at latitude 50°N has a single,

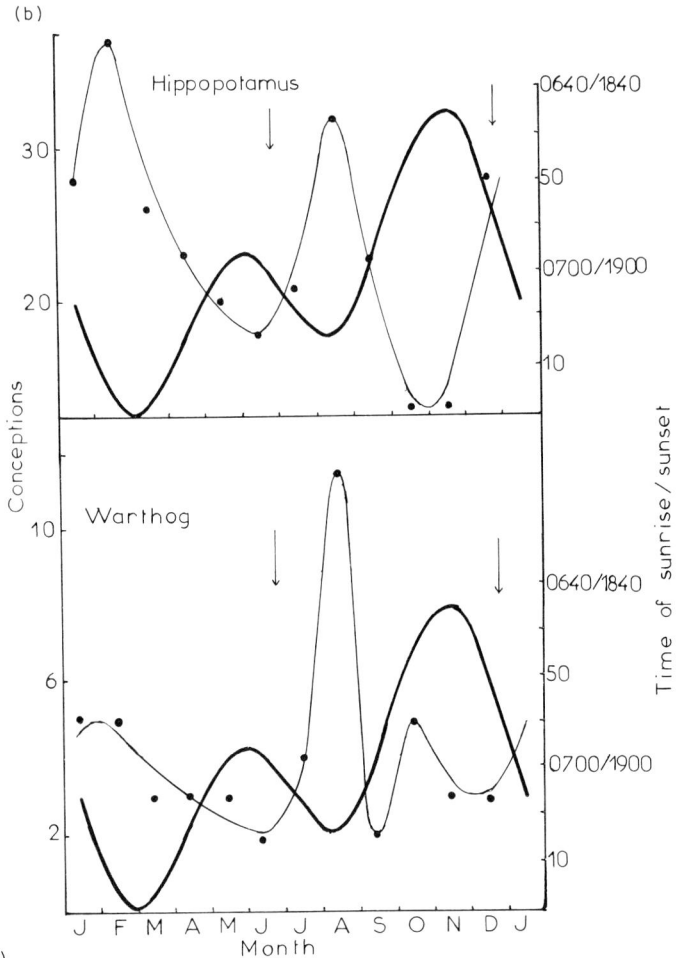

Fig. 26(b).

restricted mating season from October to November. On the Equator, in Kenya, it breeds throughout the year, but has marked peaks of increased activity in April and October (Anderson, 1964), suggesting an extension of its northerly rhythm into a continuous breeding activity. We can see that the waterbuck follows a similar tendency at the Equator, although not in phase with the sheep, which has a 4 months' shorter gestation period. Ducker *et al.* (1973) consider that the sheep has an inherent rhythm, which controls its oestrus activity in the absence of adequate changes in daylength.

Cloudsley-Thompson (1961) suggested that the breeding season was dependent upon the interaction of environment and internal rhythm acting together, and that no internal rhythm could ever act so completely as to exclude the animal from environmental control. But if environmental cues are lacking, there seems to be no reason why the internal rhythm should not be uppermost in expression, as Lofts (1970) has suggested, and which was confirmed by Ducker *et al.* (1973) in their experiments on sheep. Lofts proposed that the photoperiodic stimulus, in mammals generally, served to time a well-developed endogenous rhythm; which would result in the eventual development of the reproductive condition, even if the animal was isolated from its environmental cues. At the Equator, where the environmental cue of change in daylength is almost absent, and in the absence of a rigorous environment, a predominantly bimodal pattern, reflecting the seasonal behaviour of congeners from further latitudes, might be expected. In the same way, the loss of a photoperiodic response in temperate zone domestic species is usually only partial; the original pattern persisting in part (Clegg and Ganong, 1969). It should not be surprising, therefore, to find that the waterbuck, and other Equatorial ungulates, may be reacting to some hardly discernible astronomical cue.

Sinclair (1977) has postulated that, in the Serengeti of northern Tanzania, lunar cycles might provide the cue for ovulation in both buffalo and wildebeest. He found that 49% ($n = 90$) of buffalo conceptions occurred between the first quarter and the full moon, while in the wildebeest 82% ($n = 33$) occurred in approximately the same period. In my study, however, the waterbuck showed 64% ($n = 39$) of conceptions between the last quarter and the first quarter (the darkest period), and only 15% between the first quarter and the full moon; but conception took place during all phases. A lunar cue would, therefore, appear not to be universal.

Maternal Nutrition

Blaxter (1964) has pointed out that, during lactation, food intake in terms of metabolizable energy, can increase tremendously in the dam; yet for the grazing animal metabolizable food is minimal at the end of the dry season so that, in empirical terms, there is an advantage in producing the young in the late rains. The waterbuck shows some relationship to this hypothesis, but not a very convincing one, for there does not seem to be any marked relationship between condition of the doe, as measured by the kidney fat index, and the season of parturition; although the doe is in good condition during the long rains when births reach their peak (Fig. 27).

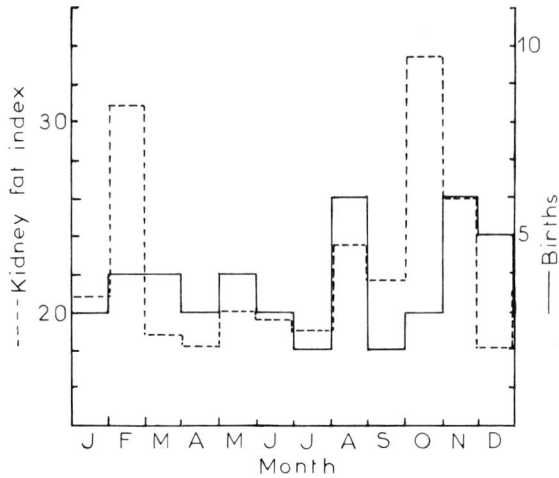

Fig. 27. Monthly births compared with the mean monthly kidney fat index.

During gestation, as Grimsdell (1969) found for the buffalo from the same area, rumen intake declines in the waterbuck, to some 5·26 kg below its average intake. Rumen contents are very variable in weight, depending upon the feeding activity of the animal prior to collection, but nonetheless the doe shows a significant decline with pregnancy, a rather abrupt drop in intake occurring when the fetus has reached about half its weight at term. Intake falls by an average of 28·3%, or 5·26 kg, to compensate for a fetal weight replacement which will equal 20 kg at term, including amniotic fluid and placental membranes (Table 27, Fig. 28).

Table 27 Change in weight of rumen contents index with increase in fetal weight.

Fetal weight (kg)	n	Mean rumen contents index	Range	SD	t	p	
10·90 — 12·73	4	9·12	7·99 — 9·72	0·81			
8·20 — 10·89	4	10·81	8·10 — 13·32	2·13	− 1·480	> 0·7	< 0·8
5·50 — 8·19	8	9·42	8·03 — 10·28	0·79	− 0·617	> 0·5	< 0·6
2·70 — 5·49	6	12·43	9·11 — 15·07	2·34	− 2·672	> 0·95	< 0·98
0 — 2·69	21	11·91	8·44 — 15·28	1·97	− 2·749	> 0·98	< 0·99
0	17	13·48	9·66 — 16·83	2·19	− 3·86	> 0·99	< 0·999

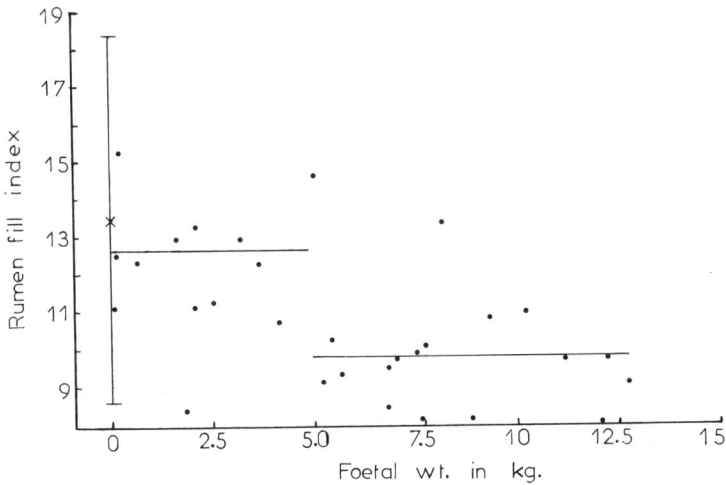

Fig. 28. The change in rumen contents with change in fetal weight. Horizontal lines show the mean rumen fill index from 0 to 5 kg fetal weight, and from 5 kg to term. The vertical line shows the range in the rumen contents index in non-pregnant animals, and "x" shows the mean.

The mean weight of rumen contents of the non-pregnant doe was 25·07 kg (SD 8·91), equivalent to 13·48% of total body weight. This was significantly different to that of the buck, whose mean content weight was 26·47 kg (SD 3·74), equivalent to 11·24% of total body weight ($t = 2·039$, df 50, $p = > 0·95 < 0·99$). Thus the doe appeared to be ingesting 2·24% more per unit body weight. Analysis of the buck showed that there was no significant difference at the 95% level of probability in the amount of rumen intake with season, so it is assumed that the same holds for the doe, and that all changes in the latter's intake relate to pregnancy. In the pregnant doe, when fetal weight has reached the region of 5 kg, rumen intake falls to an average of 9·78% of total body weight; 1·46% less than the buck's intake. When the mean intake of pregnant and non-pregnant does is calculated, 11·2%, no significant difference is found at the 95% level of probability between buck and doe intake ($t = 0·115$, df 93, $p = \ll 0·1$). Thus the doe appears to fortify herself for pregnancy well in advance of this occurring, so that what happens in the last half of gestation would appear to be less important, in terms of season and food quality, than what takes place in the first half.

Lactation

The doe has two pairs of inguinal teats, and a mammary gland which weighs approximately 40 g in the adult virgin doe, reaching a weight of nearly 2 kg when full of milk. Colostrum is produced at least 20 days before estimated parturition. In the does studied, regressed glands due to fawn loss were common, but in no case ($n = 24$) was mastitis observed. In a continually breeding animal it is expected that the adult doe will be permanently lactating, except when she stops feeding the fawn at heel in order to produce colostrum for the next birth. Hence, in the continuously breeding waterbuck, the calf should normally be weaned at about 255 days.

Doe Y120 appeared to be unable to maintain lactation after parturition. When this animal was caught on 6 April 1965, a thin, non-milky fluid was expressible from the teats. She gave birth about 10 days later, and the fawn was subsequently killed by a leopard on 10 May, 24 days after its birth. The leopard ate all but the fawn's stomach, and I found this to contain only mud (the fawn was living by the lake shore) and hair, a single hairball measuring 3 cm in diameter. There was no sign of milk or vegetation. The same doe calved again on 11 February 1966, and the fawn disappeared between 18 and 25 days later. When this doe was caught again on 23 April 1966, 17 to 23 days after the estimated date of the fawn loss, a watery fluid was again expressible from the teats. She gave birth a third time, on 9 December 1966, but the fawn was not seen. On 9 February 1967 she was killed, and found not to be lactating. She was hyperthyroid, the thyroid gland weighing 50·8 g, and of above average body weight with a weight of 192 kg.

Of 61 does examined, older than 3 years of age, 27 were lactating (3 producing colostrum are not included), hence we can calculate that the average length of lactation was:

$$\frac{27}{61} \times 280 = 124 \, \text{days.}$$

Or, putting it another way, fawn mortality up to weaning was:

$$\frac{260 - 124}{260} \times 100 = 52\cdot3\%.$$

The average interval between pregnancies is given by the number of non-pregnant does, divided by the number of pregnant ones, multiplied by gestation length, as follows:

$$\frac{9}{56} \times 280 = 45 \, \text{days.}$$

Extending this analysis, the average calving interval is equal to: the average interval between pregnancies, plus gestation length:

$$\frac{9}{56} \times 280 + 280 = 325 \text{ days.}$$

Events Leading to Mating

The buck, solicitous and imposing, attending his meek dam with her fluffy fawn, makes a charming family picture. But the buck is not a family man, there to protect his charges. His sole interest is in the willingness of the doe to mate. If the fawn gets too much in his way, then he becomes antagonistic towards it. The localism of the fawn in its first few weeks of life (see Maternal Behaviour), means that the dam is restricted to the fawn's locale. This will almost certainly be within a buck's territory, as the territories of the bucks are co-incident with the home ranges of the does. The *post partum* doe is therefore soon found by a territorial buck, who at first attaches himself closely to her; presumably until she has come into her first *post partum* oestrus. Her chances of being mated in the first *post partum* oestrus are therefore very high. I was not able to determine the length of the *post partum* anoestrus (given by Heape as 21 days) from the time passed with the doe by the buck; the buck was not always seen with the doe, and the doe sometimes moved out of the area, perhaps if the fawn was lost. One such period of attachment, however, appeared to be 27 days (Plate 21).

When any doe, or group of does, enters the territory of a buck (see Chapter 11), his invariable reaction is to investigate one or more of them, and attempt to mount one or two. If he is spending time with a group he may, from time to time, abruptly stop feeding or ruminating, and attempt mounting. Such behaviour was not usually accompanied by erection of the penis. Unreceptive does never showed aggression towards persistent bucks, and often rubbed themselves on the buck's flanks. An unreceptive doe avoids the buck by walking away, or by twisting round, flank to flank. In contrast, the oestrus doe, usually stands and awaits the buck, lifting her tail to one side when mounting attempts are made. She shows considerable olfactory interest in the buck, sniffing his inguinal region, and often stretching up to sniff at the base of the horns. When excited the doe may even attempt to try and mount the buck.

Broadly speaking, copulatory behaviour is as described by Kiley-Worthington (1965). The usual sequence of events is for a buck to approach the doe fairly quickly, eagerly licking its philtrum — the bare

Plate 21. The buck follows the post-partum dam closely in anticipation of her first post-partum oestrus.

black nose patch — and with head down and neck extended, in a posture which signifies appeasement among bucks, the horns being backwardly directed. Despite a number of occasions, which I closely observed, I never saw the buck actually lick the doe, and the licking would appear to be for moistening the sensory surface of the philtrum, making it more receptive to odours. Some does avoid the buck when they see him coming. Those which he can approach, he sniffs in the perineal region, and they lift their tails and urinate; often continuing to graze as they do so. This is a characteristic reflex, but not necessarily the result of tactile stimulus. A doe, particularly a young one, may urinate as soon as a buck approaches from behind, without any encouragement from his nose.

Curling back his philtrum in characteristic manner, the buck then samples the doe's urine, by letting it run over the curled philtrum. Some species take the urine into the mouth, and then eject it, but any imbibing was not obvious in the waterbuck. The buck stands for a few moments with lips curled, the *flehmen* of German authors, looking from side to side, and may salivate copiously (Plates 22, 23, 24). This behaviour

Plate 22. Buck urine-sampling.

Plate 23. The buck grimaces in a characteristic manner.

Plate 24. Urine sampling.

develops early, and I have seen a 9-month-old buck conduct it with a young doe of the same age. Having performed the behaviour the young buck does nothing further; whether he would do so if the doe was in oestrus is unknown, for it is unlikely that a young buck would have access to an oestrus doe. Estes (1972) has suggested that the function of this behaviour is to test the hormonal content of the urine with the vomeronasal organ, which lies in the roof of the palate just posterior to the nares, with accessory ducts opening immediately behind the dental pad. Oestrogens occur in increasing concentration in the urine with the approach of ovulation, but it seems in the waterbuck to have become more of a reflex action than an action having a physiological significance; for whatever the reproductive status of the doe, the activity of urine sampling makes no difference whatsoever to the subsequent behaviour of the adult buck, which is to try to mount the doe, whatever her state. He may even neglect the preliminary of urine sampling, ignoring her urination and attempting to mount as soon as he approaches.

It is interesting that urine sampling has not been recorded in the hartebeest and topi, but is found in almost all other groups; even in such diverse genera as giraffe and rhinoceros. I suggest that it is a primitive social behaviour which has largely become replaced in some species by more advanced behaviour patterns based upon visual appreciation and an understanding of the female response. I do not think it is used by the anoestrus doe, as Geist (1971) suggests, to distract the buck while she makes good her escape; at least, such a ruse does not work in the waterbuck.

An unreceptive doe, and the majority is unreceptive, may signify her unwillingness to mate by an action which can best be termed as "champing". Extending her neck low, and shaking her head up and down, the doe makes a rapid biting or "champing" movement with the jaws, walking away and often inclining her head towards the buck as she does so (Plates 25, 26). This seems to be employed both as a submissive and as a greeting gesture; young bucks do it to adult bucks, and a captive male fawn regularly greeted me with it. It did not appear to me to be a threat gesture, signifying an intention to bite, as one worker has suggested. The waterbuck does not bite anyway.

The chances are against any doe in a group being receptive as the majority will be pregnant. We have seen that the average interval between one fawn and the next is 325 days; so that a doe will, on average, only receive the buck for between 12 to 24 h once in every 10 to 11 months. Oestrus is probably brief. One doe observed to be in oestrus at 1400 h, had reverted to normal group behaviour by 0800 h the next day. Another, observed coming into oestrous at 1900 h, was captured at 0800 h the next day, when she was still being pursued by bucks. She did not give birth until 11 months later, so that this mating, and several successive ones, must have been unsuccessful. In the case of a *post partum* doe the buck appears to know that she will come into oestrus soon, and this is why he sticks to her until his importuning is rewarded.

When a doe is not receptive, the buck may cajole her by rubbing his chin on her rump or on her flank, investigating her inguinal region in a manner suggestive of a suckling calf, or he may paw her backlegs with an extended foreleg. This latter movement, termed *laufschlag* by German authors, was always of a rather clumsy, ill-directed nature, without any precision in positioning the leg. I never witnessed the "pincer-grip" as described by Kiley-Worthington (1965); neither could it be termed the "mating kick", for the waterbuck does not "whip" its leg up, as does, for instance, the mountain sheep (Geist, 1971). In the waterbuck it is a clumsy, undirected, jerky movement (Plates 27 and 28).

If the doe does not respond to these solicitations the aggrieved buck may push her violently from behind with his forehead between her backlegs. The doe always remains docile, rarely dodging with ears down, an attitude of dislike which is also seen in heavy rain. If a buck is over-persistent the does may try to stand parallel to him, facing the opposite

Plate 25. Doe champing in response to a buck.

Plate 26. Doe champing in response to a buck.

Plate 27. Buck cajoling a doe by rubbing his chin on her rump.

Plate 28. Rubbing the doe's udder region.

direction; what Geist (1971) has termed the "reverse parallel" position in respect to fighting rams. Sometimes a doe even appears to be apologetic, giving the buck a quick, friendly rub along the flank, before twisting away. But the buck is not always so friendly, and sometimes individuals will charge unwilling does which they appear to have taken a dislike to; and sometimes prod them with their horns. Nothing violent ever ensues, but a buck might become aggressive if he appears to be being teased.

As an illustration of this seemingly anthropomorphic behaviour, an apparently pregnant doe walked up to a territorial owner and sniffed him. She then placed her forehead against his and pushed, like a buck about to spar. The buck pushed her in the shoulder until she moved away; but she did not stay away for long. Seeing the buck lying down a little later, she stopped grazing and suddenly walked up to him again, sniffed him, and when he stood up, sniffed his hind end. Then she moved round and sniffed his forehead, again pushing against him, forehead to forehead The buck was now clearly getting annoyed, and taking the initiative he turned and pushed her violently with his head between her hind legs. Pushing her round and round like this for about 5 min, he made repeated attempts to mount. Abruptly, he stopped and walked away as if indifferent, only to suddenly turn, rush at her, and then stop. Less than 15 min later the doe walked up to him again and sniffed him. This time the buck was having none of it, and walked off towards another doe, ignoring her completely. How can one avoid the anthropomorphic interpretation, that the doe appeared to be "teasing" the buck?

If a buck is lucky, and finds a doe in oestrus, then she stands and awaits him, making no attempt at avoidance (Plate 29). In her rising excitement she may sometimes try to mount him, and the buck for his part may try to mount the doe from whatever position he may be standing in. This is generally attributed to inadequate mating experience, the actual consummatory act being half instinctive and half learned. Thus a simple table-top arrangement is said to release mounting responses in many domestic ungulates, and the reaction solely to a female in the correct position is thought to come from experience. When I witnessed one of the Peninsula bucks behaving like this, trying to mount a doe from the side, he was already an adult in the prime of life (Plates 30, 31). As the buck mounts, the doe invariably walks forward, and this may help to guide the buck into the right position. The characteristic white rump pattern may also be designed to this end, but it could only act as an accessory releaser, otherwise a buck would never try to mount from the side.

The buck mounts onto his hind legs, and copulation is effected with a single, quick thrust, the doe immediately moving forwards. The true act of copulation can be identified by the upward carriage of the doe's head, and the contracted conformation of the buck's hindquarter musculature

Plate 29. Mounting an oestrus doe, the doe stands with tail held aside.

(Plate 32). Due to its rapidity, true consummation is often difficult to detect in many wild ungulates, and casual observers often speak of copulation when full intromission has not in fact been achieved, leading to much confusion in the literature over the time of mating seasons. As we have seen, circumstances make it a relatively uncommon act, and I witnessed only three sessions in 30 months of observation. Penetration by the buck may be sufficient to cause bruising of the cervix, for slight haemorrhage was observed in the cervix of a doe which was killed and found to have just ovulated; and hence had probably just mated. Geist (1971) has argued that cautious courtship is selected for, as if a male courts roughly then the female will move into another's territory. My impression was that all waterbuck bucks were equally uncautious, a state perhaps brought about by their Equatorial existence, in which they are in season throughout the year, but does are rarely available.

After a successful union both partners may graze or rest for a while before repeating the act. The buck is usually the first to recover interest, after a period which may last for up to half an hour.

Due to its spatial distribution, a territorial buck is more often than not

Plates 30 and 31. Buck attempting to mount from the side.

Plate 32. Successful copulation. Note the carriage of the doe's head, and the contraction of the buck's hindquarter musculature.

undisturbed by others when mating. But when a wandering doe comes into oestrus, and by wandering I mean either a virgin, or one who has lost her fawn and perhaps already undergone several unsuccessful cycles, then she may attract several bucks together, by her movement from one territory to another. One February evening, not long before it was dark, I spotted a doe apparently coming into oestrus in the territory of the buck Y7. When I first saw her she was with a group of does, but the others soon wandered away, leaving her behind. As might be expected, the buck Y7, finding her alone, repeatedly tried to mount her, but she was not yet receptive and merely stood passively. By now four other bucks had appeared, three of them were territorial, and one was a 6-year-old bachelor. Y7 spent most of his time watching the doe, who grazed apparently unconcerned by the attention that she was attracting. From time to time, one of the watching bucks would try to approach her, only to be confronted by Y7, causing him to beat a hasty retreat. As I watched, quite a pantomime ensued. When one of the trespassing bucks apparently thought that Y7's attention was diverted, he sneaked round a bush with every appearance of a conspirator and, appearing suddenly, almost succeeded in mounting the doe, but quickly absented himself when Y7 spotted him, and came galloping back to the doe's side. Although considerably harassed by the trespassers, Y7 did not engage any of the envious bucks in a fight; to do so would have meant losing the doe to one of the others. Darkness closed in with the doe now lying down, the centre of attention, with Y7 standing guard, and the other four would-be suitors waiting hopefully in the bushes.

This buck always appeared to be having trouble in his area, and the following May I came across him with an old territorial buck, Y9, a young one, and a 5-year-old bachelor. Y9 had a 1 cm diameter wound in his shoulder, with blood streaming from it, while Y7 was limping from a hind leg injury. As I watched, Y9 engaged Y7, but did not persist and both broke off. It subsequently transpired that the cause of all the trouble was a doe coming into heat. No mating activity was seen, but it may well have taken place the following night.

These observations indicate that, if other bucks become aware of a doe coming into oestrus, then they tend to neglect their respect for territorial boundaries. The owner is still treated with reserve, suggesting that territorial inhibitions remain, but he is not allowed to mate without interference. As with some other species, it may be that as the doe nears her oestrus peak, so the buck becomes more excited and more aggressive towards others. When I witnessed one buck mating he was sufficiently aggressive even to chase away two young does who came to watch. I shall have more to say about this lack of territorial respect in the last chapter.

Among aberrant sexual behaviours, anoestrus does were very occasionally seen to mount one another, and one doe was seen to go through all the motions of urine sampling and lip-curling, in the typical buck manner. On another occasion, one doe following another rubbed her head on the latter's rump, inducing the urine reaction. But the doe who had induced it merely ignored the response and walked away.

Male fawns as young as one month of age may attempt mounting if they are following closely behind a doe. Those up to 3 years of age have been seen trying to mount one another, but no penis erection took place and such behaviour was not common. It was sufficiently rare as to suggest that such behaviour was not associated with the establishment of hierarchy among the young bucks, contrary to such behaviour in species such as the mountain sheep.

A buck following a doe was seen to erect his penis and stand with back arched as if in the act of copulation, but no ejaculate was seen. On another occasion a buck mating an oestrus doe was seen to discharge a fluid when following her; while yet another buck was seen to try to use a bush as a mounting object.

6. Parturition and Maternal Behaviour

Introduction

There is little information on mother–infant relationships in wild animals with which to effect comparisons; activities such as parturition, suckling, weaning and the general behavioural development of the young are but poorly known, as Lent (1974) has pointed out in his review of the subject. Most ungulate species fall into one of two categories with regard to the type of mother–infant relationship which occurs after birth; the infants are either "hiders" or "followers". The former spend the whole of their time resting in hiding between maternal nursing visits, which may be as infrequent as once in 24 h; they are the "abandoned" calves of popular belief. Followers, on the other hand, accompany the parent from birth, perhaps the most extreme example of this being provided by the wildebeest, the young of which can follow the mother as little as 5 min after parturition. But both types are regarded as precocial, although the former are probably unable to withstand sustained activity. However, the physiological difference between the two has yet to be elucidated, but such a difference must surely exist. It is in the former category, as a hider, that the waterbuck is found.

Parturition

About 2 days before she is due to give birth, the pregnant doe takes herself off on her own to a suitable patch of thicket which will provide concealment. This is an important choice for the primiparous doe, for it probably determines her home range for the rest of her life. Records of marked Peninsula does suggest that the doe always returns to the same spot to give birth; one marked doe returned to exactly the same spot three times in succession (Chapter 11). Concealment during parturition obviously has anti-predator advantages, but Gosling (1969) has shown how, in the Coke's hartebeest, which gives birth in the open, conspecifics attack the cow in labour if she does not separate herself from the herd.

Parturition seems to take place, in the waterbuck, usually early in the morning. There is an obvious advantage in this timing in that the morning sun can dry the wet fawn, reducing the risk of hypothermia that nocturnal births might pose. Gosling, however, found that the hartebeest tends to give birth most frequently between 1300 and 1500 h; but five observed cases of birth in the waterbuck, in which the fawn was still wet and had not, or had only just, risen, were at 0815, 0830, 0840, 0900 and 0915 h. Actual expulsion of the neonate was not seen by me, but *post partum* behaviour from an early stage was witnessed on one occasion. My diary for 7 March 1966 reads as follows:

0900　Parent lying with fawn wet and immobile beside it.

0910　The dam rises and licks the fawn. About 45 cm of fetal membranes are hanging from the dam's vulva.

0922　Fawn stands and tries to find the teats, but falls over. Stands up again and the dam licks the fawn's anus. Tail-raising reflex observed in the fawn.

0926　Fawn tries to suckle again and makes butting movements in the dam's groin.

0928　It finds the teats and suckles.

0932　The dam proceeds to eat the membranes which are lying on the ground.

0935　Fawn stops suckling and lies down.

0940　The dam clears up traces on the ground.

0941　The fawn gets up again and suckles intermittently.

0947　Fawn lies down after trying to shake itself.

0952　Dam takes the membranes hanging from the vulva into her mouth and pulls them out, ingesting them as she pulls.

1012　The dam examines the ground, then she pulls the last remnant of the membranes, about 60 cm long, from the vulva, and ingests it. Then she sits by the fawn.

One must be cautious in making interpretations from a single observation, and also I am unable to say with certainty how much time had passed between actual parturition and the time when I found the couple, but I believe it to have been very short. Lent (1974), however, provides figures from a wide range of ungulates, which suggest an average time for the neonate to stand after birth is about 30 min, and to first suckle is about 68 min. This waterbuck fawn probably stood for the first time about 30 min after birth, suckling at about 40 min; which compares with Gosling's (1969) observations on the hartebeest.

In the single case which Elliott observed, of early *post partum* behaviour (Elliott, 1976), the dam did not ingest the membranes, which were left protruding for 6 h. They were suddenly expelled when she lunged at an oxpecker on her back. But this dam appeared to be very weak, and is presumed to have died a few days later.

The licking of the fawn, especially of its anus, are all a part of the reciprocal stimulation which strengthens the mother–infant bond; and these activities having been completed I thought that now was the time to mark the fawn, before it became uncatchable. If the couple are approached too soon after the birth, before the maternal bond is fully forged, there is a risk of the dam fleeing and abandoning the fawn. But if the bond is well forged, then quite the opposite can happen, as I now proceeded to find out.

At 1014 I started the Landrover, from which I had been observing events, and drove slowly towards the pair from about 25 m distance. As I narrowed the gap, the dam suddenly looked up at me in surprise, her surprise quickly changing to alarm. Leaping to her feet, she came storming towards me and, halting a metre's length before the vehicle, challenged me to advance further. Snorting threateningly, standing rigidly before me with hair erected, she looked a good foot taller. It is hard to imagine the normally docile doe looking aggressive, but this one had suddenly transformed herself into a sufficiently threatening-looking object as to make me stay in my Landrover! Discretion being the better part of valour, I went to enlist the help of my assistant.

We were back within 10 min, and when I approached this time the dam got up and trotted off rapidly, with the 90-min old fawn following. They had moved about 100 m when, by driving suddenly round some bushes, I was able to separate them. My assistant and I leapt from the vehicle and seized the fawn, which was standing bewildered by the sudden disappearance of its dam, who had continued on her way unawares. In seconds I had clipped a plastic streamer to its ear, the brief pinch of the pliers causing the fawn to utter a little protesting bleat. The next moment its dam came snorting back through the bushes towards us, as we leapt ignominiously into the Landrover and drove off. But the indignant doe was not prepared to let it rest there, and gave chase. Fortunately there was no one to witness our cowardly flight from the enraged doe. A Landrover pursued by a waterbuck doe? It seemed impossible, but it was true; the doe, persistently trying to butt the rear of the vehicle, followed us for a good 75 m, before giving up and returning to her fawn. I am pleased to say that I saw them together again shortly afterwards, but the fawn did not survive for long.

This doe showed herself to be exceptionally courageous, for the customary reaction of a dam, when I marked her fawn, was to run off to a safe distance and then to run around me, snorting her disapproval. Fawns, once caught and tagged, did not run away when released, but sat down. I watched one dam return about 10 min after I had left her fawn; she sniffed it, licked it, let it suckle, and then tried to lead it to a safer place.

One dam that I came across, lying with her new, wet fawn, appeared to be exhausted by her efforts. Although she licked the fawn occasionally, she seemed too weak to stand. The fawn was first to rise, walking round her and trying to find a place to suckle. I left them, intending to mark the fawn later, but when I returned she had hidden it somewhere else and I was not able to find it.

This behaviour of the dam in apparently selecting a hiding place for the fawn should not be confused with the normal "lying-out" of the fawn. Under ordinary conditions the fawn chooses its own hiding-place, as will be described later, but where a potential danger threatens, the dam evidently leads the fawn away to a new site, even if disturbed while with it when it is older.

Behaviour of the Fawn

The waterbuck fawn as a "hider", remains hidden during the first 2 to 4 weeks; 3 weeks being the average period which I observed. This concealment is in a circumscribed area, but the fawn does not have a regular "form". It simply hides itself in the nearest long grass, or thicket, when the dam leaves it. During this period of lying-out the fawn is visited only once in 12 h, and probably only once during the whole 24 h. I was unable to determine whether there was a second visit at night, but I think it unlikely. In this respect the waterbuck fawn seems to be one of the least frequently visited species (Lent, 1974), but we have little reliable information on others. Zoo records are of little value in this respect as they probably reflect over-solicitous behaviour resulting from the abnormal conditions of confinement.

Fawns visited infrequently in this way are fed small amounts of rich milk, which they digest slowly. As time passes the maternal milk becomes weaker, and its relative constituents of protein and sugar change in proportion, so that either more frequent feeding may become necessary, or the fawn must turn to substituting its diet with vegetation. Elliott (1976) records a fawn which survived to at least 8 months of age, despite its dam dying when it was only 3 weeks old; but it was smaller and slower growing than the other fawns.

The fawn was never seen to suckle on demand, and I never recorded more than one feeding period a day, whatever the age of the fawn. In two consecutive days' observations of a 6-month-old fawn, which had suckled on the second day at 0800 h, I saw it trying to suckle several times between 1800 and 1900 h that evening, perhaps anticipating a later, nocturnal feeding period.

The doe's teats become pink and extended at parturition, and for up to about 3 weeks afterwards they are visible between the hind legs as she walks away. At other times they are not visible in this manner.

The fawn suckles like any other bovid; standing, or kneeling when it is large, it suckles from the side with a vigorous wagging of the tail, making thrusting movements at the udder. If one is feeding a captive fawn on a bottle, and it does not wag its tail violently, then one can be sure that something is wrong. I timed a 6-month-old fawn suckling its dam for almost 10 min, but shorter periods of half this time are more common. The dam stands patiently while her fawn is actively suckling and then starts to move or graze, shying if the fawn still tries to reach the udder, but never becomes aggressive or tries to push it away (Plate 33).

Feeding time is usually before 1100 h in the morning, and new fawns were usually visited at about 0930h ; this latter appears to correspond with a drop in feeding activity at this time in the doe. During a visit to the

Plate 33. Typical suckling attitude.

lying-out fawn the dam grooms the fawn and licks its anus frequently. Defaecation, which is usually induced by anal stimulation, was never seen, and it would suggest that the dam ingests the faeces, as is the case with other species. I never saw any evidence of urine ingestion, as has been reported for other species, but it may well occur. In this way no traces are left of the fawn's presence which might lead predators to search for it. The very strong scent of the waterbuck does not develop in the fawn until it is about 14 weeks of age, based on observation of a captive specimen; but I found that even the newborn animal had enough rubbed on to it from the dam to give it the characteristic waterbuck smell. The ubiquitous nature of the waterbuck's scent in the habitat probably means that it does not matter much whether the fawn smells or not, and its late development is simply related to the maturing process, rather than to an anti-predator defence measure.

After the essentials of feeding and grooming are complete, the fawn then walks or plays with the dam for an hour or so, the period lengthening as it grows older. Playing by the fawn comprises what are clearly, in the adult, alarm gaits. Commencing by dashing round madly in circles, the fawn may then break into a hackney-type gait, prancing in an erect stance. At times it may also "stott", bouncing along like a rubber ball, with all four legs held rigid. This latter is an alarm gait common among small gazelles, such as Thomson's, but uncommon among larger animals. In its second week the fawn's play might extend to facing the dam and feinting from side to side, with head lowered. If the dam lowers her head in response to the "threat", this will send the fawn scampering off in mad circles. The head-down threat is an instinctive pose, present at an early age. A fawn which had been captured on its day of birth, when presented with a small dog 2 weeks later, immediately made a head-down assault on it.

Fawn boldness was demonstrated on other occasions. One day I saw a doe, followed by three curious bachelor bucks, running to her 2-week-old fawn which had appeared from cover. The doe turned on the bucks and tried to drive them away, attacking them forehead to forehead. The rather ungallant young bucks retaliated by fighting back. Seeing this attack upon its mother, the tiny fawn ran up and unhesitatingly challenged the biggest of the bucks, who answered the fawn's head-down challenge with a blow which sent the fawn running back to its mother. Undaunted, it then took on one of the others, but by now the young bucks could see that they were not wanted and departed, leaving mother and fawn in peace.

Adult bucks were not normally aggressive towards small fawns. After a fight between two territorial bucks, one of the contestants was walking back to his does when he passed a small fawn, which walked boldly up to him. The buck showed no aggression, but simply lowered his head in what appeared to be a gesture of shepherding the fawn aside. Unfortunately, in so doing, one of the buck's horns accidentally struck the fawn on the head, sending it staggering under the blow. The buck simply walked on.

If a territorial buck followed a doe to her meeting with her fawn in its lying-out place, then she showed no aggression but merely tried to avoid him as best she could. The fawn would become quite bewildered by the bickerings which sometimes ensued, trying to stick close to its dam, while she twisted and turned to avoid the attentions of the buck. Sometimes a frustrated buck got sufficiently annoyed as to make a rush at the fawn, without following through, and the dam showed every sign of apprehension.

At 3 weeks or more the fawn follows the dam continuously. Prior to this, when the dam wants the fawn to follow her she holds her tail out stiffly in a horizontal to vertical position, keeping it held thus as long as she is walking (Plates 25, 26 and 34). I have called this the "follow-me signal". It has not been recorded for any other ungulate in the same context, but a photograph in Murie (1944, page 158) suggests that it may occur in the caribou, while in the sitatunga tail wagging is used (Walther, 1964).

Although the dam gives the young fawn a signal when she wants it to follow her, it will of course follow her without any such signal. This can pose problems when the dam wants to leave the fawn, for she does not

Plate 34. The "follow-me" signal of the doe to her young fawn, the tail held stiffly in the air. See also Plates 25 and 26.

actively induce a lying position as has been recorded in several other species (for example; elk (Altmann, 1963), red deer (Perry, 1963) and roe deer (Bubenik, 1965)); but there is no information on other African ungulates except for the hartebeest, in which species the fawn always leaves the dam of its own accord. The waterbuck dam abandons the fawn as best she can. Sometimes the fawn itself goes and lies down, in which case the dam chooses that moment to make her exit. At other times she may resort to suddenly dashing in and out of the bushes to shake it off; but this may take several attempts before she can rid herself of the fawn, for when it sees its mother suddenly run off, its natural instinct is to follow close to her. Sometimes the dam succeeds in abruptly trotting away when the fawn is looking in the other direction. Whatever the means adopted, when the fawn suddenly finds itself alone, it runs into the nearest thicket or long grass and lies down. Here it usually remains until the next visit. The dam may graze a good three-quarters of a kilometre away, and during this time the fawn is very susceptible to predation if it does not remain hidden. It tends however to be attracted to animals passing close by, perhaps more so as it gets older than in the first one to two weeks; and also perhaps more so when it is near to feeding time.

I saw one fawn come out of a bush to investigate a group of passing does. It went from one to the other looking for milk, but each doe merely sniffed it gently, and moved away when the fawn tried to suckle. Shortly after, the dam appeared on the horizon, and seeing what was happening advanced cautiously. When she saw her fawn amongst the group of does, she walked straight up to it and let it suckle. This observation seems to suggest that it was the approach of feeding time which had caused the fawn to expose itself, but it also suggests that when it is hungry the fawn is not concerned with where it gets its milk from, or it is unable to recognize its parent. But in this case the dam appeared to recognize the fawn by sight.

If the fawn remains in hiding, it is only found by a predator by chance; the cloying scent of the waterbuck, distributed everywhere by the animals' passage, would keep a predator very occupied indeed if it investigated the source every time it came across the scent. Hence it seems that it is only movement which gives the fawn away; but, as we have seen, a calculated 52·3% never reach weaning age, and of this percentage almost all disappear in the first two months of life.

The abrupt disappearance of all of the fawns which I marked suggested to me that the majority probably fell prey to hyaenas during this early period of lying-out. Lions are not a significant predator. In view of this heavy mortality a question which some might pose is why does the dam leave the fawn unattended for such long periods? The answer is not because she has a "weak" maternal instinct, although observations suggest that the instinct is weaker in primiparous does compared with multiparous. However, my adventures in fawn tagging showed just how strong the mother's instinct can be when she feels her fawn to be threatened. Further evidence of this is provided by the dam whose fawn was taken by a leopard. Although she could obviously smell the leopard's presence and the blood of her fawn, she hung about the spot for 3 days, continually going back to see if the fawn might reappear again. Geist (1971) reports a ewe whose lamb was killed by a grizzly bear also remaining in the spot where the lamb was killed for 3 days after its death. He notes that it might just have been painful milk pressure in the ewe's udder which kept her there; but in the case of the waterbuck the dam was not lactating (see Chapter 5).

I have already suggested that the reason for the necessity to lie-out may rest in the animal's physiology, the fawn perhaps lacking the glycogen reserves at birth that many other species have, so that the dam, in effect, has no other recourse open to her than to abandon the fawn and not attract attention to it. The dam has no offensive weapons with which to ward off predators; she is not very good at biting, and a kick might be nasty for a man, but not for a carnivore. Butting with the forehead is more of a gesture than anything else, and certainly no deterrent to a carnivore. Thus the fawn's best chance of survival is to remain hidden until it is strong enough to follow the dam. As Geist (1971) says of the mountain sheep, "The best a ewe can do is supply milk abundantly so that the lamb can develop normally . . .", but he goes on to say that the mother's presence appears to protect the lamb from trauma in dangerous situations unfamiliar to it. Here the waterbuck differs, but we must remember that African ungulates appear to have evolved under high and diverse predation pressure, unlike temperate zone species, so that either such trauma does not exist, or the young have little chance of escape from dangerous situations posed by predators. When the waterbuck fawn is with the dam, if alarmed it flees, with the dam trying to follow it, rather than looking to the dam to lead it from danger.

On two occasions fawns of about one month of age, which were lying apart from their mothers, although no longer lying-out in the strict sense of the term, ran to a termite mound when they were disturbed, and climbed it to look for their mothers. One of these, disturbed by me, ran to a mound, climbed it, saw its mother, apparently recognized her by sight, and ran to her. The other was disturbed by a herd of buffalo. It also climbed a termite mound to scan the horizon, and was rewarded by seeing its mother about 500 m away, coming to look for it. It ran to her, and rubbed its face eagerly on her forequarters before suckling. On these occasions recognition certainly seemed to be effected by sight, but perhaps the fawns would have run up to any doe which happened to be in the vicinity, and it was just chance that it happened to be their mothers on both occasions.

The dam recovers her fawn by calling it with low, penetrating bleats, for, despite popular belief, the waterbuck is not mute. I was first made aware of this when driving on the Peninsula and heard an intermittent, low, bleating noise, similar to a sound made by swifts on the wing. At first I put it down to just that, as there were many of these birds flying among the *Capparis* clumps. Suddenly I realized that there were no longer any birds there, and it was the waterbuck doe which I was following who was making the noise! Sure enough, she soon met up with her fawn. Once I knew what I was listening to, I could stand in the morning outside my house on the Peninsula at about 0800 h, and hear the dams in the distance calling to their fawns. I have watched a dam call her 6-month-old fawn away from a group to suckle her, with bleats audible at 100 m distance. The fawn may answer its mother with a more high-pitched call, which I can only describe as sounding like a rather squeaky, penny tin trumpet. Sometimes the wrong fawn answers a call, but after mutual sniffing no further interest is shown by either party. On one occasion I watched a 4-month-old fawn stand and stare, before running 30 m to its solitary dam. The wind was blowing from the fawn to the dam so that it could not have scented its mother. Thus it seems that a combination of sound, scent and sight, may be used for recognition.

Sometimes, when they have reached the following-at-heel stage, two fawns may team up together. This often results in the two following one doe; but I was not able to confirm whether both suckled her. Where waterbuck densities are high, "nurseries" or peer groups may be found; I have recorded as many as six small fawns resting together, some distance from the adults. These were not nurseries in the strict sense of the term, for they appeared to be associations of choice on the part of the fawns and were quite unguarded. Lent (1974) has suggested that the concept of sharing maternal duties, with one female guarding such peer groups, appears to be unfounded in the species for which it has been suggested (Plate 35).

Due perhaps to its lying-out period when newborn, the waterbuck fawn is an independent creature. Often it does not stick close to its mother when it is in the at-heel stage. During grazing, adults show no protective formation, such as is shown by some ungulates which herd the fawns in the centre of the group. But if a dam knows that danger is present she will try to lead the fawn away from it, just as she does when it is in the lying-out stage, as I was able to confirm on one occasion on the Peninsula.

I had come across a 3-month-old fawn standing quite motionless near to the path which two lionesses had taken about 3 hours beforehand. The

Plate 35. Two fawns may sometimes follow one dam together.

fawn, which must have smelt their passage, literally looked too terrified to move. Soon the dam appeared, and stalked towards the fawn extremely warily, frequently standing motionless and staring intently for long periods at a nearby bush. I knew the lionesses were no longer in the vicinity but the waterbuck did not. The fawn continued to stand as if hypnotized until the dam reached it. She licked it once, lightly, on the forehead, an unusual gesture among waterbuck, and then turned and walked slowly away with the fawn following. Here perhaps was an instance of the dam's presence protecting the fawn from trauma, as Geist postulated for the mountain sheep.

Another incident which I witnessed, was that of a hyaena which ran near to a group of does, who scattered at its sudden appearance. The hyaena was intent upon some business of its own, and kept on running without so much as sparing them a sideways glance. But one of the does trotted after it, to be followed by another. Although the hyaena soon disappeared into the distance the does continued to watch intently in the direction in which it had gone. Driving round, I found a fawn lying immobile in the grass, which the hyaena had passed within a few metres of.

Weaning and Separation

If the fawn is successful in surviving the rigours of life in the bush, it is not weaned until the dam's milk dries up in preparation for the production of colostrum for the next birth. There is no such thing as a prolonged lactation anoestrus in this species, at least not at the Equator, other than this brief period, estimated at 20 to 25 days. Average figures suggest that the fawn is weaned at about 276 days, but the oldest fawn which I saw still taking milk was one of 7 months of age, at which age it spent just as much time in grazing as the adults. When she is dry the dam turns on her offspring and tries to drive it away. This is of a relatively gentle nature, for waterbuck does are never violent. A harmless butting of the fawn when it approaches is usually sufficient.

Weaning and separation are two different things, and the young doe maintains her relationship with the dam for some time after weaning has taken place, but does not accompany her when she leaves to calve. The young buck may be separated from the dam at an earlier stage by territorial bucks, who show antagonism to other bucks accompanying a doe too closely. But this is not until the horns have made their appearance, at 9 months, which is just about the time that a full weaning period would finish. By this time the young animals are relatively well fitted for an independent life (Plate 36).

Plate 36. A young doe and a 9-month buck using a termite mound as a vantage point.

7. Population Structure and Factors Affecting Survival

Introduction

Individual growth and formation in terms of reproductive biology and behaviour have been considered. We now come to consider the structure of the population, and those factors which mould it by affecting the survival of individuals.

When I started my study of the waterbuck there were few life tables of mammal populations, the table from which the survivorship curve is derived, and in which the probability at birth of reaching any given age, is related to each age value in turn. Related to this are the number of deaths within each age group, the survivors remaining, the rate of mortality, and the expectation of further life. This useful analysis derives from Edmund Halley, better known as the discoverer of Halley's Comet, who introduced it to human demography in 1694. Since that time it has been widely developed by actuaries, but less so, until very recent times, by biologists.

For many years we had only Leslie and Ranson's studies, on captive voles and rats in cornricks (Leslie and Ranson, 1940, 1952), and Deevey's analysis of Murie's data on the Dall sheep (Deevey, 1947). Towards the end of my study Caughley (1966) produced a critical analysis of more recent life tables which were emerging, together with his own for thar (Himalayan wild goat). Since that time a number of life tables for African ungulates has been produced (Spinage, 1972), few, however, attaining the stringent demands of Caughley.

The difficulties of deriving life tables for wild mammal populations are usually those of obtaining a large enough sample on which to base a cohort, and of the accurate determination of age. Although methods of more precise age determination are now available to us, there can still be a large amount of error involved (Spinage, 1973, 1976). Generally speaking, it would appear that the robust nature of the survivorship curve is not likely to be significantly affected by lesser errors, especially in the determination of age of older individuals. This is contended by Caughley (1977); but would seem more likely to hold true if the analysis of vital statistics is attempted in depth. We are now in possession of enough data to know what type of survivorship curve to expect, in an undisturbed large mammal population, and what may influence deviations from this expected curve.

There are two ways in which a survivorship curve can be constructed. The first, the age-specific survivorship curve, is obtained by following a cohort of animals, all born at the same time, through to the death of the last member. There can be no doubts about the validity of such an analysis, provided that emigration is not confused with mortality. Unfortunately, few workers with large mammals can use such a method, for although some large populations, such as the Serengeti wildebeest, produce all of their young at the same time, and in sufficient numbers to permit a meaningfully large cohort to be marked at birth; with possible life spans of 15 to 20 years, or more, it becomes almost a lifetime's occupation to observe the cohort.

The alternative to this daunting prospect is the time-specific survivorship curve. In this analysis the members of the cohort all have different birthdays, but are treated as if they had all been born at the same time; the deaths being added from the last survivor backwards. Such a curve can be derived from a sample of a living population, or from a sample of found skulls, the ages of which can be determined. It was the latter method which I used to derive my survivorship curves for the Park. However, unless certain restrictive assumptions are met, a time-specific survivorship curve, or life table, can represent no more than a crude generalization of the probable structure of a population at the time that the data were collected. It cannot tell us how that structure was attained, nor what its future pattern might be. A requisite of such a life table is that the population is stationary in size, natality equalling mortality. If the population is increasing or decreasing in size, the data must still be forced into a model in which births are equal to deaths, so that the distribution of the age classes is distorted in pattern, and the life table can no longer represent a true rendering of the age distribution of the population. My studies in the Park suggested that the population was stationary at the time (Chapter 8), so that this requirement was met and tentative life tables could be produced.

It is still not easy to obtain such data from wild populations. I found only 134 skulls in 3 years, representing about 4% of the total population. This was similar to the estimated proportion which I collected in another study in the Akagera National Park of Rwanda, which was 3·8%. It is little wonder that if adult skulls are so difficult to recover, then those of young animals are even more so.

Waterbuck Survivorship

In animal demography attention has frequently been drawn to a lack of data on the mortality of young animals (Bourlière, 1959; Caughley, 1966), which results from the easy destructibility of the delicate remains. But there are approaches to this problem. The weaning ratio (Chapter 6) gives an approximation of losses up to weaning, and in a continuously breeding, monotocous species, the state of lactation of the doe enables us to construct a tentative survivorship curve for the fawn in its pre-weaning period (Spinage, 1968). In the waterbuck this is for the first 8 months after birth. A multiparous doe could be considered to be lactating up until at least the eighth month of pregnancy, and if she was not lactating, to have lost the fawn. If a doe was not lactating the elapsed duration of pregnancy, determined from the age of the fetus by the method of Huggett and Widdas (1951), suggested that the fawn was lost some time prior to that age. If we take the interval between lactation ceasing, plus the interval to the next conception, as 2·17 months (lactation anoestrus 20 days, interval between pregnancies 45 days), then this should provide us with our starting point. The age of gestation, plus 2·17 months in pregnant, non-lactating does, then tells us the maximum age at which the fawn must have died. Thus if a doe is 4·5 months pregnant and not lactating, the preceding fawn must have died some time during the preceding 6·7 months. We cannot, of course, say exactly when, and only by viewing the total period can we derive an approximate rate of survival. In so doing we must disregard the first two months of pregnancy, for a lack of non-lactating does revealed in this period may simply mean that they had lost their fawns but had not yet dried up. Likewise, the last month of pregnancy must be ignored, as during this period the doe dries up in preparation for the production of colostrum for the next birth.

A survivorship curve based upon pregnant, lactating does, has less inherent error, and in this case we know that the fawns have definitely survived to each stage. Thus a lactating doe which is 4·5 months pregnant, must have a fawn at heel of approximately 6·7 months of age.

But the same limitations exist for the first two months and the last month of pregnancy.

Tentative survivorship curves were constructed in this way and found to be similar in shape for both samples (Table 28 and Fig. 29). Bearing in

Table 28 Lactation state of multiparous does expressed as fawn survival in months.

Time from last parturition (months)	Age of fawn	Lactating does	As fawn survival	Non-lactating does	As fawn survival
2·17	0				
3·17	1	3		0	
4·17	2	3	20	4	28
5·17	3	7	16	4	23
6·17	4	1	9	1	19
7·17	5	4	8	6	18
8·17	6	2	4	4	12
9·7	7	2	2	8	8

See text for explanation.

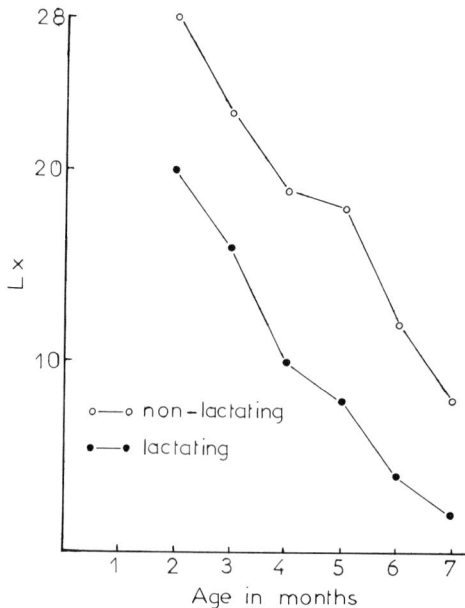

Fig. 29. Survivorship curves of fawns based on lactation state of the doe.

mind the constraints of the method, and the small sample size, these curves suggest that fawn mortality in the first 7 months of life is almost random with respect to age, although there was some hint of an increase in rate of survival at 4 months of age. The independent nature of the two samples, which both exhibited this inflexion, lends support to this hypothesis.

The ratio of non-lactating does to lactating does was 54·9%, which was close to the weaning ratio of 52·3%.

On the Peninsula, 16 assumedly parous does were kept under observation for 27 months from the time that they were captured and marked. In this period, one gave birth three times (Y120), and all potentially could have given birth at least twice. But only 22 births out of a possible 32 were recorded. Of these 22 births, eight fawns are known to have died within the first 2 months of life; a total of 36%. Assuming that all of the does produced at least two fawns during the period of observation, this suggests a total mortality of 18, or 56% in the first 2 months. Interpolating this value into our lactation ratios gives an estimated mortality of 58% in the first 7 months of life. Whether one considers these extrapolations too tenuous or not, it is clear that fawn mortality up to weaning is very high.

From 1963 to 1967, a total of 104 buck skulls, and 33 doe skulls, was found in the Park; a ratio of 3·2 : 1. The ages at death of these skulls were determined by stages of tooth eruption, degree of tooth attrition, and cementum line counts. Higher rewards were offered for the recovery of doe skulls, which makes it seem unlikely that the preponderance of bucks was due to selectivity in collection. More likely it was due to the adult buck skulls being less manageable by hyaenas, who could not carry them away to their burrows. It is unlikely that the buck skulls survived weathering better, as the doe skulls did not show any more marked deterioration after death than those of the buck. In the Akagera Park of Rwanda, I collected 1·6 times more buck than doe skulls ($n = 47$) (Spinage, 1972); but up to 6 years of age, five times more doe skulls were recovered than buck, an unexplained anomaly. In the Park, the ratio was approximately equal up to this stage (13 buck to 11 doe). In her study, Elliott (1976) found 1·5 times more buck than doe skulls ($n = 64$), but during the study 1·3 times more does died than bucks ($n = 52$).

To facilitate comparison only the first 99 buck skulls collected were used in analysis. The numbers of animals dying in each age group, when compared in equal cohorts, showed an apparent increased rate of mortality at ages 4 to 5 years in the doe, which was not shown by the buck. Mortality then declined until senescent mortality showed a sharp peak at age 8 to 9 years. The buck showed a steady increase in the rate of mortality, peaking at 8 to 10 years, which could be associated with an increasing struggle to maintain its territory; old bucks usually finally being driven out at the age of ten (Chapter 10) (Fig. 30). The Akagera collection was insufficient to demonstrate any trends.

As the counts within the study areas, other than the Peninsula, suggested that the population was stationary in size (Chapter 8, Table 33), a tentative life table was constructed from the found skulls, assuming 50% mortality in the first year. But we should bear in mind, following Caughley's caution (Caughley, 1966), that a sample size of less than 150 skulls is unlikely to produce a valid life table. But since the life table is a robust analysis, at least as far as survival and mortality are concerned, it can be considered as a crude approximation to the probable age structure of the population. Such a life table gives a mean length of life for the buck

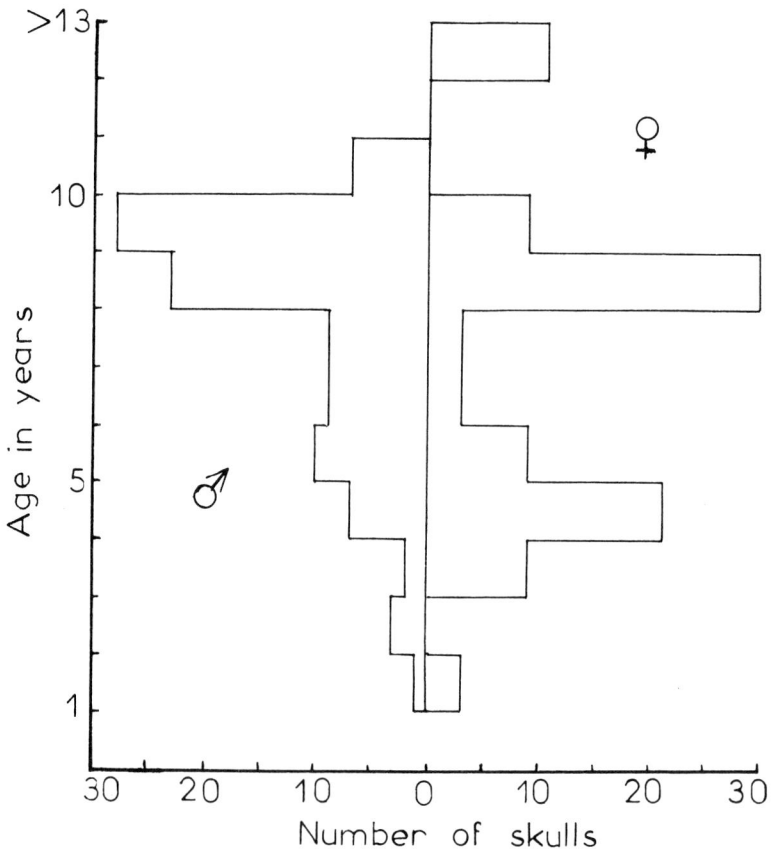

Fig. 30. Frequency distributions of the numbers of found skulls.

of 4.6 ± 1.7 years, with an expectation of life at birth of 4 years. For the doe the mean length of life is 4.4 ± 3.5 years, with an expectation of life at birth of 3.8 years. Thus there is no significant sexual disparity in mortality ($t = +0.025$, df 20, $p = \ll 0.1$, for mean length of life), and the values are of the expected order of magnitude (Table 29). If, indeed, first year mortality was higher, say 60% and not 50%, this would make a decrease of only 0.3 years in the mean length of life. But if it was of the order of 80%, then this would decrease mean length of life by almost a half.

Table 29 Tentative life tables for buck and doe waterbuck in the Rwenzori National Park, Uganda.[a]

Age	Frequency	Survival	Mortality	Mortality rate
x	f_x	l_x	d_x	q_x
Buck				
0	(198)	1.000	0.500	0.500
1	(99)	0.500	0.005	0.010
2	1	0.495	0.015	0.031
3	3	0.480	0.010	0.021
4	2	0.470	0.035	0.075
5	7	0.434	0.050	0.116
6	10	0.384	0.046	0.118
7	9	0.339	0.046	0.134
8	9	0.239	0.116	0.397
9	23	0.177	0.141	0.800
10	28	0.035	0.035	1.000
11	7			
Doe				
0	(66)	1.000	0.500	0.500
1	(33)	0.500	0.015	0.030
2	1	0.485	0	0
3	0	0.485	0.046	0.094
4	3	0.439	0.106	0.241
5	7	0.333	0.046	0.136
6	3	0.288	0.015	0.053
7	1	0.273	0.015	0.056
8	1	0.258	0.152	0.588
9	10	0.106	0.046	0.429
10	3	0.061	0	0
11	0	0.061	0	0
12	0	0.061	0.061	1.000
13	4			

[a] Standard demographic notation is used and the values expressed as probabilities.

The survivorship curves (l_x) (Fig. 31), and the force of mortality curves (q_x) (Fig. 32), derived from these data, take the form for the buck that we might expect for an undisturbed large mammal population, the survivorship curve being convex in shape, indicating an environmentally resistant, mature population. The force of mortality curve is "J"-shaped, indicating that the rate of mortality declines from birth to the second year, but then, in the buck, increases again continually throughout life. In the doe it probably more or less levels out in the prime of life, and then accelerates in maturity to the termination of life. More specifically, the doe rate of mortality appears to show an acceleration at 3 years, which could be a result of dystocia or *post partum* death in the primiparous doe, rather than a spurious result of inadequate sampling, as this acceleration is also shown by the much larger sample of collected does $(n = 87)$ (Fig. 33).

As time-specific life tables represent a stationary population, the net reproductive rate (R_o) is unity, and the intrinsic rate of natural increase (r_m) is zero. Thus we can proceed no further with our doe life-table calculations, for we know what the answer should be, and any departure from the expected answer would simply be a reflection of the inaccuracy of the mortality data.

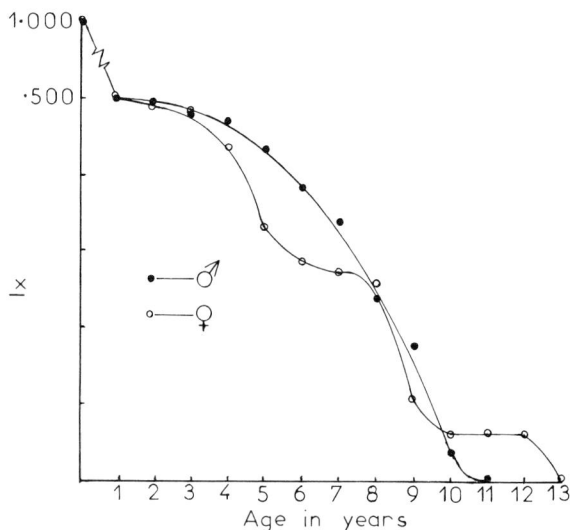

Fig. 31. Survivorship curves for bucks and does in the Rwenzori Park.

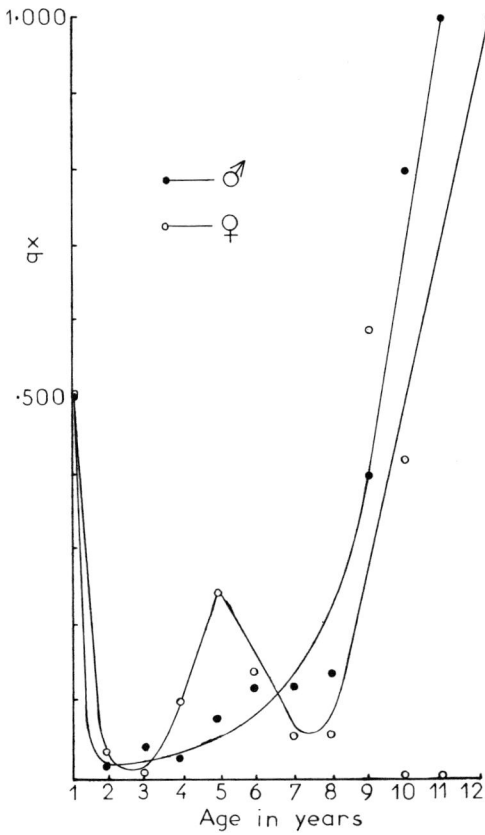

Fig. 32. Smoothed force of mortality curves for bucks and does in the Rwenzori Park.

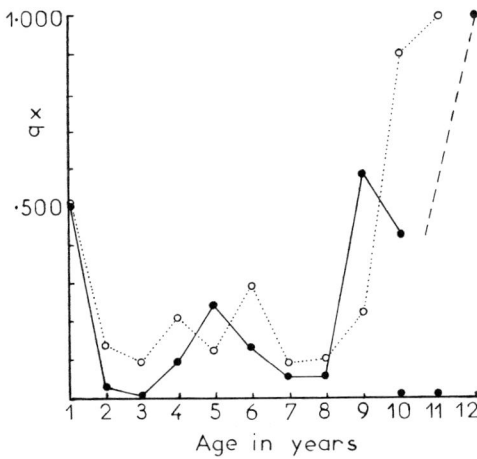

Fig. 33. Force of mortality curves for cropped does, dotted line, and natural mortality, continuous line.

By means of tooth impressions I was able to determine the ages of the entire Peninsula buck population ($n = 22$). This age structure suggested a somewhat erratic recruitment to the population, with a tendency to recruit groups of animals of approximately the same age (Fig. 58). A "survivorship" curve was constructed in order to illustrate how the Peninsula population differed from the Park mean. Although such a derivation is representative of the exact age structure of the population, absent from all errors other than those possible in age determination, it is not, of course, a real "survivorship" curve. This is because its form is determined by a large unknown emigration factor, rather than by mortality. But it does suggest that an apparent age-constant survival curve could, in fact, be shaped by a large degree of emigration rather than by mortality alone (Fig. 34).

I succeeded in determining the ages of only two-thirds of the Peninsula doe population ($n = 29$), which when expressed as survival, possessed the shape of an age-constant survivorship curve, resulting, apparently, also largely from emigration (Fig. 35).

Factors Limiting Increase

Survivorship curves aid in telling us how a population is behaving, but give us no intimation as to what limits the numbers in a population. Since Malthus first focussed attention on it (Malthus, 1798), ecologists have long puzzled over the fact that populations do not continually increase exponentially in size, until exhaustion of environmental resources cuts them down. But when we consider all of the chance factors which militate against an animal's survival, I am more inclined to the

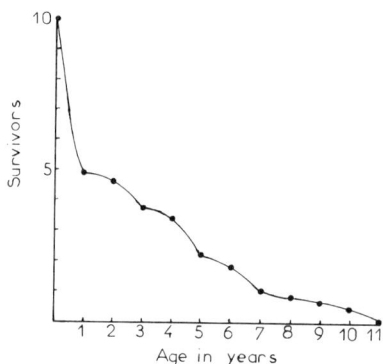

Fig. 34. "Survivorship" curve of Peninsula bucks.

view that it is a wonder that any survive at all, let alone increase in number. As we have noted, apart from a localized increase on the Peninsula, the waterbuck population of the Park appeared to be stationary. There are many factors which could contribute to this.

A doe may have her first fawn at a little under two-and-a-half years of age, and is then capable, on average, of producing another every 10·8 months up to a maximum of 10 years; a possible lifetime production of 11 offspring. Clearly not every doe is up to this rate of performance, but more importantly, only a very small number will live out their full reproductive life. The mean length of life was calculated as 4·4 years, allowing a mean production of only 2·1 offspring, which is just sufficient to keep the population constant in size, if all does contribute.

Infertility was uncommon in the animals which I examined; only one in eight was not pregnant, and this did not necessarily imply infertility. But there was a tendency in those of 10 years and older to ovulate more frequently without conception taking place (Table 26). These does, however, represent the smallest sector of the population, and thus contribute the least.

One of the quickest ways in which to retard population increase is to lengthen the interval between the production of one fawn and the next (the other is to delay the date of first conception but this rarely happens). Lengthening the calving interval effectively disrupts the rate of production, as the doe is unable to shorten the length of gestation in order to recover lost time. If a breeding doe is merely removed from a population, she simply makes way for others, and the rate of production is not retarded.

In their studies on the red deer, Guinness *et al.* (1978b) found that the hind's reproductive performance differed widely between different years,

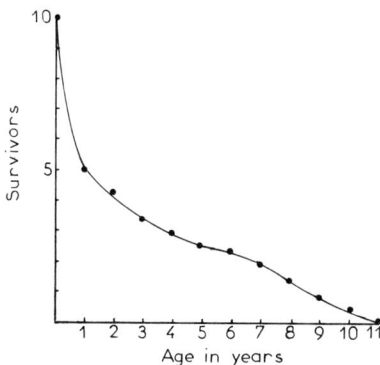

Fig. 35. "Survivorship" curve of Peninsula does.

and that within a population the dam's home range had an important effect upon most aspects of its reproduction. Those in areas of good grazing showed high fertility (especially among young hinds), high birth weights and early birth dates (important in this species in view of the winter period). Poor areas showed the opposite, while intermediate areas produced intermediate effects, such as early birth dates but low birth weights. The question of the neonate's survival is therefore revealed as a very complex affair, but I feel that, in considering African ungulates, we should be guided by my earlier reservations with respect to the difference in habitats between temperate zone and tropical zone species; tropical species probably being better adapted to their habitats, which have suffered less disturbance by man.

As with does, so with the bucks it is most unusual to find an infertile one. An exception was a territorial Peninsula buck which was accidentally killed during marking. This animal was found to be infertile due to perforation of the scrotum by a thorn, the seminiferous tubules being purulent. It will be readily appreciated how such an animal upsets production, by keeping oestrus does in his area, but not fertilizing them. Of course, such does may be mated by potent bucks in their next cycle, by which time, if they are virgin does, or if they have lost their fawns, they may have moved into another buck's territory. But the calving interval will have been irretrievably lengthened, and, as we have seen, the chances are that a doe will return to the same area to calve throughout her life; and it is there where she comes into her first *post partum* oestrus. If this has become the area of an impotent buck, then the effect will be cumulative.

Factors Affecting Fawn Survival

The above are what we might term "primary factors" which retard population increase, but there are also many secondary factors; the many accidents in life which affect the survival of the individual.

Some fawns are born with congenital defects, but it is rare that a chance occurs to examine them; although one case was reported to me of a fawn which was unable to feed because of a "twisted" oesophagus. Such causes of death are, however, undoubtedly much rarer than are the many accidents which may befall a fawn.

A captive fawn, reared to the age of 7 months, died of a perforated oesophagus, which had probably been pierced by a thorn, the wound then ulcerating. I recorded six fawns lame from sprains, fractures and dislocations of the leg. Although I have found a case in an impala of a complete fracture of the femur healing successfully (Spinage, 1971), such injured animals are vastly more prone to predation. There was the case of the fawn which, although killed by a leopard, would have died anyway as the doe was not lactating (Chapter 5). The fawn was 24 days old when killed so the dam may have lactated for a period after giving birth.

Predation, although relatively unimportant in the adult, probably outweighs all other factors affecting fawn survival. The main predator of fawns in this area was the spotted hyaena, with the leopard taking an unknown portion. Other predators, such as the lion, are probably rarely involved, and only one case was recorded. Elliott (1976) recorded a 3-month-old calf as probably dying from over-exertion after escaping predation, and another case where a jackal took a calf.

The role of disease is completely unknown, but may well be a significant factor. It is dealt with more fully in the next section. This host of natural factors operating adversely against the fawn, makes its high infant mortality a matter of no surprise.

Factors Affecting Adult Survival

Those factors which may terminate the life of the adult are: senescence, starvation, accidental death, fighting, sickness, disease and predation. It is popularly held that wild animals almost never die from the first of these factors, senescence, predation overtaking them first. Death from senescence is, however, more common than opinion would generally suppose, and is often found among the largest of the mammals, such as the elephant, hippopotamus and buffalo, but is also not uncommon in mature populations of other species in areas where there are few predators.

Senescence

To determine such a cause of death requires that very close observations be kept of old animals, and I was able to record only one instance of a very aged doe which apparently succumbed to senility, the body being found in a thicket, untouched by scavengers. This is the one which I estimated to be 18·5 years of age (Chapter 4). I also recorded a case of death from dystocia, the remains of a doe being found with those of her neonate. As this doe was estimated to be 13 years old, it would seem likely that this accident of parturition resulted from senility.

Starvation

No deaths were recorded that could be attributed to starvation, but a poor diet is probably a contributory cause of death in old bucks in the dry season, as they tend to be forced to inhabit the least favourable areas, where the preferred species of grass may be in short supply. But in general, in this region, since the waterbuck appears to balance its protein-deficient dry season graze with protein-rich browse (Chapter 8), death from starvation is likely to be rare. There are contrary records from other areas. During a bad drought in Kenya in 1960 to 1961, the waterbuck population of the Nairobi National Park decreased by approximately one third (Foster, 1968). Herbert (1972) reports 12 deaths allegedly from starvation in a private game reserve in the eastern Transvaal during a 9-year period; while Stevenson-Hamilton (1929) recorded waterbuck as dying in "considerable numbers" at the end of a drought in the Kruger National Park of South Africa in 1916. Hirst (1975) reports a high mortality rate as "commonly seen in waterbuck populations" in poor years in South Africa, but this seems to relate to only six deaths observed during the winter of 1964 (Hirst, 1969). Conditions are undoubtedly different in South Africa, the winters approaching the limits of the waterbuck's ability to maintain thermoneutrality, thus placing greater demands upon its bodily resources (Chapter 9).

Enterotoxaemia

If a herbivore survives starvation, then it is next liable to succumb to enterotoxaemia. The condition of the waterbuck in the Park, as expressed by kidney fat deposits, was at its lowest after the commencement of the short rains, co-incident with what would be expected from a green-grass-loss. No cases of scouring, indicative of enterotoxaemia, were seen at this time, and it is possible that, with its high water and high protein intake, the waterbuck is not as affected by the sudden exuberance of green grass as are some other species.

If a waterbuck is suddenly approached, it removes itself to a safer distance, regards you, and then invariably defaecates. Payne (1964) has pointed out the importance of faecal water as a measure of water intake; so making use of the waterbuck's reaction, I collected fresh faecal samples from two known bucks of the same age, each month of the year for 6 months, including the month of April when fat reserves were lowest. The samples were recovered immediately upon voidance, wrapped in foil, and weighed in the laboratory within a few minutes of collection. They were then dried to constant weight and re-weighed. This showed that the waterbuck maintained its water balance within rather narrow limits, although faecal water content was significantly higher, at the 95% level of probability, in both animals during April ($t = -3.99$, df 4, $p = > 0.98 < 0.99$, and $t = -4.8$, df 4, $p = > 0.99 < 0.999$), with an average increase of 3·23% (Table 30). It is interesting to note that one of the two animals had a consistently higher water content in the faeces than did the other, averaging 3·5% more ($t = -2·88$, df 10, $p = > 0·99$ $< 0·999$). That this difference between the two was only slightly greater than the seasonal increase, suggests that the latter is insufficient to represent any physiological disturbance, supporting the hypotheses that the waterbuck has an efficient method of coping with increased water intake, and is not liable to suffer an excessive green-grass-loss, nor to develop enterotoxaemia as readily as some species.

Table 30 The percentage of moisture in fresh faecal samples per month, January to June 1966.

Month	Jan.	Feb.	Mar.	Apr.	May	Jun.	Mean
Rainfall (mm):	36	44	81	149	46	88	
Buck Y7 (7 yrs) n	3	4	1	3	1	3	
% water	72·0	72·6	73·4	76·4	71·5	73·5	73·23
Buck Y11 (7 yrs) n	3	3	3	3	2	3	
% water	75·3	76·7	75·2	80·0	76·7	76·4	76·72

n = number of samples collected during the month.

Accidental death

A waterbuck died in the Murchison Falls Park of northern Uganda by falling into a gulley when fighting with another. No such fatal accidents were recorded by me. One adult buck and one adult doe were seen to be lame; the doe eventually recovered, but the buck disappeared. Such accidents obviously make the animals more prone to predation, especially as the does tend to reject lame members from the group (Chapter 10).

Fighting

Death from fighting is not uncommon, as we might expect from a territorial antelope, and a popular tendency to regard wild animals as non-intentional killers, I find to have no basis in fact. Statements such as that of Matthews (1964): "the more I have sought examples of such intraspecific overt fighting in mammals the less I have succeeded, and I doubt that it normally occurs in nature", are made in the absence of detailed studies of life histories in the wild.

This view is supported by Geist (1971) who deals at some length with the misconceptions that have grown up from the writings of Lorenz (1966), Eibl-Eibesfeld (1963, 1966), Barnett (1967) and Tinbergen (1969). He points out that "It is a popular game today, as in the past, to scrutinize the actions of animals and exalt examplary conduct, thereby justifying it for humans". The apogee of this approach was reached in mediaeval theology, when animals were endowed with virtues representative of God's meaning. Geist refers to combat deaths in several species of temperate ungulate; "These figures . . . suggest that ethology's conventional wisdom, that animal rarely kill or maim each other in intraspecific fighting, has been overstated. Ewer (1968) has also cautioned against its overenthusiastic acceptance." He goes on to state that "Intraspecific combat tends to be very dangerous in the majority of ungulates since these are large, agile animals often armed with lethal weapons and quick to fight. Such fights end more commonly in serious injury or even death of one opponent than is generally appreciated . . .". Lorenz (1966) considered that animals with dangerous weapons rarely used them against conspecifics, but this is exactly what they are used for, and quite frequently! In a recent study in South Africa, Wilson and Hirst (1977) considered that intraspecific aggression was a very important source of mortality in both sable and roan populations, species with large and conspicuous horns.

I recorded three deaths from fighting in the Park, resulting from abdominal horn wounds. Herbert (1972) records one instance in his South African study, and I have even witnessed a case in the north of the Central African Republic in which a buck, suffering from an abdominal horn wound, apparently voluntarily drowned itself by thrusting its head under shallow water, and keeping it there. Seeking water as a balm for abdominal wounds is not uncommon, and I witnessed this behaviour in the Rwenzori Park (Plate 37); obviously this habit could lead to a lack of recovery of carcases of animals that have died from fighting.

Elliott (1976) records one buck killing another on Kenya's Crescent Island in Lake Naivasha, but no deaths from fighting were recorded in her high density study area.

Sickness

It was seldom that persons with a knowledge of veterinary pathology were present at autopsies, but when they were the following pathological disturbances were noted:
   (i) perforated ulcer in the omasum of a 12-year-old doe;
   (ii) caseous cyst under the skin of a 10-year-old doe;

(iii) pleuritic lung in a 2-year-old doe, probably parasitic bronchitis resulting from lungworm infection;

(iv) small vesicles in the meninges of the brain.

In addition, a waterbuck was found in Tanzania with numerous kidney stones. Elliott (1976) records a 6-year-old buck dying from suppurative pneumonia, another from glomerulonephritis and a rumen ulcer, while a 10-year-old doe appeared to have died from prussic acid poisoning.

Plate 37. A buck dying of a horn wound in the abdomen.

Viral diseases. No epidemics affected the population during my study, and no records exist of any past ones. Rinderpest, the most virulent of wildlife diseases, does not seem to affect the waterbuck to any great extent. When Lugard was witness to some of the effects of the great rinderpest plague in this area at the end of last century, he thought that the waterbuck must be immune (Lugard, 1893). Following up this supposition, Mettam (1936) inoculated one with the virus, inducing a severe reaction, but the animal survived. He concluded that waterbuck in Uganda were seldom infected naturally. Later he wrote,

> "they appear to fear the disease for they have been observed to trek away from rinderpest infected places. One herd which was observed moving north from Singo (where rinderpest existed) towards Lake Wamala (where it did not) numbered well over the 280 animals counted" (Mettam, 1937).

Unlikely as Mettam's conclusion seems to be of waterbuck "fearing" rinderpest, this is the only account, to my knowledge, of waterbuck migrating. If indeed this was really what they were doing. They have not always shown immunity to the disease, and were reported as slightly affected by an outbreak of rinderpest in Kenya in 1960–1962 (Stewart, 1964). In the Central African Republic, following an alleged outbreak of the disease in 1969, Thal (1972) found 8 out of 13 waterbuck sampled, positive for rinderpest antibodies, but no animals were recorded as dying. There have been no outbreaks of the disease in the Rwenzori Park since 1944–1945 (Plowright *et al.*, 1964), so it is unlikely that any of the present population would have had contact with it.

In the Kruger National Park one waterbuck is reported to have died of foot-and-mouth disease (Pienaar, 1961), and in Kenya, one out of three animals sampled showed a positive antibody response to this disease (Fay, 1972). The Kenya study also showed positive responses to bluetongue and to bovine virus diarrhoea.

In the Central African Republic, Thal (1972) obtained positive antibody responses to several of the mosquito-borne arboviruses: Group A (Sindbis), Group B (Yellow Fever, West Nile, Uganda S, Zika, Wesselbron) and ungrouped Zinga. Several of these viruses are fatal to man.

Bacterial diseases. No bacterial diseases affecting waterbuck were identified in the Park. I saw one adult doe scouring and in a weak condition. She disappeared, but I was unable to locate a carcase. In the Kruger National Park, during an outbreak of anthrax in 1959–1960 and 1962, 75 waterbuck were reported to have died from the disease, representing 7% of all deaths recorded from it (Pienaar, 1967). One positive brucellosis isolation was made in Herbert's study (Herbert, 1972), but results in a Kenya study were negative for this disease (Fay, 1972).

Protozoal diseases. Tsetse flies, the vectors of trypanosomiasis, like the waterbuck least of all, and the anomaly of this is that the waterbuck has been reported as fairly heavily infected with trypanosomes, namely *Trypanosoma vivax,* and the *T. congolense* — *T. brucei* group (Wenyon, 1926). Other protozoal parasites recorded for the waterbuck are *Theileria* sp. (Brocklesby and Vidler, 1965; Thal, 1972), and *Anaplasma* sp. (Fay, 1972), while positive antibody results have been obtained for *Theileria parva, Anaplasma marginale* and *Babesia bigemina* (Fay, 1972). These produce the diseases in domestic stock known as East Coast Fever, Gall-sickness and Redwater.

In South Africa, Wilson and Hirst (1977) reported "cytauxzoon-like parasites" in smears from the waterbuck, *Cytauxzoon* being a parasite of the erythrocytes which can cause a high mortality in some antelopes.

I found one heavy infection of the little-known sporozoan parasite *Sarcocystis,* in the hind leg muscle of a 9-year-old buck. This parasite is not uncommon in some antelopes, but its effects are unknown.

Endoparasites. No heavy infestations of endoparasites were noted in the animals which I examined. On the whole waterbuck appeared to me to be remarkably "clean" animals in this respect. But detailed investigations in Kenya of the frequency of nematode worms in the gastro-intestinal tract, showed the waterbuck to be the third most heavily infested of the eight species of plains herbivore examined. However, it came eleventh out of 12 animals examined for the quantity of strongyle eggs passed in the faeces, but conversely, had a relatively heavy infection of strongyloides (Fay, 1972).

I occasionally found tapeworm cysts in the liver, and a single specimen of the tapeworm *Moniezia benedeti* was found in the rumen of a 7-month-old fawn. Fay (1972) recorded five of 26 waterbuck examined to be infested with "intestinal tapeworms". In Central Africa, Thal (1972) found heavy infestations of *Stilesia hepatica* in the biliary canal, and he also found the liver fluke *Fasciola gigantica,* which was absent from all the specimens which I examined. I found two paramphistomes, *Paramphistomum phillerouxi* Dinnik and *Carymerius* sp. occurring together in the rumen; numerous in some animals, and absent in others. Paramphistomes were also reported as common in the Kenya study of Fay, and in Thal's study, the latter noting "several species". Thal also found the larvae of *Linguatula nuttali* in the heart chambers. I found whipworms, *Setaria* sp. to be common in the peritoneal cavity; these are probably the source of the microfilariae recorded in the blood by Brocklesby and Vidler (1965). In animals from the Selous Game Reserve of southern Tanzania, *Setaria castroi, Haemonchus* sp., *Cooperia* sp., *Longistrongylus* sp., and *Bunostomum* sp. have all been recorded (Sachs, R. in Gainer, R.S., 1979). In Kenya, Elliott (1976) recorded larvae and eggs in the faeces of *Eimeria aubumensis, Bunostomum trig-*

onocephalum, Trichuris ovis, Cooperia sp., *Moniezia* sp., *Paramphistomum* sp., and *Trichostrongylus* sp., the highest numbers occurring in calves.

In South Africa, Herbert (1972) recorded *Paramphistomum microbothrium* in the rumen, *Calicophron calicophron* in the body cavity, and *Taenia* sp. (probably *Stilesia* sp.) on the liver surface. He also recorded *Linguatula serrata* (sic) from the heart.

Ectoparasites. The most commonly recorded external parasites were Ixodid ticks, of which only two species were found: *Amblyomma cohaerens* and *Rhipicephalus tricuspis*. The former was by far the most common, its main sites of attachment being the legs and axillae, and, in does, the teats. The presence of ticks on the teats was an indication that the doe was not suckling a fawn. Large numbers of *Rhipicephalus* sp. were found on fawns of only a few hours *post partum*, so they must transfer almost immediately after parturition, either from the doe or from the grass.

Herbert (1972) recorded *Amblyomma hebraeum, Rhipicephalus appendiculatus* and *R. evertsi* on waterbuck in his study, with infections apparently heaviest in January to June. Other ticks which have been found on waterbuck are given in Table 31.

I treated almost entire skins of waterbuck in a solution of 5% caustic soda, which dissolves hair and flesh, but not the chitinous exoskeletons of arthropods (Spinage, 1969). By this means I found that numbers in excess of 4000 ticks, comprising mainly the young stages of nymphs and larvae, were not unusual on a healthy host (Table 32).

When an animal is sick, or about to die from some debilitating cause, its ticks appear to engorge very rapidly, probably because of a breakdown of the host's immunological resistance. This tends to give the erroneous impression that the ticks may have been responsible for the animal's death. Hutchins (1917), for example, reported a waterbuck as having died of anaemia from an infestation of *R. appendiculatus,* a most unlikely event, in my opinion.

Another common ectoparasite is the biting louse *Damalinia (Bovicola) hilli,* which probably occurs on most waterbuck, although it is often overlooked. It is not numerous, and using the same technique of dissolving the skin, I estimated up to 500 to occur on one host, with as few as 11 on others (Table 32 and Plate 38).

Undoubtedly the parasite which causes the waterbuck the most distress in the Park, is a parasitic mite of the sub-order Mesostigmata. This rather strange mite, *Railletia hopkinsi*, lives in the eustachian tube of its host, and was first described by Radford (1938) from specimens collected from a defassa waterbuck on the Aswa River of northern

Table 31 Tick species recovered from waterbuck.

Species	Country[a]	Authority
Amblyomma cohaerens	Uganda	Spinage (1968)
	Zaïre	Doss *et. al.* (1974)
gemma	Kenya	Walker (1974)
hebraeum	South Africa	Herbert (1972)
pomposum	Mozambique	Doss *et al.* (1974)
splendidum	Central African Republic	Thal (1972)
	Portuguese Guinea	Doss *et al.* (1974)
variegatum	Mozambique	
	Portuguese Guinea	
	Zaïre	Doss *et al.* (1974)
Boophilus decoloratus	Mozambique	Doss *et al.* (1974)
Haemaphysalis aciculifer	Sudan	Doss *et al.* (1974)
	Kenya	Elliott (1976)
Ixodes cavipalpus	Kenya	Walker (1974)
Hyalomma nitidum	Central African Republic	Thal (1972)
rufipes	Central African Republic	Thal (1972)
truncatum	Cameroons	Doss *et al.* (1974)
Rhipicephalus appendiculatus	Tanzania	
	Zaïre	Doss *et al.* (1974)
	Kenya	Fay (1972)
	South Africa	Herbert (1972)
cliffordi	Central African Republic	Thal (1972)
evertsi	Tanzania	Doss *et al.* (1974)
	South Africa	Herbert (1972)
kochi	Mozambique	Doss *et al.* (1974)
lunulatus	Central African Republic	Thal (1972)
maculatus	Mozambique	Doss *et al.* (1974)
muehlensi	Mozambique	Doss *et al.* (1974)
pravus	Mozambique	Doss *et al.* (1974)
pulchellus	Tanzania	Doss *et al.* (1974)
	Kenya	Fay (1972)
punctatus	Mozambique	Doss *et al.* (1974)
reichenowi	?	Doss *et al.* (1974)
sanguineus	Mozambique	Doss *et al.* (1974)
simus	Mozambique	Doss *et al.* (1974)
tricuspis	?	Doss *et al.* (1974)
zumpti	Mozambique	Doss *et al.* (1974)

[a] Most of the species undoubtedly have a wider distribution than that which is indicated from the records.

Table 32   Numbers of ectoparasites recovered from four waterbuck skins.

Collection date	Sample	Ticks			Lice
		Adult	Nymph	Larva	
Nov 1964	10-year buck, legs only	4	406	462	25
	Fore half of body minus legs	0	169	132	82
Apr 1964	9-year doe, complete skin	700	1780^a		16
Jly 1965	10-year buck, right half of body with legs		2033^a		247
Jun 1966	6-year doe, complete skin	140	612^a		11

[a] Stages not separated

Plate 38.   *Damalinia hilli*, a biting louse common on waterbuck.

Uganda, in April 1936 (Plate 39). It was very common in the Park, its presence being easily identified by the occurrence of black (from the dried blood), scab-covered swellings at the bases of the host's ears, which result from the host's persistent scratching. Examination of histological sections of these swellings confirmed Hinton's report (Dollman, 1931) that they showed no sign of glandular cells, contrary to what is sometimes suggested. The swellings showed multiple extravasation and considerable pathological disturbance of the tissue. On old animals, as much as half of the cheek may have been rubbed bare by the host. Scabs were found on 3-year-old animals, and the incidence in mature animals was over 50% in most areas. An exception was the Peninsula, where the infection was apparently benign.

By contrast, in a higher rainfall area in the north of the Central African Republic, I saw no scars whatsoever in a region where waterbuck were

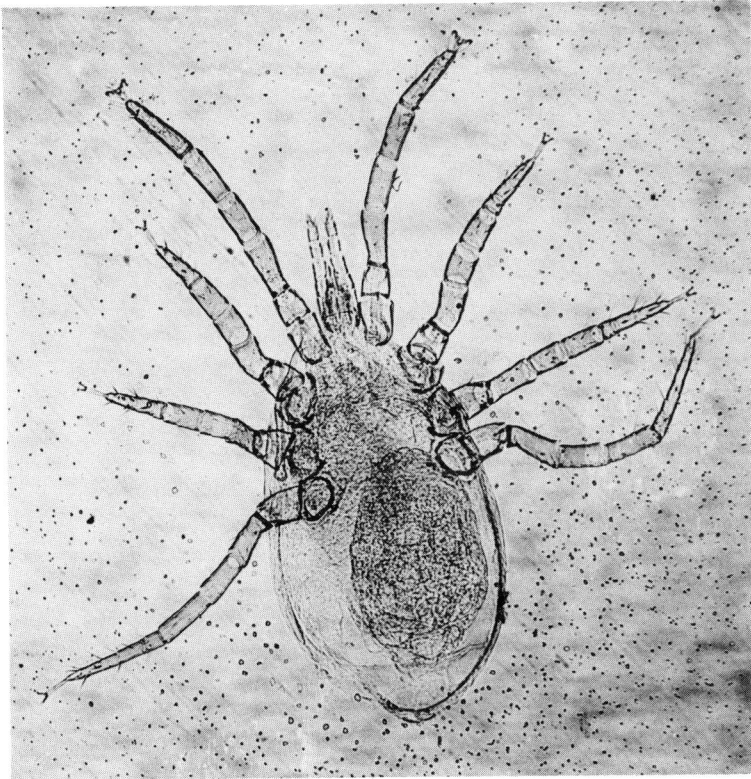

Plate 39. *Raillietia hopkinsi,* the ear mite of the waterbuck.

common, but noting some mud at the bases of the ears of an adult buck, which had just been killed by a lion, I examined the eustachian tubes and recovered one mite from each ear. The absence of scars could suggest that the normal state is an infection low enough that irritation is not excessive, and such low infection must be attributable to an absence of ready transmission. In the Rwenzori Park it would seem that conditions are exceptionally favourable for transmission, much to the waterbuck's disadvantage. Strangely enough, the mite may not be a parasite in the true sense, for in cattle, one of its few known hosts, it is said to feed upon debris in the ear. No blood meals were present in the specimens that I recovered. Clearly, however, it can cause an unbearable irritation, which is unusual in an ectoparasite, but as it is protected within the ear, no selection can act against its survival.

The only other known hosts of *Railletia* are domestic cattle for *R. auris*, and, of all creatures, the Australian wombat for *R. australis*. But observing similar patches to be present under the ears of Uganda kob, I came to the conclusion that this animal was also a host. With the aid of a colleague, a new species, *R. whartoni,* was found to infest the Uganda kob, which seems to be closely related to *R. hopkinsi* (Potter and Johnston, 1978).

Mites were found, but not identified, adhering to a specimen of *Siphona minuta* from a waterbuck in the Park (Kangwagye, 1965, personal communication), which suggests that the mode of transmission from one host to another may be phoretic.

Two species of *Siphona*, *S. minuta* and *S. latifrons*, were common on waterbuck in most areas of the Park. An exception was the Peninsula, a rather dry area not favoured by the humid-loving flies. These blood-sucking flies swarm around the head and over the back of the host, *S. minuta* seeming to more or less live on it, while *S. latifrons* takes a blood meal and then drops into the grass to digest it. Both lay their eggs on the fresh droppings of the host. The flies are particularly attracted to the ear scabs, where they can easily penetrate the skin (or it may be the smell of blood which attracts them), and it could be this fact which aids the transmission of *R. hopkinsi*, for it is significant that both *Siphona* and ear scabs are absent from the Peninsula.

One or two hippoboscids, *Hippobosca hirsuta,* were usually found in the long hair of the neck. Other biting flies feeding on waterbuck were probably tabanids and tsetse flies, but as we have noted, it is not a preferred host of the latter. At night, buck which were lying or ruminating were seen on several occasions to suddenly leap up and gallop away 30 m or so, only to lie down again. This appeared to be in reaction to biting flies, probably mosquitoes.

The differences in parasite loads recorded above, and the occurrence, for example, of many arbovirus infections in waterbuck in the Central African Republic (they are also common in several other species of plains herbivores), can be attributed partly to locality such as predominantly wet areas, to close contact with domestic cattle, and so forth. But they are due in part also to a limited knowledge, few detailed systematic parasite surveys of the larger game animals having been conducted. In the case of the arboviruses, for example, workers have concentrated on small mammals and birds. For the most part we know nothing of the effects of these and other diseases upon their wild hosts, but statements such as that of Glasgow (1963): "The trypanosomes concerned have little or no pathogenic effect on those indigenous African vertebrates which are fed on commonly by tsetse," are most unlikely to be true. The theilerioses and babesioses are both fatal to domestic stock, but immunity can be acquired; the same is probably true of their wild hosts. How many stock farmers, for instance, tolerate a calf loss of over 50% in the first 2 to 3 months? The stock farmer nurtures his calves against these diseases, the wild animals either simply succumb, or develop immunity. Disease epidemics usually affect all ages, but once a zoonosis is endemic in a population the primary effect may be confined to the calves. Plowright (1965) and Plowright *et al.* (1969) have carried out some interesting preliminary studies in this respect on wildebeest and warthog, but unfortunately such research is extremely limited. We still have a great deal to learn.

Predation

Conjuring up violent visions of nature "red in tooth and claw", perpetrated by Victorian naturalists, predation is the factor that many think of as most affecting the survival of an animal. In the waterbuck this is probably true with respect to predation on the calves, but the seizure of a helpless calf by a hyaena is hardly a dramatic event to witness.

In the Park three potentially important predators of waterbuck were absent: the crocodile, hunting dog and cheetah. Cott (1961) recorded waterbuck remains in 25% of those crocodiles which had fed on antelopes in Zambia, suggesting that this predator may play a significant role in some regions. Herbert (1972) states that findings in the Kruger National Park rate the waterbuck there in terms of numbers as the most important mammalian prey species of the crocodile.

Among minor predators of the waterbuck, Pienaar (1969) lists the brown hyaena and the python in the Kruger National Park.

Lion. Rüppell, when he first described the defassa waterbuck, noted that it was preyed upon by lion (Rüppell, 1835), although it is often asserted that it is not favoured by lion (e.g. Kiley-Worthington, 1965). I myself have found no cause to challenge Rüppell's view: my observations in this Park, and elsewhere, confirm the view that a lion's prey depends upon the relative frequency of prey species. In the northern part of the Park buffalo are the most common animals, and consequently they fall prey more often than any other species. In the southern sector it is the topi which is the most numerous animal, and thus in that region is the species which figures largest in the lion's diet. The waterbuck is not a common item on the lion's menu, but if a lion has the opportunity to take one, then it will. My experience has always been that the lion is a facultative, rather than a selective, predator.

I recorded 12 deaths from lions, and the real number was undoubtedly greater than this. Of the total, one was a calf, one a doe of about 4 to 5 years, one doe's age was not determined, and nine were bucks. Of these bucks, two were aged $2\frac{1}{2}$ years, the others were aged approximately 5, $5\frac{1}{2}$, 6, $7\frac{1}{2}$, $9\frac{1}{2}$, 10 and 11 years; a very even distribution of mortality. Twelve is admittedly an insufficient number for meaningful analysis, but it is interesting to note that of this total only one was a calf, and two were does. Of the nine bucks, four, perhaps five, would have been territorial. But on the basis of this sample there was no apparent selection for either territorial or for old bucks. The chances of being taken by a lion appeared to be random with respect to age. The preponderance of bucks compared with does may be partly due to solitary animals presenting less aware targets, but it may also be partly attributable to a greater boldness on the part of the buck, as I discuss later (Chapter 9). In her study Elliott (1976) recorded 40% of waterbuck which fell prey to lions as being about 6 years of age, and 20% about 8–9 years, from a sample of eight bucks and nine does. Three sub-adult does were also taken.

In the Akagera Park of Rwanda I recorded three buck deaths attributable to lions.

Lion were found to be important predators of waterbuck in the Kafue National Park of Zambia (Hanks *et al.*, 1969), confirming an earlier report by Mitchell *et al.* (1965). Hanks and his co-workers found that 1·67 more adult bucks were taken than adult does, while some 21% of the total number of kills ($n = 68$) were of animals less than 18 months of age. This study showed that apparently little of the carcase was devoured, and this has fortified a belief that lions do not like the taste of waterbuck. I find this point of view hard to accept, especially when lion will feed upon putrefying hippopotamus carcases without any appearance of fastidiousness. I think it is more likely that the lions in this area were killing so frequently that they took only the choicest parts, or that it was single lions which were doing the killing, and a complete carcase was more than enough for it. To kill a buffalo, for example, requires the concerted efforts of several lions, and all take part in the meal afterwards. Hanks *et al.* (1969) also quote an observation of an adult buck with claw marks showing that it had been killed by a lion, but the carcase was untouched. I think it probable that the buck escaped the lion and died afterwards. It is not unusual to find adult bucks carrying the claw marks of lion, and in Spinage (1962) I illustrated a well-known case in the Park of a buck which escaped from a lion in 1959. The animal's back had been badly clawed, and it healed over leaving the animal deformed (Plate 40).

Plate 40. A buck whose back healed into this deformed shape, after attack by a lion.

Stevenson-Hamilton (1912) commented upon the lion's preference for waterbuck in South Africa, and Pienaar (1969) stated that it was the preferred prey species in the Kruger Park, and had apparently been so for as long as records were available. Herbert (1972) also found that waterbuck were taken in good numbers by lion, but did not relate the numbers to the lion's relative feeding preferences. Hirst (1975) states that in South Africa the waterbuck is subject to "heavy losses" from lion predation in poor years; generally he considered that lion predation was fairly heavy (Hirst, 1969). All in all, therefore, I think there is no evidence to show that lions exhibit any lack of preference for waterbuck flesh.

Leopard. Leopards are always much more common than people think; their nocturnal and secretive habits mean they are rarely seen, although often heard. This makes it difficult to assess their role as a predator. I have referred to one taking a fawn on the Peninsula, and I once came across one hunting yearlings at night. I expect that young waterbuck figure quite largely in their diet, as was found by Herbert (1972) in his study. In the Akagera Park I recorded one case of a 3-year-old doe which died of wounds inflicted by a leopard.

Spotted hyaena. The important role of the spotted hyaena as a predator on fawns has already been discussed, but this animal seems to pose little, if any, threat, to adults. Does showed fear of hyaenas, snorting and running away in a "hackney" gait, neck held stiffly erect, legs brought sharply straight up and down, flexing the joint as they trot. But bucks gave chase if a hyaena came too close to them, although one night I watched a buck allow a hyaena to approach within 2 or 3 m without showing any interest in it. Herbert (1972) records five cases of predation on adults in 10 years, compared with 131 from lions.

Man. Man is the remaining, and at the time of my study, not very important predator in the Park. I think that as a predator he was significant in only two areas which were adjacent to heavy settlement, one of these being the Kayanja area. Due to the taint of the flesh which occurs if the fat is not properly removed, poachers do not go out of their way to take waterbuck if other game is available. But they eat it readily enough if there is no alternative. There is a myth current among many people in Africa that the meat is unpalatable, but, as indicated, this is only if it is allowed to become tainted with the fat, which imparts a strong, musky flavour. I must confess to having eaten a considerable amount of waterbuck meat, and my tastes are not catholic. When served to unwitting guests they were unable to distinguish it from poor quality beef. I say "poor quality" because the muscle fibres are rather coarse.

A small number of adult bucks was taken by sporting hunters in the adjacent hunting reserves to the north and south of the Park. In the former one of my marked bucks was taken shortly after the end of the study.

8. Population Density, Food Supply and Habitat Preference

Introduction

A considerable part of any population study relates to determining the number of individuals in a population. By itself, such a number may tell us little, except that we have a lot, or not many individuals. But it becomes meaningful if it can be compared with other estimates, or if the census can be repeated, for then we may be able to determine whether a population is stationary, or in a state of change. Such deductions may require a high degree of precision in censusing dispersed populations, which due to the generally cryptic appearance of wild animals, or simply to their numbers, is not often attained in census methods. The wide confidence limits which may derive from sample counts, often leave one unable to determine whether changes are taking place or not.

Although subject to the same constraints, numbers expressed in terms of density are generally more meaningful, comparison between areas giving us some insight into the habitat requirements of a species. A number of critics has suggested that density itself has little meaning, and that it is biomass, the mass of animals, which is important to an understanding of animal–habitat interrelationships. Biomass figures, however, usually introduce further errors, and many published ones are probably highly erroneous, being derived simply from adult weights. The only true estimate of biomass is derived from the size frequencies in a population, and if the population does not have a stable age distribution, the size frequency distribution may change rapidly. Other critics have taken this population aspect further, and offered the opinion that the only meaningful figure is that of energy exchange, expressed as the number of kcal/m² of land occupied. It is only this figure which can take into account the different rates of metabolism between young and old animals, and hence their different feeding requirements.

I believe that all of these determinations can be useful: numbers, density, biomass and energy exchange. It all depends upon what we are trying to define. I have found the most useful determination to be density, because it is the most widely used, and therefore offers the greatest scope for comparison. Before we can compare biomass, we need to know precisely how different workers have derived their estimates. For example, Field and Laws (1970), by using a mean weight, quote a biomass figure 27·6% higher than I have calculated for the same area. Yet this is still lower than that of other workers.

Animal counts were conducted by driving back and forth throughout a study area in a Landrover, all species being noted by an observer standing in the back of the vehicle. In the intensive waterbuck studies only the waterbuck was counted, and I found that I could perform this adequately from the driver's seat.

Waterbuck Densities

From March 1963 to March 1967, a total of 419 counts was made of waterbuck in 11 study areas (excluding the Peninsula) by various workers. This gave a total of 14 489 waterbuck in 7177 km², an overall mean density of 2·0/km² (standard error as % = 170) (Table 33). Assuming 518 km² of the Park to be uninhabitable to waterbuck due to swamps, dense forest, human habitation, or too great a distance from water, the remaining 1461 km² held an estimated 2922 waterbuck. This compares with park-wide aerial counts conducted in 1971 and 1972, which gave a mean density of 2·3/km² (standard error as % = 54), an estimated grand total of 3563. At the 95% level of confidence limits the population size was between 2335 and 4791 (Eltringham and Din, 1977). These results are comparable, especially when we consider that some increase in numbers may have taken place due to the higher rainfall in the area between 1967 and 1971. In this analysis I have continued, therefore, to use my original estimates rather than those of later workers.

Throughout the Park, densities ranged from 0·15/km² along the Ishasha River in the south, to 17·8/km² in the lacustrine Kayanja region (Table 34). Analysis of the figures presented by Field and Laws (1970) of the Park game counts, reveals no apparent interspecies competition between waterbuck and the other dominant herbivores, when the areas are viewed as a whole. There is no significant correlation between waterbuck density and the total large herbivore density, between

Table 33 Waterbuck population means in count areas for each year of study.

Area	Year 1	Year 2	Year 3	Year 4
Ogsa	39	45	41	53
Nyamagasani	83·5	94·5	118	111
Crossroads	17	23	16	25
Royal Circuit	26·4	28·5	16	9
Katwe	24	16	27	39·5
Kikorongo	41	56	53	42
Craters	11	10	13	9
Lion Bay[a]	76	84	86	—
Yearly means	39·8	44·6	46·3	41·3
Total mean = 43				

[a] Incomplete data not used in analysis of variance (see p. 162).

waterbuck biomass and total large herbivore biomass, nor between waterbuck density and the densities of Uganda kob, buffalo and hippopotamus. However, if we select the four areas of highest waterbuck concentration, we find that a significant inverse correlation exists between waterbuck density and hippopotamus density ($r = 0.97$, df 2, $p = < 0.05 > 0.02$). This selection of areas is justified by the fact that these are the main areas of hippopotamus concentration, close to water

Table 34 Comparison of waterbuck densities in different areas of the Park.

Area	Size in km²	Number of counts	Density/km²
Kayanja "A"	9·6	2	17·8
Kayanja overall	52	2	9·9
Peninsula	4·4	45	10·5
Nyamagasani	14·6	45	7·0
Ogsa	12·2	43	3·7
Lion Bay	24·4	29	3·4
Katwe	8·8	43	3·1
Kikorongo	22·6	40	2·2
Katunguru	16·1	41	1·2
Royal Circuit	20·5	45	1·0
Craters	22·1	45	0·5
Kikeri	9·6	44	0·3
Ishasha	17·2	44	0·15[b]
Mean[a]	18·0		5·5

[a] The mean density is calculated from the total figures and not the individual area means.
[b] Assuming this population to use only up to 0·6 km from the river the density would be 3·1/km².

(Table 35). In other areas waterbuck density seems to be determined by physical habitat factors, rather than by interspecies competition. The reality of the waterbuck–hippopotamus interspecies competition has been highlighted by the Peninsula hippopotamus removal experiment.

Table 35 Inverse correlation of waterbuck density with hippopotamus density in four areas.

Area	Waterbuck density/km²	Hippopotamus density/km²
Peninsula[a]	10·46	0·91
Nyamagasani	6·73	15·15
Ogsa	3·52	20·53
Katwe	3·16	27·99

$r = 0.97$, df $= 2$, $p = < 0.05 > 0.02$
[a] Originally the hippopotamus density was 29/km² before reduction and the waterbuck 4·6/km². Figures are from Field and Laws, 1970.

Analysis of variance of the data on waterbuck counts (Table 33), show that the population numbers in the study areas did not change from year to year, during the four years of counting ($F = 0.062$, df $3:24$, $p = \gg 1.0$), and thus we can confidently assume that the overall Park population was stationary during this period.

At the level of stocking recorded here, the biomass, derived from the survivorship curves, and the mean sex ratio, relating to the age class size frequency, was of the order of 332 kg/km²; equivalent to an energy exchange of roughly 10·12 kcal/m²/year. According to Field and Laws (1970) waterbuck represented 2·33% of the total herbivore biomass.

Compared with other published figures, a density of 2·0/km² is high (Table 36), with the exception of Lake Nakuru National Park in Kenya, where, in 1979 the density was 78/km². The difficulty in comparing waterbuck densities is that figures may refer to a clumped distribution near to water, rather than to some arbitrary national park surface, although the boundaries of national parks are often real enough for the animals. This difficulty can be overcome by using the contour density method, derived from surface trend analysis, as described by Norton-Griffiths (1975). But this method of aerial analysis has yet to be applied to waterbuck.

The very high density recorded for the Lake Nakuru Park seems to have been largely artificially induced. In 1972 it was estimated by Kutilek (1974) to be 31·1 km², and had risen to 78/km² in 1979; with an estimated density in one part (Baharini) of 106/km²; but towards the end of 1979 there was a heavy mortality (Wirtz, 1980). This increase appeared to result from an absence of all predators, except the leopard and hunting

Table 36 Some approximate densities of waterbuck.

Area	Density/km²	Authority
Lake Nakuru, Kenya	{ 72–106	Wirtz (1980)
	31·1	Kutilek (1974)
Nkala River, Zambia	7·0	Hanks *et al.* (1969)
Toro Game Reserve, Uganda	4·0	Leuthold (1966)
Tarangire Reserve, Tanzania	2·6	Lamprey (1963)
Rwenzori Park, Uganda	2·1	Spinage (1970)
Kivu Park, Zaïre	1·3	Bourlière and Verschuren (1960)
Akagera Park, Rwanda	0·80	Spinage *et al.* (1972)
Nairobi Park, Kenya	0·73	Foster and Kearney (1967)
Sabi-Sand Wildtuin, South Africa	0·62	Herbert (1972)
Ngorongoro Crater, Tanzania	0·54	Estes and Goddard (1967)
Mara Game Reserve, Kenya	0·50	Darling (1960)
Safarilandia, Mozambique	0·46	Dalquest (1965)
Murchison Falls Park, Uganda	0·43	Laws *et al.* (1975)
Henderson Ranch, Rhodesia	0·27	Dasmann and Mossman (1961)
Kruger Park, South Africa	0·23	Pienaar (1966)

dog, which are both rare, and a greater destruction of other herbivores than was suffered by the waterbuck. National park status was given to the area (land area 29 km²) in February 1961, and the land area was increased in 1973 to 157 km², which has allowed the waterbuck to recover its numbers more quickly than have the other relict species, and thus to dominate the habitat. In the Solio Reserve of Kenya, Elliott (1976) found densities ranging from 0·76/km² to 3·35/km², while a part of the area, which had been fenced in in 1971 (55 km²), contained a concentration of 82·4/km² in one region. This dropped to 55·9/km² in 1974, after a die-off resulting from excessively dry conditions. The 1962–1968 mean annual rainfall, however, was 777 mm, approximating to that at Mweya, but the highest density that I recorded in a circumscribed area in my study, was 17·8/km² at Kayanja.

Sex Ratio

The total number of animals sexed in the counts which I conducted was 12 433. Of this total, 4785 were bucks, and 7648 were does, giving a mean sex ratio for the Park of 1 : 1·6 bucks to does, within the range of 1 : 0·5 to 1 : 4·3. Bourlière and Verschuren (1960) gave ratios of 1 : 2·3 to 1 : 4·6 for the Kivu Park in Zaïre, probably having overlooked the bachelor groups. Kutilek (1974) gave a ratio of 1 : 2·3 for the Lake Nakuru Park, but this had changed to 1 : 1 by 1979 (Wirtz, 1979). Other authors give ratios of

1:1·9, 1:1·8, 1:1·5, 1:1·1 and 1:1, for the Solio Ranch, Nairobi National Park, Umfolosi Game Reserve, the Sabi-Sand Wildtuin and the Kafue National Park, respectively (Elliott, 1976; Foster and Kearney, 1967; Mentis, 1970; Herbert, 1972; Hanks *et al.*, 1969). The experience at Nakuru would seem to suggest that high densities may result, initially, in fewer bucks. With the exception of the Kikeri and Ishasha areas, this hypothesis is borne out to a large extent by the Park density–sex ratio relationship (Table 37), which showed a highly significant positive correlation ($r = 0.89, p = < 0.02 > 0.01$), approximating to a loss of one buck for every increase of 5 animals/km². Of course this is not a real loss, but only a redistribution of bucks, and we cannot make predictions of density from the sex ratio or vice versa. The mean Park value shows a much higher number of bucks than would be predicted. Kikeri and Ishasha were omitted from the regression analysis because the former appeared to be only visited by waterbuck from time to time, and had no resident bucks; while Ishasha's density figures were rather eclectic, as there were insufficient observations to determine with accuracy the range of waterbuck away from the river.

Table 37 The relationship of sex ratio to density in the study areas.

Area	Density/km²	Ratio $\overset{\curvearrowright}{\male} : \underset{+}{\female}$
Kayanja "A"	17·8	1 : 3·7
Peninsula	10·5	1 : 1·8
Kayanja overall	9·9	1 : 3·2
Nyamagasani	7·0	1 : 1·7
Ogsa	3·7	1 : 1·5
Ishasha (within 0·6 km of river)	3·1	1 : 1·4
Royal Circuit	1·0	1 : 0·5
Kikeri	0·3	1 : 4·3

Changes in the Peninsula Population

One of the interesting features of the Peninsula population is that it has undergone considerable changes in size in recent years, unlike the remainder of the Park population, and records documenting these changes date back for some time. In this study I was fortunate to have the, admittedly isolated, observation of Worthington and Worthington (1933), and the counts of Drs G.A. Petrides and W.G. Swank for the years 1956–1957. This enabled me to follow the albeit artificial, dynamic changes of the population over a time span of 44 years, taking into account observations which have been made since the end of this study.

From November 1956 to May 1957 the population was counted at monthly intervals, and group positions plotted, by Petrides and Swank, who kindly placed their data at my disposal. Their results for this period are summarized in Table 38, together with subsequent counts, including

Table 38 Summary of waterbuck counts of the Peninsula and Ogsa areas, from 1956 to 1973.

Date	Number of counts	Mean total	Range	Mean population numbers					Density /km²	♂ : ♀+
				Adult ♂	Yg ♂	Adult ♀+	Yg ♀+	Fawn		
Peninsula										
Nov. 1956– May 1957	8	20	9 — 31	4	0	16	0	0	4·5	1 : 4
Jly 1957– Nov. 1957	5	16	13 — 20						3·6	
May 1962	1	37							8·4	
Mar. 1963– Oct. 1964	46	53	35 — 71						12·0	
Nov. 1964– Mar. 1967	488	48	30 — 72	6·8	9·1	22·7	5·4	3·7	10·9	1 : 1·8
Feb. 1968– Dec. 1968	39	43	14 — 64	11·2		28·4		1·6	9·9	1 : 2·5
Jan. 1969– Dec. 1969	50	37	18 — 53	10·5		25·1		1·9	8·5	1 : 2·4
Jan. 1970– Dec. 1970	44	39	29 — 47	10·8		26·1		2·2	8·9	1 : 2·4
Jan. 1971– Dec. 1971	40	37	25 — 55	9·4		25·7		1·9	8·4	1 : 2·7
Jan. 1972– Dec. 1972	45	33	16 — 54	9·4		22·2		1·8	7·6	1 : 2·4
Jan. 1973– Sep. 1973	34	38	20 — 44	11·8		24·1		1·6	8·5	1 : 2·0
Ogsa										
Nov. 1956– May 1957	7	44	20 — 66	5·3	0	28·4	9·7	0	3·1	1 : 6·8
Jly 1957– Nov. 1957	5	65·4	49 — 80						4·6	
Mar. 1963– Mar. 1967	43	45·2	15 — 76						3·2	
Jun. 1965– Mar. 1967	36	44·2	14 — 83	8·7	7·8	22·2	2·7	2·9	3·1	1 : 1·5

those of Eltringham (1979, and personal communication). From this information it has been possible to obtain a fairly accurate idea of the population changes which took place subsequent to the elimination of hippopotamus from the area.

Since at least 1955, the population of waterbuck on the Peninsula appears to have been stationary at about 18 animals (4·0/km²), remaining at this level until about 1960, after which the number began to increase exponentially. This increase appears to have been a direct result of the removal of the hippopotamus population in 1958, which has already been referred to (Chapter 2). This action removed the intra-specific competition created by the hippopotamus which, in large numbers, has the facility to modify the vegetation to its own advantage, changing the sward by its grazing behaviour to the low-lying, mat-forming grasses which it favours. As we have seen, consequent upon the removal of this grazing pressure, these mat-forming grasses were replaced by tussock grasses, particularly *Sporobolus pyramidalis,* which is favoured by waterbuck and buffalo. The year in which *S. pyramidalis* replaced *Chrysochloa orientalis* as the dominant grass, was the year in which probably both waterbuck and buffalo began their exponential increase in numbers, namely 1961. Unfortunately counts are lacking between December 1957 and April 1962, but the calculated rise is in accordance with 1961 as the year of increase.

Mean population size, accompanied by considerable fluctuation in numbers, continued to increase until 1964, when the population size started to decline gradually. This would probably have resulted in a levelling-off at a density of about 10·9/km², but the cessation of hippopotamus control in 1967 meant that the hippopotamus rapidly recolonized the area, almost achieving its former density level with 111 occupants by 1973. A number of disturbances also took place in the area, such as the construction of an airstrip and a large, ditched, plant study area, clearance of bush, etc. The result was that the waterbuck population declined a little, stabilizing from 1969 to 1973 at about 8·4/km². This suggests that the hippopotamuses had not as yet modified the habitat sufficiently to act as significant competitors. Neither did there seem to be any competition with the buffalo, for their numbers declined by 43% from those present in 1968; at that time the buffalo averaged 109 head. The numbers of hippopotamuses seem to have remained relatively stable since 1973, but by the end of 1977 apparently only some 11 waterbuck remained, and about 68 buffalo, suggesting that by this time the hippopotamuses were exerting effective competition. That the waterbuck numbers have declined to half their original total seems to be illustrative of the oscillatory nature of disturbed populations.

Since a number of writers has claimed that mixed herbivore

populations are probably intra-facilitative rather than intra-competitive (e.g. Gwynne and Bell, 1968), these population changes are rather important in demonstrating that quite the contrary situation may occur, supporting the conclusion already derived from density figures (Table 35). All the evidence points to the increase in waterbuck numbers being a response to lack of competition from the hippopotamus in a favoured habitat; thus certain *habitats* appear to be competed for by different species.

Let us consider other possible factors which might have influenced waterbuck numbers. In mid-1963 there were two grass fires on the Peninsula, and although some preference for burnt areas by waterbuck has been claimed (Field and Laws, 1970) these burnt regions had no obvious effect in attracting more animals. Bere informs me (personal communication, 1966), that prior to the removal of the hippopotamus resident hyaenas numbered about 20 animals, "favouring the long grass of the northern point", which in 1964 was a short grass area, probably resulting from buffalo trampling. Bere considered that the hyaena population eliminated most of the fawns at, or soon after, birth. This was due not only to predator density, but also to the fact that the grass was kept so short by the hippopotamuses that there was very little cover for the fawns. However, the fawns usually lie up in thickets, which, if anything, might tend to increase rather than decrease in size in the presence of hippopotamus grazing; the hippopotamuses avoiding woody seedlings and also reducing the risk of fire by overgrazing. But in January 1964, 16 hyaenas were eliminated from the area, and the only ones seen thereafter were occasional nocturnal visitors. This elimination had no apparent effect on the downward trend in the waterbuck numbers, which had already commenced, suggesting that the hyaenas had played little part in these population changes.

Rainfall on average was 19·5% higher from 1963 to 1967, than it was from 1959 to 1962, but the increase in waterbuck density had already started in 1961, so that an increase in available food could not be attributable to rainfall alone. Furthermore, although rainfall continued to increase from 1968 to 1973, waterbuck numbers declined.

Using Dean and Gallaway's (1965) population model, a theoretical increase was constructed, assuming emigration and survival rates calculated according to the Park data, and commencing with the population present in 1959. If we adopt the assumption that the habitat was not conducive to population increase until some 16 months after the hippopotamuses were eliminated, when in fact the grass composition had changed significantly in the waterbucks' favour, then we find that the calculated increase matches the observed increase very closely; and indicates that the increase could largely be accounted for by reproduction rather than by immigration. Evidence from the adjoining Ogsa suggests, however, that the former was not the case. Peak numbers were reached in January 1965 when 72 animals were counted, but this marked the beginning of the decline, probably resulting from reduced immigration and increased intra-specific competition, all available individual niches having been occupied.

Jungius (1971) quotes figures from the Kruger Park in South Africa of an estimated 100 waterbuck in 1903, rising to an estimated 4000 in 1916, a figure of acceptable magnitude since this would imply an "r" (the intrinsic rate of natural increase) of approximately 0·265, or a finite rate of increase of 1·300; which would result in a population of 3025 waterbuck by 1916, or 3933 by 1917, produced from the original 100. The same finite rate of increase applied to the Peninsula's 20 waterbuck in 1960, gives a total of 57 in 1964, compared with an observed mean of 11 monthly counts in that year of 55·6 (SD 5·2) (Fig. 36). Elliott (1976)

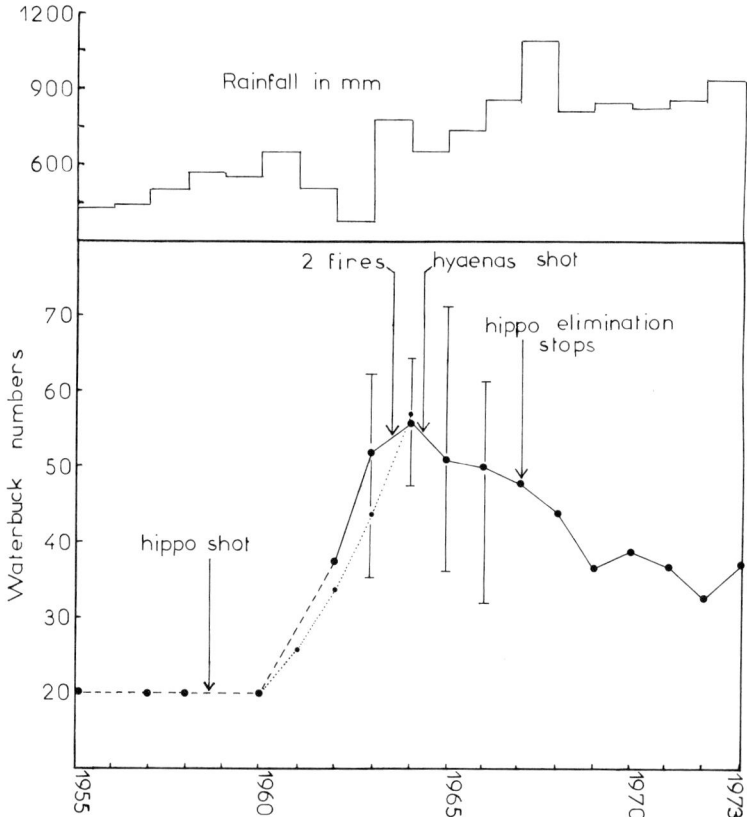

Fig. 36. Changes in the numbers of waterbuck inhabiting the Mweya Peninsula from 1955 to 1973. Dotted line shows the calculated increase from 1960, vertical lines show the range.

reports that, on Crescent Island, there were eight waterbuck in 1964 (one buck and seven does), rising to 133 in 1973, a slightly faster finite rate of increase than is given by 1·300, which would have produced a total of 85 animals.

The Overgrazed Study Area (Ogsa)

This area, adjoining the Peninsula, was also counted by Petrides and Swank from November 1956 to May 1957. They covered a slightly larger area than the one that I chose, and so only those animals recorded within my boundaries were taken into account when analysing their data. The results of these counts, and of subsequent counts, are given in Table 36. They indicate a relatively stationary population with a mean density of 3·5/km² (SD 0·73).

Prior to the commencement of my work in October 1964, two adult bucks were unfortunately irresponsibly shot, and a further two died during drug capture. The real number of bucks should therefore have been slightly higher, and I estimated it to be about 12 adult territorial bucks, giving a sex ratio of 1 : 1·3. The population seems to have remained virtually stationary for 10 years, although the sex ratio would appear to have undergone a marked change, a bachelor herd apparently being absent from the area in 1957.

As the Ogsa and the Peninsula are adjacent to one another, and a part of the doe population is common to both, movements between the two areas would be expected to be in inverse ratio. The monthly population counts were thus compared (Fig. 37), but whereas peaks in the Ogsa population were indicative of immigration from the Peninsula, dips in the Ogsa population usually meant that animals were on the steep, inaccessible slopes along the lakeside or Channel, and had been overlooked in counting. Taking the two areas together, the overall increase in population density in the 10 years 1956 to 1966 was 1·34 animals/km², with the mean population size showing only minor fluctuations, and the increase being confined to the Peninsula. If the initial increase in Peninsula numbers was attributable to immigration from the Ogsa, one would expect the Ogsa population to show a decline at this time. It does indeed show this, the density falling from 3·8/km² in 1956–1957, to 3·1/km² in 1963–1967; a mean loss of 10 animals. This does not account for all of the Peninsula increase, which was undoubtedly reinforced by increased recruitment, producing the real increase in size in the overall population.

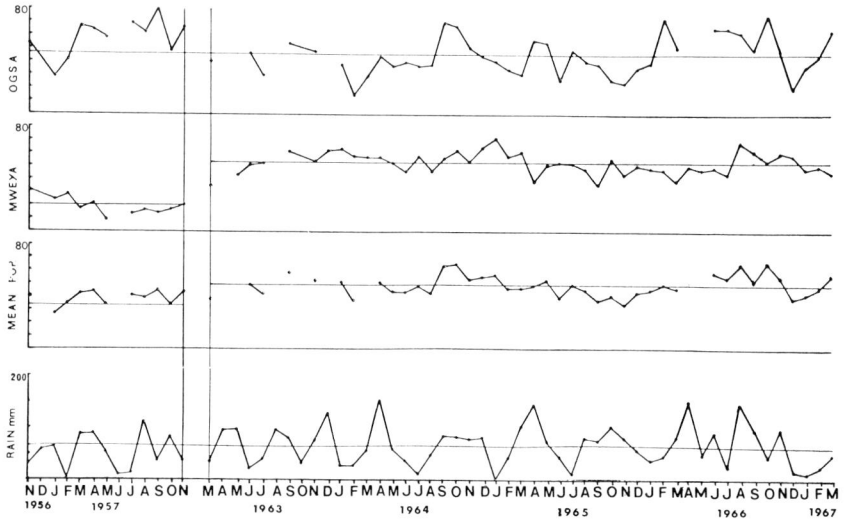

Fig. 37. Monthly changes in the Peninsula and Ogsa waterbuck numbers from 1956 to 1967, with mean monthly rainfall. Horizontal lines show the means.

Densities in Other Study Areas

Kayanja

No previous records were available for this area, and circumstances did not permit of monthly counts. A two-day count was made of the entire area in October 1956, and repeated in July 1966. These counts gave a total of 131 bucks and 365 does with fawns, and 113 bucks and 421 does with fawns respectively. This represented a mean density for the area of 9·9/km², with a sex ratio of 1 : 3·2 bucks to does. Area "A" (see Fig. 16) had a mean density of 17·8/km², showing how the population concentrated towards the lake.

Nyamagasani

Although adjacent to Kayanja, this area had a much lower density, 7·0/km², with a sex ratio of 1 : 1·7, close to that of the Park mean.

Royal Circuit

Five counts were conducted in the Royal Circuit area during the dry season of February to March 1966, which gave a mean density of 0·54/km², comprising 5 adult bucks, 1·2 young bucks, and 4·6 adult does, a sex ratio of 1 : 0·7. These results differ slightly from the overall counts (Table 33), but established that the area was generally avoided by does.

Ishasha River

In July 1966 three counts were made along 8 km of river bank, giving a mean population of 47 animals. Assuming that an area of only up to 0·6 km from the river is used by the population, this gives a density of 3·1/km². The routine Park counts, however, produced a much lower density of 0·5/km² since they did not specifically take in the riverine habitat, largely counting only those animals seen outside it. The Park-wide densities are summarized in Table 34.

From these densities we can conclude that both the Peninsula and the Ogsa showed a marked increase in the number of bucks in the areas during the 10 years 1956 to 1966. This suggests that, when areas become available for colonization, they are first occupied by adult bucks seeking territories, and later immigration and reproduction by the doe sector adjusts the sex ratio. We have seen that this took place on Crescent Island, which was colonized by a solitary buck in 1964; later followed by seven does. By 1973 the sex ratio had adjusted itself to 1 : 1·3 (Elliott, 1976). But in areas of very high stocking rates the number of does then tends to exceed the number of bucks by a greater proportion than expected, probably due to increased intra-specific competition among the bucks resulting in higher mortality rates. In contrast, in a poor area such as the Royal Circuit area, which has a high proportion of sword grass *Imperata cylindrica*, which is tough and unpalatable in all except its very youngest stages, there is a high ratio of bucks to does, as the does avoid the area.

It is the topography of an area rather than its vegetation which seems to produce high stocking rates, for the Peninsula and the Ogsa have virtually identical grass swards, at least with regard to the dominant species, but the Peninsula had roughly three times the waterbuck density of the other area. Food supply must, however, be the ultimate factor determining density.

Food and Feeding

In September 1966 I conducted an analysis of the grasses on the Peninsula and in the Kayanja "A" and "A1" study areas. A selected number of compass-bearing transects was chosen and the vegetation along these transects was examined at every 100 paces. Bushes and thickets were avoided since it was the grassland composition which interested me. I used a quadrat of 50 × 50 cm, divided into four squares each of 25 × 25 cm. The presence or absence of a grass species rooted in a square was noted, and the occurrence expressed as a percentage of the total number recorded. The Ogsa was analysed by Dr C.R. Field by means of random quadrats (Field and Laws, 1970). Selected grasses were collected throughout the year by Dr Field for analysis of their organic chemical content. I collected samples of *Acacia* and *Capparis* browse for organic chemical analysis.

As already noted (Chapter 2) the Peninsula was relatively rich in grass species; the 21 recorded by me are given in Table 39. The analysis showed

Table 39 Plant species identified in the Peninsula and Kayanja study areas.

Grasses (species)	Peninsula	Kayanja	Shrubs and trees (species)		Kayanja
Leptochloa obtusiflora	+		*Capparis tomentosa*	+	+
Eragrostis tenuifolium	+		*Maerua triphylla* subsp.		
Harpachne schimperi	+		pubescens	+	−
Dactyloctenium aegyptium	+		*Pavetta albertina*	+	−
Chloris gayana	+	+	*Abutilon guineense*	+	−
pycnothrix	+	+	*Erythrococca bongensis*	+	−
Cynodon dactylon	+	+	*Euphorbia candelabra*	+	+
Microchloa kunthii	+	+	*Acacia sieberiana*	+	+
Chrysochloa orientalis	+	+	*Teclea* sp.	+	−
Sporobolus pyramidalis	+	+	*Cordia ovalis*	+	−
robustus	+		*Hoslundia opposita*	+	−
spicatus	+	+	*Azima tetracantha*	+	−
stapfianus	+	+			
Panicum maximum	+				
repens	+	+			
Brachiaria decumbens	+				
Urochloa panicoides	+				
Digitaria sp.	+				
Cenchrus ciliaris	+	+			
Bothriochloa sp.	+	+			
Hyparrhenia filipendula	+	+			
Heteropogon contortus	+	+			
Themeda triandra		+			

− Indicates that records were not made.

that *Sporobolus pyramidalis* was the most dominant (Table 40), with the fire-induced red-oat grass *Themeda triandra,* common throughout the greater part of the Park, conspicuous by its absence.

Table 40   The dominant grass species with a frequency of occurrence of $> 5\%$ in four study areas.

Peninsula (248)		Ogsa[a]	
Sporobolus pyramidalis	31·0	*Sporobolus pyramidalis*	39·5
Bothriochloa sp.	12·7	*Bothriochloa* sp.	16·5
Chloris gayana	12·2	*Chloris gayana*	12·5
Cynodon dactylon	8·9	*Sporobolus stapfianus*	9·0
Cenchrus ciliaris	8·6	*Chloris orientalis*	6·0
Kayanja "A" (496)		Kayanja "A1" (512)	
Heteropogon contortus	19·0	*Hyparrhenia filipendula*	28·8
Hyparrhenia filipendula	16·0	*Themeda triandra*	23·0
Microchloa kunthii	14·0	*Microchloa kunthii*	14·0
Bothriochloa sp.	10·0	*Heteropogon contortus*	13·0
Themeda triandra	10·0	*Sporobolus stapfianus*	10·0
Sporobolus pyramidalis	10·0	*Bothriochloa*	7·8
Sporobolus stapfianus	7·5		

[a] Data kindly supplied by Dr C.R. Field.
  Numbers in brackets refer to the number of quadrats examined.

The Ogsa had identical grass dominants to the Peninsula (Table 40). Of other grasses *Cynodon dactylon* was much less frequent here as it favours shore lines, whereas *Brachyaria decumbens* and *Sporobolus stapfianus* were more frequent. The former is a grass which tends to be localized in its distribution, while *S. stapfianus* is indicative of overgrazing.

Kayanja was less rich in species than was the Peninsula, with only 14 recorded (Table 39). In the regularly burned area "A1" adjacent to a main road, the track separating it from "A" served as a mild firebreak (Fig. 16), *S. pyramidalis* disappearing altogether as a dominant species, *T. triandra* reaching second place (Table 40). Table 41 shows how the composition of the main grass species changes as one moves east from the lake shore into the more and more frequently burned area.

Seasonal changes in the percentage of crude protein were determined

Table 41   The change in the occurrence of dominant species of grass moving inland from west to east from the lake shore at Kayanja.

Species	Kayanja "A"								Kayanja "A1"							
H. filipendula	1	4	3	11	12	9	10	12	12	11	13	13	14	14	15	14
H. contortus	1	7	11	12	14	10	8	9	12	10	8	4	3	4	2	2
T. triandra	0	1	2	2	5	7	8	6	10	6	9	10	13	13	15	14
S. pyramidalis	9	14	10	2	0	1	1	2	3	2	0	2	0	0	0	0
Bothriochloa sp.	0	5	8	6	4	6	3	6	6	9	6	3	0	0	0	0

Figures are expressed as the frequency of occurrence in four quadrats.

in a number of grass species (Fig. 38a, b). This showed that although the highest protein values were mostly attained in the wet seasons, the lowest were not necessarily found in the dry seasons. This apparent anomaly depends upon whether the onset of the dry weather is rapid, curing the grass as standing hay, or whether it is slow, allowing the organic status to decline. The mean values showed *Chloris gayana* to be the most nutritious of the species examined, a grass which was not dominant at Kayanja, but which was in both the Peninsula and Ogsa areas. The fire-induced *Hyparrhenia filipendula* and *Themeda triandra* ranked second in nutritional content, although the latter species is anomalous in that in the dry season it is too tough to be palatable. *Sporobolus pyramidalis* was shown to be a relatively poor species, nutritionally speaking.

Feeding preferences of the waterbuck in the Park were investigated by Dr Field, by means of rumen samples (Field 1972). He studied samples from three areas, which did not coincide with my study areas, but the vegetation was not greatly different. Rumen analysis confirmed the importance of *S. pyramidalis* in the waterbuck's diet (Table 42), figuring largest in the diet of those animals from the Kamulikwezi area to the north-west of Lake George. Surprisingly, *T. triandra* was the most dominant item of diet in the other two areas, Nyamagasani and Lion Bay, although it was not the most available species. This apparent preference may, however, have been a seasonal effect related to plant growth. In both the Kamulikwezi and Nyamagasani areas Field found a significant difference between wet and dry season intake, although the dominants remained the same for both seasons, but Lion Bay did not show a significant difference. The main overall species preference was: *S. pyramidalis* 97·2%, *T. triandra* 95·8%, *Heteropogon contortus*, *Hyparrhenia filipendula* and *Bothriochloa* sp. 91·7%, *C. gayana* 87·5% and *C. dactylon* 50% out of the 20 species recorded (Table 42).

Field concluded that waterbuck could be considered as intermediate

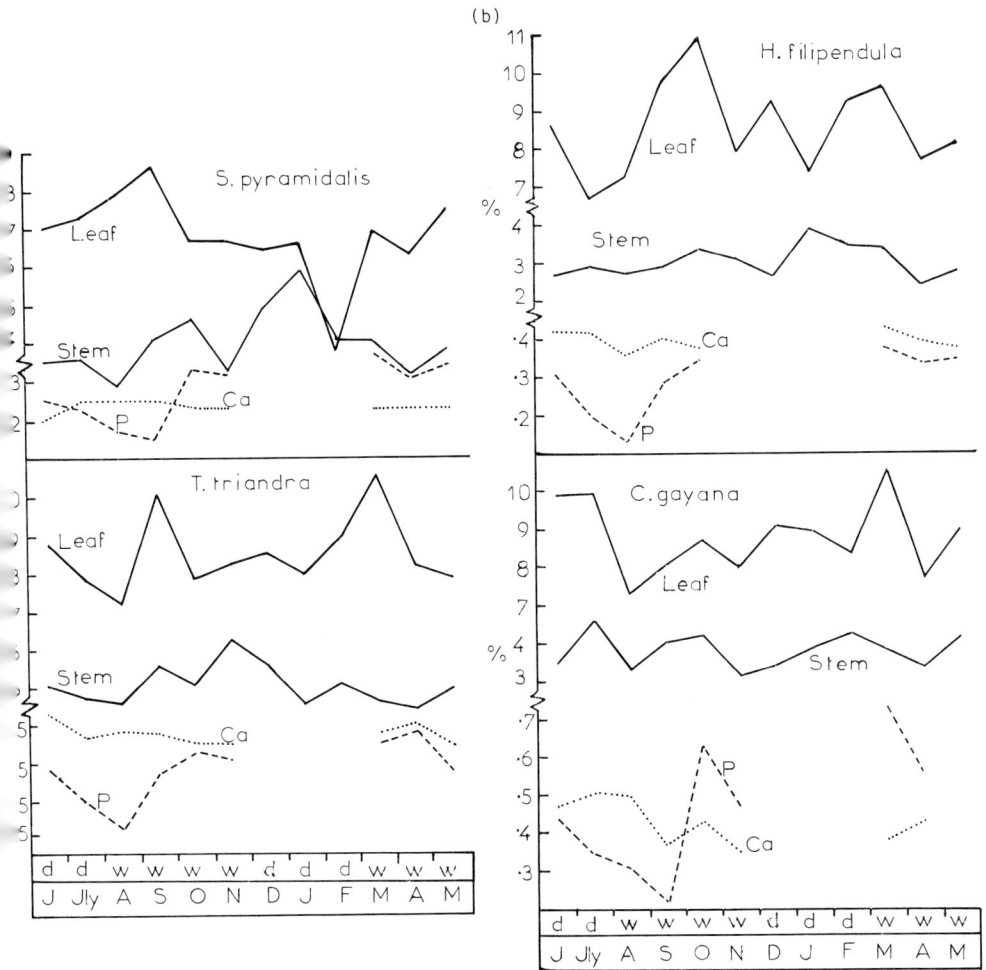

Fig. 38. Monthly changes in the crude protein values in the stem and leaf of four principal grass species. Changes in calcium and phosphorus for 9 months are also shown; "d" — dry, "w" — wet.

grazers, along with buffalo, eating both medium height pyrophilous grasses (mostly *Andropogoneae*) and short grasses (mostly *Chlorideae, Paniceae* and *Sporoboleae*). He considered that they might compete with buffalo for *Chloris* in the Kamulikwezi and Lion Bay areas, for *Heteropogon* in Kamulikwezi, and *Themeda* in Lion Bay. At Nyamagasani and Lion Bay, they might compete with hippopotamus for *Sporobolus*, and for *Heteropogon* at Nyamagasani. The greatest range of

Table 42 Rumen analysis of grass species ingested by waterbuck, occurrence
> 10% (adapted from Field, 1972).

Area	Ingested		Availability
	Wet season	Dry season	> 5%
Kamulikwezi	*S. pyramidalis* *H. contortus* *I. cylindrica* *Bothriochloa* sp.	*S. pyramidalis* *H. contortus* *C. gayana*	*S. pyramidalis* *C. gayana* *Sporobolus homblei/spicatus* *Hyparrhenia* sp. *I. cylindrica* *Panicum repens* *H. contortus*
Nyamagasani	*T. triandra* *H. contortus*	*T. triandra* *H. contortus*	*M. kunthii/S. stapfianus* *Hyparrhenia* sp. *H. contortus* *Bothriochloa* sp. *T. triandra* *S. pyramidalis*
Lion Bay	*T. triandra* *S. pyramidalis*	*T. triandra* *S. pyramidalis* *C. gayana*	*S. pyramidalis* *Hyparrhenia* sp. *C. gayana* *Brachiaria brizantha/decumbens* *T. triandra*

Species are shown in declining order of occurrence.

grass types was apparently eaten in the October–November wet season
(Fig. 39).

Kiley-Worthington (1966) investigated 14 faecal samples collected in
the Park in the dry season, and found that the species most consistently
encountered were *Brachiaria decumbens/dictyoneura*, *Chloris* sp.,
Panicum maximum, *S. pyramidalis* and *T. triandra*, all of which were
found in Field's study. She concluded that waterbuck were able to subsist
on relatively poor quality vegetation, but this was not confirmed in my
study. On the contrary, captive waterbuck were found to require about
four times more protein in their diet than did other bovids in captivity,
namely the Hereford steer, African buffalo and oryx. This necessitated
feeding them with lucerne in order for them to survive (Taylor *et al.*,
1969). The high protein intake accorded with their high water intake and
consequent high urine output. Kiley-Worthington reached her conclusion from the predominance of species such as *Aristida* sp. and *H.
contortus*, to the exclusion of available species such as *C. dactylon* and

Table 43 Species of grasses recorded eaten by waterbuck in the Park (after Field, 1972).

Species	% frequency[a]
Sporobolus pyramidalis	97·2
Themeda triandra	95·8
Heteropogon contortus	91·7
Hyparrhenia filipendula	91·7
Bothriochloa sp.	91·7
Chloris gayana	87·5
Cynodon dactylon	50·0
Imperata cylindrica	44·4
Brachiaria brizantha/decumbens	31·9
Panicum maximum	31·9
Eragrostis sp.	25·0
Sporobolus stapfianus	15·3
Microchloa kunthii	12·5
Setaria sphacelata	11·1
Sporobolus homblei/spicatus	8·3
Leersia hexandra	6·9
Hyparrhenia dissoluta	4·2
Panicum repens	4·2
Chrysochloa orientalis	1·4
Brachiaria platynota	1·4

[a] Percentage frequency is expressed as a percentage of the total number of samples (72).

Brachiaria sp. in the faecal samples. This may have been a spurious result deriving from the greater digestibility of more palatable species. From samples from two waterbuck collected in South Africa, Wilson and Hirst (1977) found a considerable preference for *H. contortus*, *Trachypogon spicatus* and *C. dactyylon*.

In the dry season, I observed the waterbuck to balance its protein deficiency in the graze, by browsing for up to almost 21% of its feeding time (Table 44); browsing at night apparently being much higher in the buck than it was during the day. This may have been related to a greater need to drink in a hot environment, compounded by the need to remove the increased nitrogenous waste resulting from browsing. During the day a mean time of 10·2% was spent in browsing by the doe, and 11·1% by the buck, values sufficiently close as to suggest that the doe might also double her browsing time at night. This would give a mean browse period of 15·7% in 24 h. But we must remember that browse is much richer in protein than is grass; the mean value for *Capparis tomentosa* in the dry season is more than three times that of the richest grass at this time (*C.*

	annual	Pyrophilus short	Perennial short medium	Pyrophilus medium tall
Dec				
Jan				
Feb				
Mar				
Apr				
May				
Jun				
Jly				
Aug				
Sep				
Oct				
Nov				

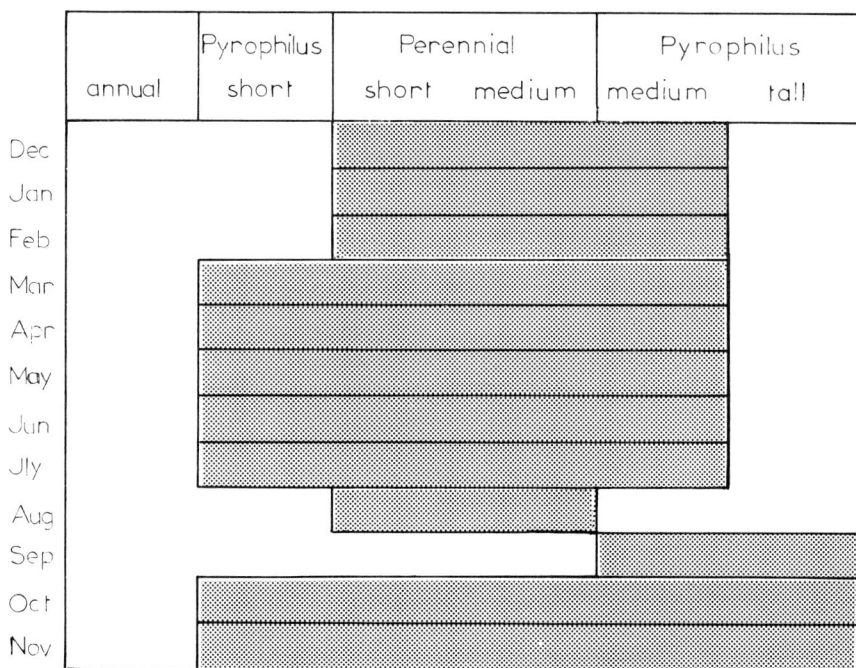

Fig. 39. The types of grass eaten by waterbuck through the year. Adapted from Field (1972).

gayana). *Capparis tomentosa* clumps on the Peninsula had a characteristic browse shape created by waterbuck. It flowers in the dry season and all of the heavily scented inflorescences within reach were usually eaten (Plate 41). In the Kayanja area the favoured browse was young *Acacia sieberiana*. Analysis of this species showed that the parts selected by waterbuck were rich in protein at the end of the dry season (Table 45). A single fawn observed over two consecutive days was seen to browse for only 0·9% of its feeding time, the protein deficiency in this case presumably being made up for by the mother's milk.

Analysis of calcium and phosphate in the principal grass species showed that calcium remained fairly constant throughout the year, as much is bound up in the calcium pectate of the cell wall, whereas phosphate declined at the beginning of the wet season (Fig. 37). A lack of phosphate would mean, of course, that the calcium would not be available to an animal.

Other studies on feeding preferences in waterbuck have not been very helpful for comparison as most give no measures of plant species'

Plate 41. Doe browsing on *Capparis*.

Table 44 Summary of time spent in browsing *Capparis tomentosa* compared with grazing in Peninsula waterbuck.

Animal	Date	Wet/dry season	Ratio browse : graze	Browse as %
Adult doe	7–8/2/67	D	39 : 200	16·3
Adult doe	22–23/2/66	D	7 : 213	3·2
Adult doe	22–24/2/66	D	38 : 306	11·1
			Mean:	10·2
Fawn, 6 month	7–8/2/67	D	1 : 107	0·9
Adult Buck	16–18/2/66	D	30 : 218	12·1
	18–20/5/66	D	22 : 197	10·0
			Mean:	11·1
	19–21/9/66	W	0 : 241	0
Adult buck	Nocturnal			
	7–8/3/66	D	15 : 64	19·0
	25–26/1/67	D	13 : 45	22·4
			Mean:	20·7
	28–29/10/66	W	0 : 30	0

Table 45   The percentage crude protein in some browse samples collected at the
end of the dry season.

Species	Date	Plant part: crude protein
C. tomentosa	11.2.67	Stem:19·1 Growing point:38·4 Flower bud:31·9 Leaf:36·0
A. sieberiana	14.2.67	Leaf:27·5 Green seed pod:11·1 Green seed:19·3

availability, and, of course, there are geographical differences in species occurrence. But Von Richter and Osterberg (1977), from studies in the Chobe National Park of Botswana, concluded that the waterbuck throughout the year seeks those species which are richest in protein. Several other workers refer to a preference for the protein-rich *C. dactylon*, but these are not quantitatively supported statements. Herbert (1972) states that the Bantu name for the waterbuck is "phiva", meaning "the one who eats the mealies", suggestive of the animal's affinity for high protein, but perhaps related to a greater indifference to man than that shown by many other antelopes. Herbert also refers to them flocking to the sites of abandoned settlements, presumably to feed on the *C. dactylon* which is found there. Since such a preference for *C. dactylon* was not observed in my study, despite its presence, this might indicate a generally protein deficient habitat. During the dry winter months of June to September, Herbert observed waterbuck standing in the river feeding on the aquatic plants *Typha capensis* and *Phragmites communis*. Child and Von Richter (1969) also observed waterbuck standing up to their bellies in water to feed on aquatics in the Chobe River in Botswana. Such behaviour was never witnessed by me; but the sandy lake edges in the Rwenzori Park were generally free of aquatic emergents. Tomlinson (1980) in a study in Rhodesia (Zimbabwe) found wading and feeding to be not uncommon during the hot dry season, and goes so far as to say that published records indicate the possibility that there is a difference in feeding habits between ellipsiprymnus and defassa, as this habit is rarely recorded in the latter. However, more studies upon defassa in protein deficient habitats may reveal this habit to be more common than has hitherto been observed. Feeding on hydrophytic plants would seem to be an opportunistic habit, and in the Central African Republic I have seen Buffon's kob standing in the middle of a river at low water, submersing their heads above the ears to feed on algae from the river bed. In the same area it is not uncommon to see roan antelope wading in ponds to feed upon unidentified aquatic vegetation (Thal (personal communication, 1981) informs me that he has identified this as a species of *Nymphae*.) Where such habits are persistent they probably indicate unduly poor conditions; Child and Von Richter noted that most adult waterbuck were in poor condition during the 1965 dry season in their study, but were in better shape where *C. dactylon* was plentiful. In the following year, when conditions were "less severe", they found waterbuck to be among the animals in the best condition at the end of the dry season.

In his study of enclosed waterbuck in Zimbabwe at a density of 4·9/km², (90 in 18·67 km² with much higher densities of other large ungulate species present), Tomlinson (1979) also confirmed the waterbuck's preference for high protein forage. In the rainy season favoured grasses were: *Sporobolus panicoides*, *Scleria rehmannia* and *Heteropogon contortus*; while during the dry seasons *Rhynchelytrum* sp., *Digitaria gazensis*, *D. milanjiana*, *Eragrostis capensis* and *Panicum dregeanum* were selected. On burnt areas *Pogonarthria squarrosa*, *Brachiaria brizanthia*, *D. milanjiana* and *H. contortus* were taken. In the hot dry season Tomlinson observed that as much as 28% of feeding time was spent in wading and feeding on sedges by territorial bucks, but only 2·6% of the time by adult does, 16·9% by immature animals, and 0·9% by bachelor bucks. At this time there was an increase in browsing, from nil in the rainy season to 5·9% of feeding time in the territorial buck, and 4·8% in the adult doe; this being almost exclusively on *Parinari curatellifolia*. Immature animals browsed only 1·1% of the time, and bachelor bucks 1·5%. This much lower amount of time spent in browsing compared to that in my study may be attributable to the use of the waterside community, where such species as *Cyperus leptocladus*, *Hemarthria altissim*, *Phragmites mauritianus* and *Brachiaria radicans* were eaten; but the use of this community may have been occasioned by overstocking of the area, forcing the waterbuck into a greater use of this habitat.

The availability of protein in the dry season in my study lends me to suppose that death from starvation is unlikely. However, in an acute drought the growing tips and buds of the bushes which provide the protein supplement, would wither, and the tannins present in the mature leaves may render them unpalatable, despite a high protein content. This remains an uninvestigated aspect of herbivore nutrition. Another factor to consider is that a high protein intake requires a high water intake to flush out the urea. Taylor *et al.* (1969) have calculated that if the waterbuck ate low quality hay, and lost amounts of urea similar to that of the Hereford steer, then its daily urine volume would be reduced by about one third, reducing its water intake correspondingly by about one fifth. Thus the waterbuck's high protein intake could result in a vicious circle developing between the animal's food and water supplies. The more concentrated the protein intake, then the more frequently must the animal drink. That the waterbuck does not select browse throughout the year, suggests that a concentrated protein intake is not preferable to a bulky diet with a more moderate protein content.

Where protein-rich browse is available in the dry season this obviously provides for a favourable waterbuck habitat, a requirement that may possibly be met in some areas by aquatic emergents. But waterbuck density is unlikely to be related to food supply alone, as is amply demonstrated on the Peninsula. Rather it is attributable to a complex of factors: food, shelter, lack of competitors, lack of disturbance, lack of predators and availability of water. Some of these factors are complementary, but experiments have shown that water availability overrides all others (Taylor *et al.*, 1969).

Sinclair (1977) has adopted a suggestion by Field (1968) that the diets of the waterbuck and buffalo in the Rwenzori Park are very similar, and this leads Sinclair to speculate that, although the two species may eat the same grasses, because the waterbuck eats less per animal due to its absolutely smaller size, it can exist where food is too widely scattered for buffalo to use, such as areas already grazed and depleted by buffalo. Sinclair thus suggests that the buffalo could create a habitat for the waterbuck, while at the same time competing with it. What both Sinclair and Field overlook, however, is that the waterbuck either, supplements its grass diet with small amounts of protein-rich browse or, it feeds extensively on protein-rich grass species, as has been demonstrated by workers in other areas. This is because it requires, as already pointed out, a diet richer in protein than most species. The correlation between basal metabolism and water turnover rates (MacFarlane and Howard, 1972) supports this latter assumption, for the waterbuck has a higher minimum water requirement than that of most African ungulates so far examined, indicating a higher fasting metabolic rate, which must be met with a richer food intake. Thus, in effect, the buffalo must have a much less demanding diet.

Habitat Preference

Lamprey (1963) was the first worker to try to define the habitat preferences of African herbivores in a quantitative manner. His observations of several ungulate species in the semi-deciduous woodland of the Tarangire Game Reserve in northern Tanzania (now a national park), indicated that the waterbuck was very selective in its requirements, with a strong preference for medium density *Acacia drepanolobium* woodland along drainage lines, but proximity to water seemed to be the main determining factor as there was also a reasonable preference for dense *Commiphora* bush where it was within 1500 m of a river, a habitat totally avoided when it was 2500 m distant. Harris (1972), also working in northern Tanzania, but in the drier Mkomazi Game Reserve bordering Kenya's Tsavo National Park, supported this concept of the selectivity of waterbuck with regard to its choice of habitat.

Most areas where waterbuck occur are riverine, and a number of observers has referred to the animal's affinity for riparian thicket and gallery forest. Kiley-Worthington (1965) writes:

"There was observed to be a daily rhythm of movement. The animals left the thicket cover of the river and wadis between 8 and 10 a.m., depending on the weather. They would stay longer in the cover on cold misty mornings, or when the percentage cloud cover was high, which during the pre-rains season, was common. After leaving the cover they would move out onto the *Themeda-Pennisetum* open grassland adjoining the riparian thickets, which is scattered only occasionally with *Acacia mellifera*. The animals would stay there throughout the heat of the day, apparently not in the least incommoded by the direct sun. At dusk they move back slowly down to the river thickets . . . it is sure that the animals remain in the riparian thickets overnight."

Although this apparent indifference to direct sunlight correlates with my own observations in the Park, it is not universal among waterbuck. Where they are hunted this rhythm is reversed, at least in part, and it is in the late evening that they emerge from cover to feed; they may also adopt this behaviour where daily temperatures are very high, although this is not so in the Saint Floris National Park in the north of the Central African Republic. Here, with a mean maximum temperature of about 34°C they can be seen passing the day on the open plains. In the Mont Fouari Game Reserve of Congo Brazzaville, close to the West African seaboard, the waterbuck shelter in the long *Cymbopogon-Eragrostis* grassland during most of the day, emerging to feed in the open short grass areas at about 1600 h.

Whether Kiley-Worthington is correct in assuming that her animals passed the night in cover, is open to question. There would appear to be no physiological reason for this, but threat of predation may have been a contributory factor in her study area.

Tomlinson (1979, 1980), in his Zimbabwe study, found that in the rainy season the waterbuck tended to feed in woodlands during the morning, but used grasslands in the afternoon. The dry season saw a marked decrease in the use of the woodlands, the most important period for their use being mid-day.

I found no circadian rhythm in the Rwenzori Park, perhaps only because the habitat structure was not one which could accommodate this type of behaviour in the majority of the area. Only west of the Kayanja region, and the Kamulikwezi area, were found wooded areas, while to the east there was, of course, the Maramagambo forest ecotone. These areas were not typical of the waterbuck's habitat here, and were not studied by me. For the most part waterbuck inhabited the open grassland of medium height, dotted with *Capparis* thicket, a habitat offering only scanty shade during the day provided by the occasional large *Euphorbia candelabra* tree.

Using grid superimposition upon the routine count distribution maps, Field and Laws (1970) concluded that the waterbuck in the Park consistently avoided *Capparis,* and preferred some water, but avoided permanent supplies, except in the dry seasons when the inland wallows dried out. These deductions would seem to be largely due to the artificiality of the method of analysis. It is true that waterbuck were seldom encountered within *Capparis* thickets, and as we have seen, only in the dry season did this shrub provide a limited, but nevertheless highly important, part of its diet. The animal's association with water is temporal, there being no inducement to remain in the immediate vicinity after drinking. An exception to this was provided by some areas, such as Kayanja, where relatively extensive lakeside areas of green vegetation of *Cyperacae* sp. and *C. dactylon* caused the waterbuck to linger feeding near the lake after drinking.

Only on the Peninsula did I conduct an analysis of habitat preference in any detail, excluding adult bucks whose distribution was determined primarily by social factors (Chapter 10). Although does occupied most of the area, at one time or another during the study period, there was a clear affinity for certain regions. Between April 1965 and March 1967, an average 33% of does was recorded on the lowlands, compared with 67% on the high plateau. Of this distribution more appeared to be recorded on the plateau in the wet season than in the dry, but I did not have equal numbers of seasonal observations to confirm this. The preference for the high plateau may have been due to the cooling effect provided by the prevailing winds, which were usually from the north-east in the morning, blowing across the Channel, and from the south in the afternoon, coming off the lake. But the absence of tsetse flies may also have been a contributory factor, especially in the wet season. Although I noticed few on the lowlands, there was much more thicket there with which they are associated.

The composite distribution map of the does on the Peninsula during the study period, shows that, on the plateau the most favoured spots were at the edges of the north-east slopes to the Channel, where there was a certain amount of thicket, and towards the centre of the plateau, an area which also possessed some scattered thicket (Fig. 40a). Completely open grassland was not the habitat most selected for, but, conversely, those areas most avoided, were the low-lying bushed areas, and the bushed slopes. It appeared that the does liked to have some cover near to them, but not too much.

Distribution of the bachelor bucks was also analysed. They also showed a preference for the high ground, with the highest occupation in the vicinity of scattered thicket, but the bachelors tended not to overlap the most favoured doe areas. Although their distribution was less evenly spread over the available habitat, the bachelors were absent from only 37 sectors (14·8 ha) occupied by does, so that the bachelors were not in any significant way denied resources that were available to the does (Fig. 40b).

When applied to the territorial bucks, the same analysis of occupation shows, as might be expected, a much more uniform coverage of almost the whole of the Peninsula area. They were absent from only one sector occupied by the does, and bachelors were absent from only 37 sectors. These were mostly sectors to which does had access, about 15 of which were in the territory of Y8. Thus there appears to be no evidence that the bachelors were excluded from the majority of areas occupied by the territorial bucks (Fig. 40c, Table 46). Of course, it must be realized that the densities given in the figures are cumulative, and the mean density per 0·4 ha would be these densities divided by 428, so that average occupation is rather low.

Analysis of the differing densities in the Park revealed that the highest concentrations of waterbuck, averaging about three times the density of the next category, were those of populations within close reach of the lake shore. When densities in this type of habitat were low this appeared to be attributable to competition by hippopotamus. Grass species composition, although varying with regard to its main dominants, was not greatly different between the areas. The distinctly low density areas were all marked by an absence of hippopotamus competition, but were characterized by relatively long distances from permanent water. They could all be termed "inland" areas. In some of these latter, for example the Craters, a dominance in the grassland composition of the coarse grass *Imperata cylindrica,* also probably exerted a negative influence on waterbuck density (Table 47). I conclude, however, that the most significant factor influencing waterbuck density and habitat selection is easy accessibility to water, a conclusion which we shall see is substantiated by the waterbuck's physiology. This does not explain why

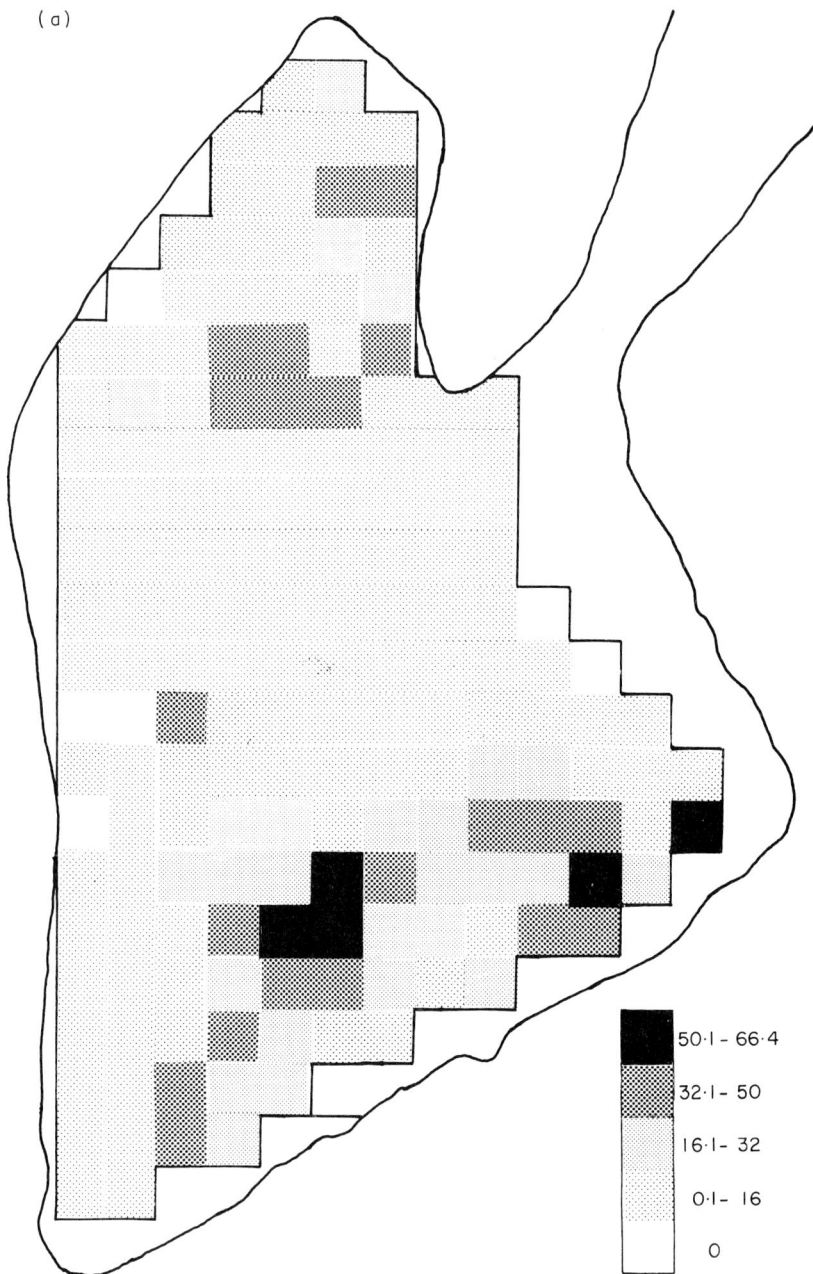

(a)

	50·1 – 66·4
	32·1 – 50
	16·1 – 32
	0·1 – 16
	0

Fig. 40. Comparative analysis of occupation of the bachelor bucks, territorial bucks and does on the Peninsula from April 1965 to March 1967, expressed as cumulative numbers per 0·4 ha. a — does, b — bachelor bucks, c — territorial bucks.

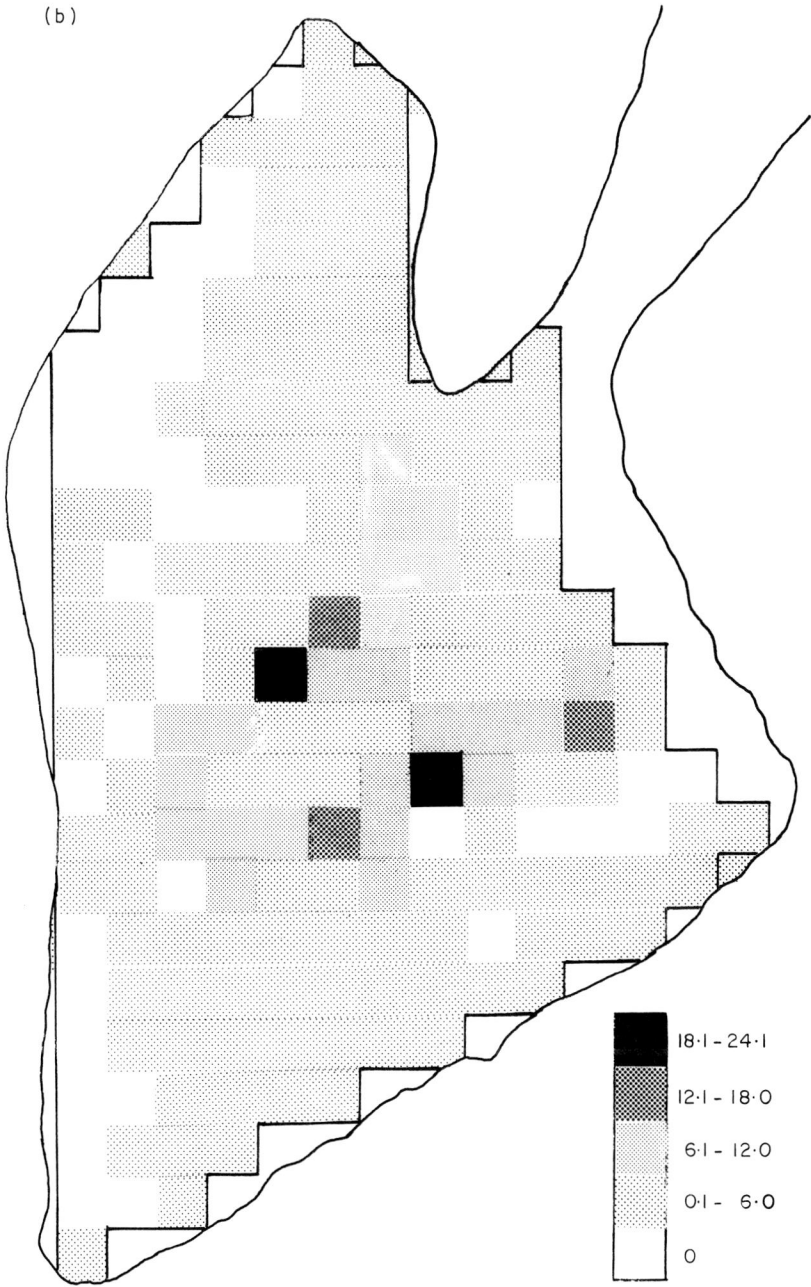

(b)

18·1 - 24·1

12·1 - 18·0

6·1 - 12·0

0·1 - 6·0

0

Fig. 40(b).

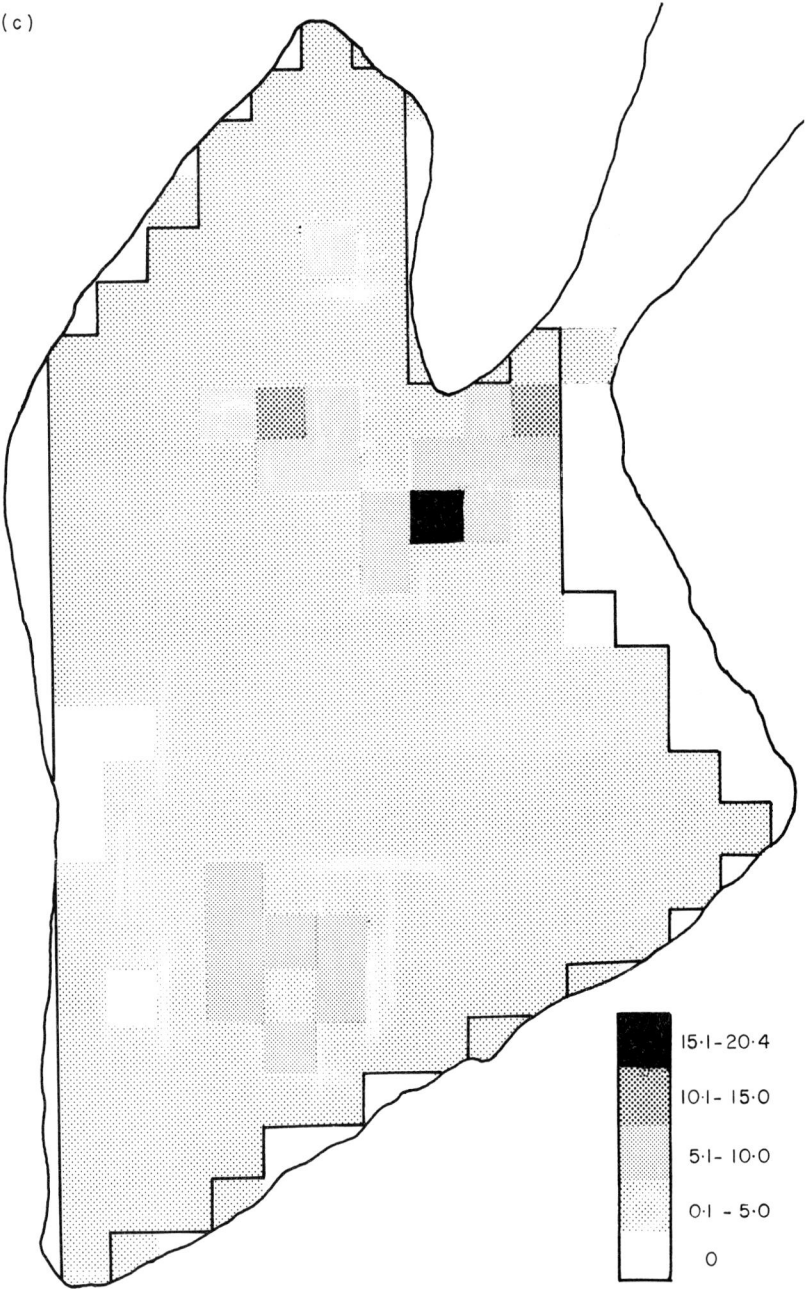

(c)

15·1 - 20·4

10·1 - 15·0

5·1 - 10·0

0·1 - 5·0

0

Fig. 40(c).

Table 46 Comparative analysis of occupation of the Peninsula by does, bachelor bucks and territorial bucks.

	Density of does				
	0	0·1–16·0	16·1–32·0	32·1–50·0	50·1–66·4
Density of bachelor bucks					
0	2	32	3	2	0
0·1–6·0	2	61	14	17	5
6·1–12·0	0	15	5	2	0
12·1–18·0	0	2	0	0	0
18·1–24·1	0	2	0	0	0
Density of territorial bucks					
0	3	1	0	0	0
0·1–5·0	3	104	21	16	3
5·1–10·0	0	8	4	4	2
10·1–15·0	0	0	0	4	0
15·1–20·4	0	1	0	0	0

	Density of bachelor bucks				
	0	0·1–6·0	6·1–12·0	12·1–18·0	18·1–24·1
Density of territorial bucks					
0	2	3	0	0	0
0·1–5·0	38	100	19	3	2
5·1–10·0	0	16	3	0	0
10·1–15·0	0	2	0	0	0
15·1–20·4	0	0	1	0	0

Cumulative densities recorded in 0·4 ha squares.

a riverine habitat, such as at Ishasha, should carry a density which probably equates with that related to hippopotamus competition. In this particular case waterbuck may suffer competition from both Uganda kob and topi, the dominant grazers in this area, with very high densities estimated at 71·1/km² and 83·2/km² respectively.

In the Akagera Park of Rwanda, approximately 53% of waterbuck were found on the short grass open plains near to the lakes; while inland open tall grassland was totally avoided. Some 42% of all animals encountered were in the same habitat as that favoured by the waterbuck. Second in the waterbuck's preference was *Acacia* scrub woodland, with 32% of encounters being in this habitat. Aerial surveys confirmed the

Table 47 Waterbuck densities grouped according to habitat factors.

Area	Density /km²	Habitat type	Grass dominants	Water availability	Hippo density	Intense hippo competition
Peninsula	7 — 11	Medium height grassland, scattered *Capparis*.	SP	BSP Lakeside	0·9[a]	No
Kayanja		,,	HC	HF Lakeside	[a]	No
Nyamagasani		,,	BSP	HC Lakeside	15·2	No
Ogsa	3 — 4	Medium height grassland, scattered *Capparis*.	SP	BSP Lakeside	20·5	Possibly
Lion Bay		,,	H	SP Lakeside	[a]	Possibly
Katwe		Overgrazed, scattered *Capparis*.	SP	BSP Lakeside	28·0	Yes
Ishasha		Heavily grazed.	HF	BSP Riverine	6·1	No
Kikorongo	2 — 3	Dense long grass, scattered *Capparis*.	HF	BSP Limited	1·0	No
Katunguru	1 — 2	Dense medium height grassland, scattered *Capparis*.	MK/SS	BSP Limited in dry season	4·4[a]	No
Royal Circuit		Medium height grassland, scattered *Capparis*.	BSP	HF Limited in dry season	13·5	No
Craters	< 1	Long grass.	IC	HF Limited	0	No
Kikeri		Dense medium height grassland.	BP	SP Limited in dry season	2·0	No

[a] Hippopotamus reduction has taken place.
SP *Sporobolus pyramidalis*, BSP *Bothriochloa* sp., HC *Heteropogon contortus*, HF *Hyparrhenia filipendula*, H *Hyparrhenia* sp., MK/SS *Microchloa kunthii/Sporobolus stapfianus*, IC *Imperata cylindrica*, BP *Brachiaria platynota*.
Vegetation data in part from Field and Laws (1970) and Field (1972).

affinity of the waterbuck for the lake regions, 39·2% of observations (mean of two surveys) being there (Spinage and Guinness, 1972). This was confirmed by Montfort (1972), working in a different area of the Park. Of the total mean density observed, 59% were found near the lakes, 27·7% in the *Acacia* woodland and *Sporobolus* plains, 7·4% on hilltops with *Loudetia*, and 5·9% on the sides of hills in *Themeda* grassland. Areas of *Cymbopogon* and *Bothriochloa* grassland were avoided completely, although I noted that the fresh flush of *Bothriochloa*, after burning, was heavily grazed by some species of ungulates.

A detailed analysis of waterbuck habitat selection has been presented by Hirst (1975). In a study of Kempiana, an area adjoining the Kruger National Park in South Africa, Hirst found that of the seven most common ungulate species which occurred there waterbuck were the most selective in their habitat requirements. Here, the habitat more or less consistently preferred was the riverine gallery forest, but the waterbuck also frequented the savanna areas which had variable tall and short grass patches. This showed no seasonal distribution pattern. Occurrence in other habitat types, away from water, was sporadic; and mixed semi-deciduous *Combretum*, and thorn woodland, was almost completely avoided. Hirst found that the waterbuck spent most of their time in the riverine vegetation, and were located most easily during the late afternoon, and when the wind was blowing steadily. Wind gusting tended to drive them into cover. His analysis of preferences and avoidances implied a preference based on abundant vegetation, especially abundant tall grass within woodland, and in dense evergreen vegetation along watercourses. Although a definite selectivity for denser habitats was evident, this appeared to be related to the grassland component, since a significant negative response was shown to dense shrub understorey, and no particular relationship to the forb layer could be found.

Hirst concluded, that a preference for tall and abundant grasses under a high woodland canopy indicated that waterbuck were indeed stenoecious, that is, limited to a narrow range of habitat types; since such a combination of habitat factors in semi-arid or subtropical Africa can only be found along watercourses where moisture is locally available, total habitat is also limited. He concluded that vegetational factors fully explained the densities and distribution of waterbuck. Such a conclusion is at variance with the results obtained in my study, where over the greater part of the waterbuck's range dense vegetation exerted no attraction to waterbuck, habitat preference relating apparently to the proximity of water. Hirst, however, appears to have analysed habitat preference solely upon vegetation type, and not vegetation type in association with water. His conclusion, therefore, seems to confuse the waterbuck's affinity for the presence of water, with the type of vegetation found in association with water.

Dependence Upon Water

I never found the animals far from water, but then most areas within the Park were within relatively easy reach of it. Apart from its physiological dependence, the waterbuck appeared to show no real affinity for water, never, for example, wallowing in pools like buffalo. I have found no confirmation of Swayne's assertion that the waterbuck, in Somalia, "delight in a mud bath" (Swayne, 1895). Perhaps his observations were made under conditions of extreme drought. Timid animals, surprised close to the lake shore, always ran inland; but if hotly pursued then a complete reversal of behaviour may take place, the waterbuck taking to water and swimming from its pursuers. The losers in fights between two bucks have been observed to swim to safety, as have also animals fleeing from hunting dogs. Wounded animals often seek water to lie in as a balm for their wounds. Once, after much lion activity on the Kazinga side of the Channel, two does, one with a fawn at heel, were seen to swim the Channel to the Peninsula. As the fawn did not follow its parent, the latter returned to it after having swum halfway across.

At 0730 h on 1 February 1966, I saw one doe with her 2½-month-old fawn drinking in the Channel, an unusually early hour for this. She then sat in the water for a minute or two, an action which was copied by the fawn which immersed itself up to its neck. The couple then emerged and shook themselves like dogs. The weather was very cool and I imagine that the water felt warmer than the ambient temperature, or perhaps I had witnessed the doe teaching its fawn not to be afraid of water?

This somewhat indifferent attitude to water as a medium, apart, the waterbuck's complete physiological dependence on water has been found to be very real, fully justifying the name of "waterbuck" which was given to the animal, when it was first encountered by Europeans in South Africa in the nineteenth century. This dependence has been experimentally verified by Dr Dick Taylor of Harvard University, using three waterbuck which I captured in the Park (Taylor *et al.*, 1969).

In various controlled laboratory experiments carried out in Kenya, measurements were made of the animal's water requirements, and the reduction in water loss which took place when water intake was restricted. Other parameters which were measured were temperature regulation and oxygen consumption, in order to describe more completely the physiological mechanisms involved in water loss. In the first of these experiments, to measure water balance, the animals were housed in metabolism cages. Each day the amount of water drunk, the amount of food eaten, the amount of faeces and urine voided, the moisture content of the food and faeces, and its weight, were measured. Water balance was measured in a constant environment of 22 °C, at which evaporation should have been almost minimal (the mean temperature at Mweya was 23·5 °C,), with a daily periodic heat load of 12 h at 40 °C, alternating with 12 h at 22 °C. It was not possible to measure minimum water requirements under a heat load, as waterbuck on a restricted water intake would die in less than 12 h in a temperature of 40 °C.

Water is lost from a body in three ways: through the faeces, through urine, and by evaporation (sweating and panting). At 22 °C, when water was freely available, the waterbuck's loss was approximately equally divided between these three pathways. But when water intake was restricted, faecal loss was reduced from one-third to about one-sixth of the total water loss. Drier faeces were formed, and almost 22% less food was eaten, but what it did eat was better digested, dry matter digestibility increasing by some 10%. When water was freely available, for each 161 g of dry food eaten, 100 g were digested and 61 g of dry faeces produced. When water was restricted, only 147 g of food was eaten, for every 100 g digested, and 47 g of dry faeces were produced. The more complete digestion, which occurred with restricted water intake, could account for one-half of the reduction in faecal water loss. Dr Taylor thus concluded that a high digestibility was an extremely efficient way in which to minimise faecal water loss. The periodic heat load had no effect on the amount of water lost in the faeces, but since evaporation increased under these conditions, the faecal water loss thus constituted a smaller percentage of the total water loss.

Restricting water intake had no effect on the daily volume of urine which was passed, except when the animal was severely dehydrated. Otherwise, the concentration remained the same (about 1100 mOsm/litre), urea accounting for about one-third to one-half of the osmotic concentration when water was restricted or available. When an animal was subjected to a periodic heat load, however, urine volume increased by about 26% when water was freely available. The waterbuck required a high protein diet in the laboratory in order to survive, about four times that of other bovids which were examined (Taylor, 1968). The high intake of protein, and the high urea excretion in the urine, help to explain the high urine volume. If the waterbuck ate low quality hay and lost amounts of urea similar to a Hereford steer, then its daily urine volume would be reduced by about one-third. Another way for the waterbuck to reduce its urinary water loss would be to have a lowered metabolism and lowered food intake, but the waterbuck did not exploit this possibility.

Most mammals increase the concentration, and reduce the volume of their urine, when water intake is restricted, but the waterbuck proved an exception to this. That it did so Dr Taylor considered to be very surprising, for the waterbuck's kidney does not lack the ability to form a concentrated urine; up to about four times as concentrated as plasma, a urine similar in concentration to that of the dehydrated Hereford steer. Thus it appeared that the waterbuck did not drink an excess of water when it had the opportunity, and always formed a nearly maximally concentrated urine. This, however, was only about one-half as concentrated as that of the eland, an animal which can make do with very little water, and one-third as concentrated as that of the camel.

The third pathway of water loss, evaporation, was not altered when water was restricted at 22°C. At this temperature about half of the evaporative loss took place from the respiratory tract by expelled air, and the other half from the skin. Under the periodic heat load, evaporation approximately tripled, the waterbuck both panting and sweating, increasing its respiratory loss by 50% and its cutaneous loss by over 76%. Under these conditions the skin accounted for about two-thirds of the animal's total water loss. The rate of evaporation was about 40% greater than that of an eland of similar size and metabolic rate, the waterbuck not exploiting any methods of containing evaporative water. In total, when subjected to a periodic heat load, it lost approximately 12 litres of water per 100 kg of body weight per day. Thus to maintain equilibrium an adult buck under these conditions would have to drink some 60 litres per day.

Rectal temperature was measured at various ambient temperatures from 25°C to 45°C. It was shown that in hot environments the waterbuck is able to maintain a rectal temperature well below that of ambient, even at 45°C; its temperature increasing by an average of only 0·6°C from 39·2 ± 0·05 to 39·8 ± 0·05°C. It used both sweating and panting for heat dissipation and saved little water by heat storage or hyperthermia. But if water was restricted, then rectal temperature continued to rise until the animal collapsed. As ambient temperature increased from 25°C to 45°C, cutaneous evaporation approximately tripled, respiratory rate increased by about ten times, and rectal temperature increased by about 2°C (Fig. 41). The waterbuck thus maintains an almost constant body temperature and must compensate for this by an increase in evaporation when the ambient temperature rises above the thermoneutral zone.

Oxygen consumption was measured at various temperatures from 5°C to 42°C, after both it and rectal temperature had been constant at a given temperature for nearly 2 h. This showed that the thermoneutral zone of the waterbuck extended from approximately 13°C to 37°C, a relatively wide zone more or less covering the range of temperature that it normally encounters (Fig. 42). Within this zone it consumed about 6·4 litres of oxygen per kg of bodyweight per day, oxygen consumption of the hydrated and dehydrated animal being approximately the same.

When dehydrated, the camel, cattle, and six other species of African wild bovids studied, all reduce their metabolism, thereby minimizing water loss. The waterbuck thus appears to be unusual in that its metabolism remains unaltered. Within the thermoneutral zone its metabolism was almost 30% higher than would be predicted by Kleiber's equation (kcals/day = 70 (kg$^{0·75}$) (Kleiber, 1961). But this is similar to eland and wildebeest (Rogerson, 1968), and is probably related to the small amounts of fat which these animals lay down compared with domestic stock.

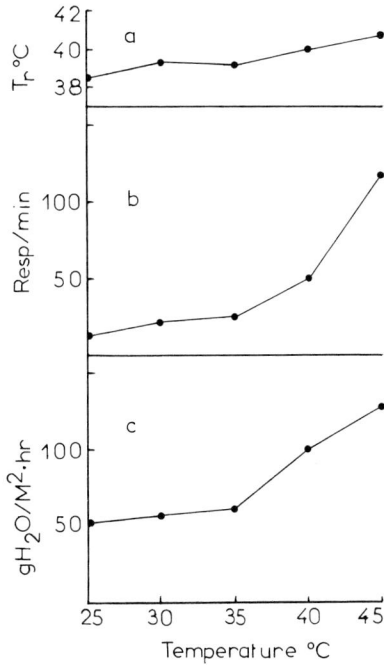

Fig. 41. Rectal temperature change, respiratory rate and evaporation in the waterbuck with increase in ambient temperature, after Taylor *et al.* (1969).

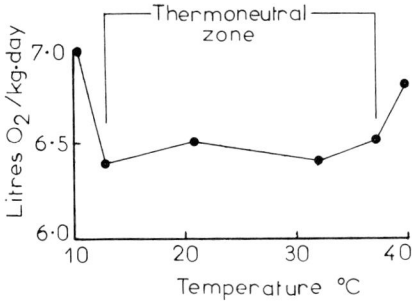

Fig. 42. The thermoneutral zone of the waterbuck, after Taylor *et al.* (1969).

The experiments leave no doubts concerning the waterbuck's dependence on water. At a temperature of only 22°C it requires some 25% more water than a Hereford steer, and nearly three times that of the arid-dwelling oryx (Fig. 43). Only faecal water loss was reduced when water intake was restricted; urine volume and evaporation were unchanged. The waterbuck lacks even the ability of a water dependent animal like the Hereford steer, to reduce its water loss in response to a shortage of water. This inability to withstand short periods of dehydration in hot environments restricts it to areas where water is readily available.

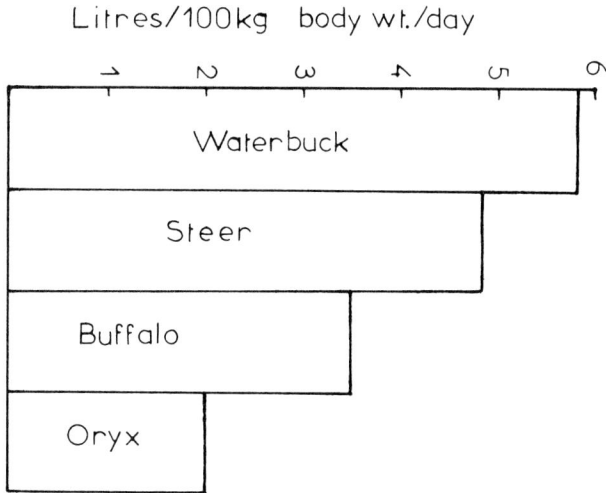

Fig. 43. Water requirements of the waterbuck compared with three other species, after Taylor *et al.* (1969).

At Mweya the lowest temperature recorded during my period of study was 12·3°C, and the highest 35·6°C, while mean temperatures were 17·6°C and 29·7°C, minimum and maximum respectively. This provided a near optimum temperature range and may have accounted for the animal's apparent success there. At Skukuza, in South Africa, close to the Kruger National Park, the mean minimum temperature is given as 12·5°C (Hirst, 1975), 5·1°C lower than it is at Mweya. Thus, if the waterbuck there exhibit the same physiology, then they may experience some difficulty in temperature maintenance during the winter. The low temperatures occurring as they do in the dry season, may explain Hirst's assertion (Hirst, 1969) that the waterbuck is particularly prone to starvation at this time, a higher than normal energy intake being required to maintain thermoneutrality.

Using a somewhat different methodology to that of Taylor, Schoen (1971) has studied some aspects of heat stress and water deprivation in the Uganda kob, the reedbuck and the bushbuck. Naturally enough, the physiology of these species differs from that of the waterbuck, but it is of interest to look at the results for the kob, the waterbuck's close relative. This animal loses roughly half of its water through evaporation, and the remainder almost equally divided through faeces and urine. When dehydrated under heat stress, it reduced its feed intake by some 25%, about the same as the waterbuck, its faecal water loss dropping from about one-fifth of the total water loss, to one-eighth, or by 62%. This is better than the camel, which reduces its water loss under heat stress by about only 70%.

The kob's urine osmolarity is similar to that of the waterbuck, about 1109 mOsm/litre compared with 1100 mOsm/litre; but there the resemblance ends. Under heat stress, with water available, the urine becomes slightly more diluted, to about 1045 mOsm/litre, indicating a greater intake of water. When water is restricted, however, it concentrates the urine, unlike the waterbuck, reaching about 1594 mOsm/litre.

The major avenue of water loss, evaporation, was almost doubled under heat stress when water was freely available, water intake increasing by about 70%. When water was restricted, evaporative water loss was reduced by over 40%.

There was little evidence of thermal storage, the highest recorded rectal temperature was 39·6°C, compared with the lowest recorded rectal temperature, between heat stress, of 38·2°C.

These results suggest that the kob is able to withstand both heat and dehydration better than can the waterbuck, mainly through its ability to restrict its urinary and evaporative water losses. Its normally high evaporative water loss, however, makes it also a relatively water-dependent animal.

9. Daily Life

Introduction

When I was able to identify individuals with certainty, a lot could be learnt about the waterbuck's daily life by following a selected co-operative individual around, and noting its behaviour. It was important to select only those individuals who were indifferent to my presence. The main thing was to keep the same animal in sight all of the time, which was sometimes difficult if it decided to enter a thicket, or to go over the edge of a steep bank.

To watch a waterbuck day, I was out before seven in the morning, observing continuously from my Landrover until the same hour in the evening, repeating this for two to three consecutive days. A paperback helped to relieve the tedium when an animal did the same thing for long periods. I became quite adept at keeping one eye on the page and the other on my subject.

Being on the Equator, sunrise and sunset varied only a quarter of an hour either side of 7 o'clock. So when I went out at sunrise it would be about 0630, not getting really light until half an hour later. It was often chilly by tropical standards, with a whisp of damp in the air, and as the mists chased through the *Capparis* clumps, the waterbuck lay dozing.

The clarion call heralding the waterbuck's day is given by the voluble red-necked spurfowl *Pternistes cranchii*, a bumptious, partridge-like bird, common in the grasslands of eastern Africa. At dawn this bird is to be found scuffling in the grass, searching for food, and from time to time it breaks off, to climb a convenient termite mound and rend the still air with its harsh "kraaek, kraa-ek, kraa-ek!" It repeats its call last thing in the evening as well.

Activity Rhythm

At this time of the morning, the doe waterbuck were usually found resting in groups, and as the spurfowl reminded them that 7 o'clock was approaching, first one, and then another, would rise, stretch, yawn, and amble away to start another endless round of grazing. There always seemed to be one, more lazy than the others, who would remain dozing until the rest were well out of sight. Only then did she slowly rise, stretch, yawn, and amble after them. I was often to be reminded that waterbuck had what Charles Elton called "real lives". Each one had a certain amount of individuality, even if it did not extend beyond that of wanting to stay in bed longer than everyone else!

The next hour was often one of the busiest for the does, the whole time being spent in grazing. The following hour, activity would decline a little, and then from 0900 to 1000 h they would take a rest whilst they ruminated. They then continued to feed throughout the day, with peaks after noon and in the last hour before dark. Surprisingly, they did not slow down their grazing activity during the heat of the day. At some point they interrupted their feeding, journeying to water, where a few minutes were spent drinking; this seemed to take place at almost any time, usually however before 1500 h. But it was, as we shall see, apparently dictated by location. When the animals reached water, they drank.

At the time of my study, as far as I was aware, there were no quantitative observations on the activities of wild herbivores; at least not in Africa. Working with cattle in Africa, Rollinson and co-workers (Rollinson *et al.*, 1956) referred to the fact that various workers had used differing time intervals for the study of activity of domestic animals, and presented data in support of a four-minute recording interval for major habits. Following their lead, I introduced this interval into the study of African herbivores, in the hope that other workers would follow suit; thus standardizing procedures and making data easily comparable. For the study, waterbuck activity was divided into feeding, noting whether the animal was browsing or grazing, ruminating, differentiating between lying and standing, lying, referring to sleeping or resting, and "other", which included all other activities such as moving, fighting, drinking, standing, etc. At every four-minute interval the appropriate activity was ticked off on a record sheet.

All observations were carried out from a Landrover, in which animals could be approached to within a few metres without causing alarm. Observation distances varied from as little as 4 m, when an animal grazed up to the vehicle, to perhaps 400 m. To observe natural activity it was essential to select an animal which was completely indifferent to my presence, and did not keep watching me. This was especially true of nocturnal observations, which had to be performed at much closer range. The behaviour of herbivores at night was completely unknown, so a number of nocturnal observation periods, extending from 1900 h to 0700 h, was included. As I carried out all of the observations myself, it was not possible to watch continuously for 24 h periods, and nocturnal studies were carried out separately. I found these more fatiguing than diurnal observation, and so did not do more than one night during any period of time. It was also necessary to have fairly clear, full-moon nights, so that the animal could be watched and followed without using any lights on the vehicle. These conditions were not often realized, which accounts for a paucity of nocturnal observations. I found 8 × 30 binoculars adequate for night observation, but it had to be restricted to bucks which I could recognize by their horn shape. It was not possible to adequately identify does at night, especially as they moved in groups.

Spot checks were carried out on completely moonless nights using an infra-red sniperscope, to ascertain whether grazing occurred then. However, I was liable to disturb the animals when moving into position, and furthermore it was easy to lose sight of the animals with the sniperscope, which only had about a 10 m range. Much improved night vision equipment is now available.

In general, I found it virtually impossible to keep more than one animal under observation at once, although I achieved this on two occasions for 2 days. If two animals are selected for study, and they separate, one must decide which of the two one is going to continue following, and the information gained from the other may be insufficient to be worth retaining. Most observations were done on a single territorial buck which could be easily followed, and whose habitat did not include much cover. The aim of the observations was to establish the basic patterns of diurnal and nocturnal activity, as revealed by prolonged consecutive periods of observation. I did not have enough time to extend the study beyond this.

A total of 7 days' observation on does in the dry season, revealed that the average doe passes approximately 62% of the day in feeding, 19% in ruminating, almost always lying down, 11% resting without ruminating, and 7% of the time in other activities such as walking, drinking, avoiding bucks, and so on. Lactating does spent almost 10% more time in feeding than did non-lactating does, mainly at the expense of resting. A 6-month-old fawn, which was observed for one day, spent only 2% less time in both feeding and ruminating than did an adult, spending more time in resting and other activities. With such a large part of the time devoted to a monotonous round of grazing and ruminating there was little excitement in watching does, which, on average, seemed to pass a very bucolic existence. Elliott (1976) found a similar pattern in Kenya.

During the day the bucks followed much the same pattern as the does. Rising at 0700 h, feeding industriously for the next two hours, but spending only 50% of each hour in actual grazing. Like the does, they then took their break from 9 to 10. After this, feeding was fairly irregular, but gradually intensified in the late afternoon, the most intense period being from 1800 to 1900 h, just before darkness.

I had 11 days and 3 nights of observations on bucks, with periods both in the wet and dry seasons. In terms of mean percentages these observations showed that bucks passed approximately 44% of the day in grazing, 21% in ruminating, 15% lying without ruminating, and 21% in other activities such as walking, drinking, importuning does, sparring with other bucks, and so on. Thus almost 20% less time was spent in feeding by bucks, than by does, during the day, yet the average adult buck weighs 21% more. It could be that does graze less at night, but unfortunately I lacked any nocturnal doe observations. If a buck was with a doe who might be coming into oestrus, then his "other activities" devoted to following and importuning the doe, could rise to as much as 30% of his daily activity. But this was found to be at the expense of resting, rather than of feeding and ruminating.

The figures given are averages for the total period of observation, the hours within which feeding and ruminating took place varying within limits from day to day, as did the length of time spent on any one activity. In the wet season, for example, the buck showed a much more even distribution of feeding throughout the day, but there was no difference in the total amount of time spent in this activity in the wet and dry seasons. In the 24-h period, roughly equal amounts of time were spent on feeding and ruminating.

In Zimbabwe, Tomlinson (1979) conducted a daylight activity study using a five-minute recording interval. His results for territorial bucks are remarkably similar to mine, in the wet season 45% of the time being spent in feeding, compared with 44·4% in my study. His "hot dry" and "cold dry" seasons showed 42·6 and 51·1% feeding time, which compares with 42·7 to 45·9% in my dry season study. The does, however, showed much greater differences, feeding falling to 35·1% of total time in the hot dry season in Tomlinson's study, with an increase in resting time, whereas I observed 60·7 to 68·4% feeding time in the dry season. Unfortunately I lack any wet season observations for the doe, but Tomlinson's 60% is close to my dry season observations. This could suggest that in his study area the does may have been suffering from heat stress in the hot dry season, for as we have seen this is acompanied by less food intake with more thorough digestion (Chapter 8), and their rumination time remained approximately the same.

Naturally there was no great difference in the amount of time spent in rumination, since feeding time was mostly similar, but for other activities there was little similarity between the two studies. In general, Tomlinson's territorial bucks appeared to be less active, spending more time lying and resting, and less time in "other activities" than did Peninsula bucks. His does also spent more time lying and resting, but yet also more in "other activities", possibly attributable to greater vigilance, "standing/vigilant" in Tomlinson's terms.

Although less happened at night, nocturnal observations were nonetheless more exciting than diurnal ones. Without lights, and in lowest gear, I would crawl through the long grass in my Landrover, hoping against hope that I would not crash into a warthog hole, or ram into a termite mound. Bushes and grass appeared in tones of silver and black under the tropical moon, but to the waterbuck, without colour vision, presumably it is only the intensity of the light which changes. Certainly there was nothing in its behaviour to suggest that its vision was impaired. More than once, the buck which I was watching would stare into the darkness, and then deliberately walk up to another which I had been unable to see. Once a buck suddenly galloped off to chase away a possible intruder from his territory, far too distant to be discernible to my myopic night vision.

A change comes over the savanna at night, when man has scuttled into his troglodyte retreats, and his harsh, discordant noises, slowly subside to stillness. Y7, the buck which I used for all of my nocturnal studies because of his indifference to me, knew this. He also knew that in the dry season there was richer grass around the native habitations, and when all was quiet he would make his way there. If any of the inhabitants of the huts heard me, they must have thought my behaviour rather strange, my vehicle stopping and starting, stopping and starting, crawling at a snail's pace without lights among the huts in the middle of the night, crunching over old tins, bottles and goodness knows what. Chasing among the dustbins was not exactly my idea of the beauties of nature revealed, but then for all their mystery, with echoing lion roars and hippo grunts, the moonlit nights revealed little of the secrets that they held. One imagines the tempo of life to increase at night, for then the majority of carnivores, large and small, is active. But my impression was that everything slowed down, and the hours of darkness were quieter than those of the day.

From 2000 to 2100 h, Y7's grazing ceased altogether. So constant did I find this that I could go and have my own supper, returning to find the buck still dozing in the same position in which I had left him. There was always the risk that he might gallop off to another spot, but this never happened. From 2100 to 2200 h he still hardly fed, but his feeding activity would gradually build up to the midnight hours, until from midnight until two in the morning, 50% of his time was spent in grazing. It would then slowly decline in intensity to almost nothing from 0500 to 0600 h, ceasing entirely in the long hours before daylight. Feeding was more intense at night, in that longer periods were spent feeding without pause, but the periods were of comparatively short duration, with long intervals spent lying ruminating. Thus there was a distinct change in the activity pattern at night, which was mostly devoted to rumination. Feeding decreased on average by some 30%, while rumination increased by 137% to accommodate the evening feeding peak. Lying without ruminating declined by almost 50%, as did other activities. The results are summarized in Table 48, and in Figs 44–47.

Table 48   Activity of waterbuck expressed as %.

Animal	Season	Date	Feeding	Ruminating	Lying	Other[a]	Moon's phase
Diurnal							
Adult doe	Dry	7–8.2.67	68·4	20·0	4·5	7·1	Last quarter
	Dry	22–24.2.66	63·3	18·1	9·9	7·6	New moon
	Dry	22–23.2.66	60·7	20·5	11·4	7·5	New moon
Mean			64·1	19·5	8·6	7·4	
6-month fawn	Dry	7–8.2.67	60·0	17·0	14·0	9·0	Last quarter
Adult buck	Dry	16–18.2.66	45·9	21·9	16·9	15·3	Last quarter
	Dry[b]	18–20.5.66	41·9	19·6	8·9	30·5	Last quarter
	Wet	19–21.9.66	44·4	21·8	18·0	15·9	First quarter
	Dry	28–29.6.66	42·7	22·5	6·9	26·6	First quarter
Mean			43·7	21·5	12·7	22·1	
Nocturnal							
Adult buck	Dry	7–8.3.66	43·7	44·2	2·8	9·3	Full moon
	Wet	28–29.10.66	16·7	48·4	12·8	22·2	Full moon
	Dry	25–26.1.67	31·9	60·4	4·5	3·4	Full moon
Mean			30·8	51·0	6·7	11·0	

[a] See text for explanation of "other activities".
[b] With doe for second and third days.

Herbert (1972) recorded feeding times in his South African study, and found a gradual increase in activity from 0400 to a peak between 0600 and 0700 h. This was followed by a gradual decrease, and then a recovery with the main peak at 1700 to 1900 h.

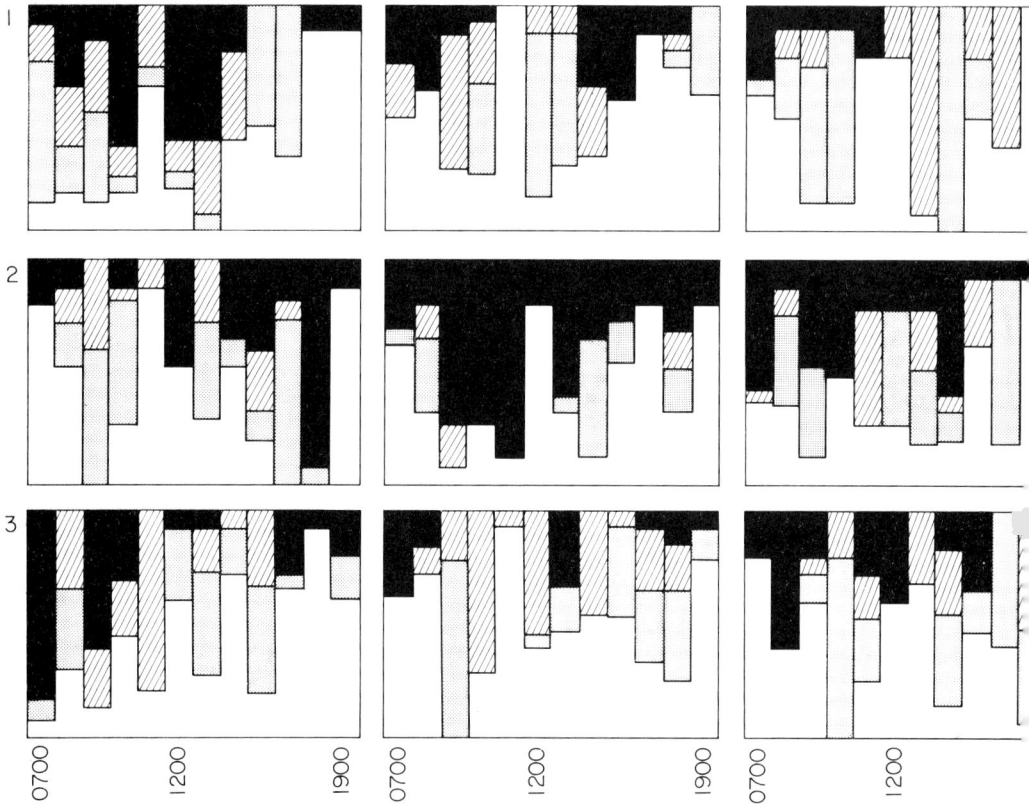

Fig. 44. The daily activity pattern of the adult buck Y7 for three periods, each on three consecutive days: (1) 16–18.2.66; (2) 18–20.5.66; (3) 19–21.9.66. Blocked areas = "other activities"; hatched areas = lying resting; medium density dots = ruminating; blank = feeding.

At night, little time was spent in resting or sleeping, and no deep sleeping was seen at all. Deep sleeping was only seen to occur during the day, and never lasted for longer than 4 min in either bucks or does. It was easily recognized, for when the buck goes to sleep in this way he stretches his neck out on the ground, and suddenly his head flops over onto its side from the weight of the horns, just as if the animal had suddenly dropped dead. The doe, in contrast, tucks her head round into her flank like a sleeping fawn. Deep sleeping of this nature was a rare occurrence, most of the time resting waterbuck merely dozed with their eyes half-closed, sitting with the head held erect.

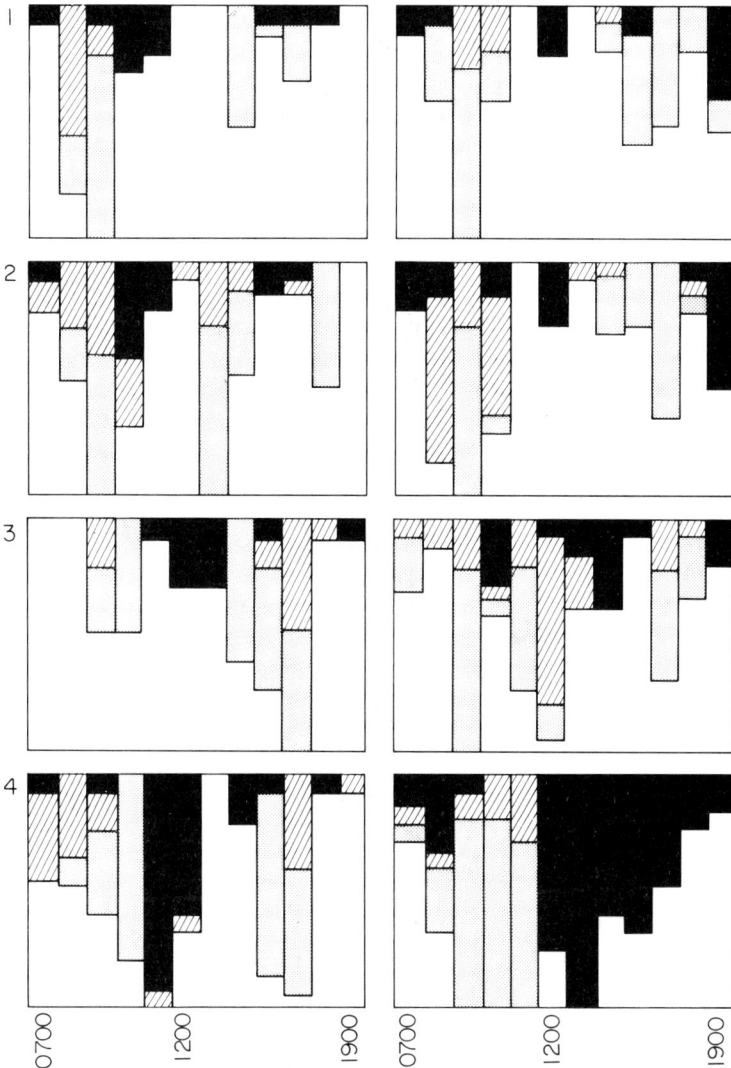

Fig. 45. The daily activity patterns of: (1) adult doe Y128 for two consecutive days 7–8.2.67; (2) 6-month-old fawn for two consecutive days 7–8.2.67; (3) adult doe Y123 for two consecutive days 22–23.2.66; (4) adult buck Y22 for two consecutive days 28–29.6.66.

Between·August 1967 and December 1968, Eltringham and Flux (1971) conducted a series of nocturnal counts on a mown grass airstrip in the Park, at times ranging from 2300 to 0300 h. In this study some 409 waterbuck were recorded in 32 nights out of 64. A significant correlation ($p = < 0.04$) was demonstrated between an increase in the numbers seen

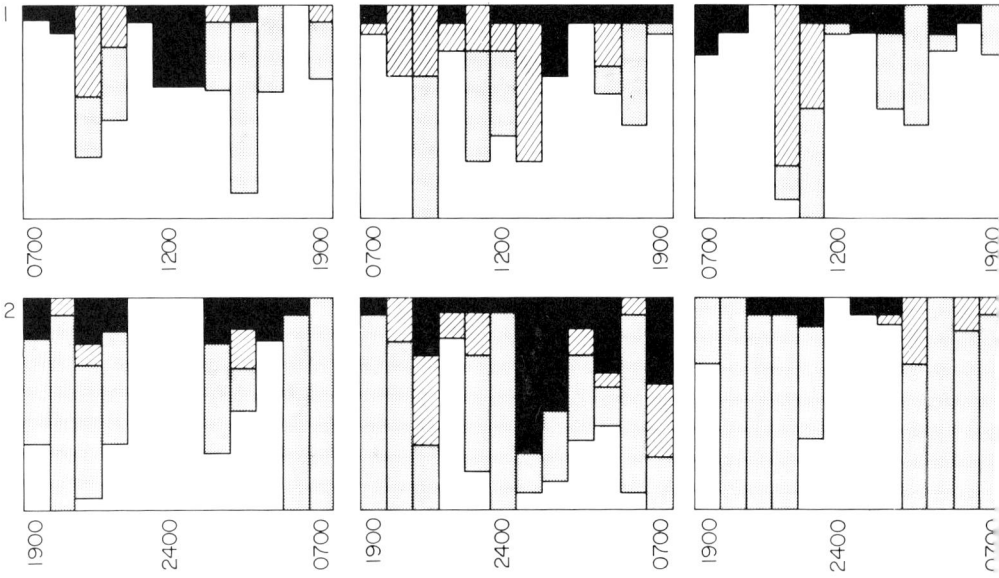

Fig. 46. The daily activity pattern of: (1) adult doe Y114 for three consecutive days 22–24.2.66: and (2) the nocturnal activity pattern of adult buck Y7 for three separate nights: 7–8.3.66; 28–29.10.66 and 25–26.1.67.

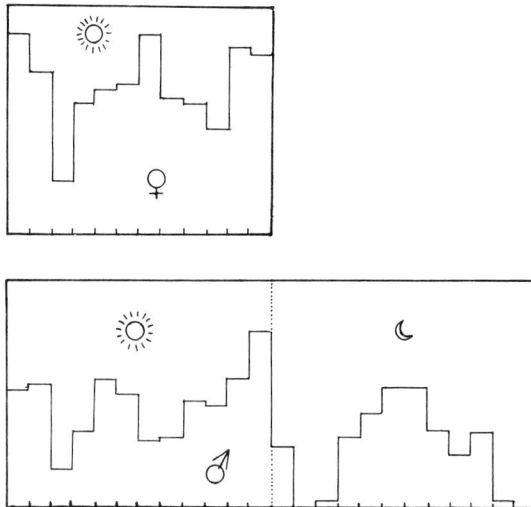

Fig. 47. The mean time spent in feeding per hour for bucks (24 h) and does (12 h), as a percentage of the total activity per hour.

on moonlit nights, compared with those seen on dark nights. On nights close to the full moon an average of 10·6 waterbuck was counted each night, compared with an average of 6 on intermediate nights, and 5·1 on moonless nights. When dark nights were compared with illuminated nights the average number counted was 2·6 for 27 observations, compared with 11·8 for 29 observations respectively. The authors concluded that this difference signified a real difference in waterbuck behaviour. Although this is possible, it is also possible that on dark nights the waterbuck moved away from the approach of the vehicle's headlights, more so than on moonlit nights.

Assuming a correlation between increased waterbuck activity and lunar activity to exist, what conclusions might one draw to explain such behaviour? Is their vision incommoded on dark nights to the extent that they remain in cover to avoid detection by predators? Herbert (1972) states that they appear to have a preference for open areas when resting at night, especially when the moon is full, and suggests that this may allow them to see approaching predators more easily. This is somewhat contradictory, in that their exposure in the open makes it easier for the predator to see them. There seems to be no valid reason why they should feed more actively on moonlit nights, unless we admit of the possibility of their behaviour being geared to a lunar activity pattern rather than to a solar one. But my observations suggested that they fed less at night than they did during the day. If they do indeed feed more on moonlit nights than on dark nights, then this should be reflected in less feeding activity the following day. Limited as they were, my observations (Table 48) did not support this. Although of the three adult doe observations I undertook, one commenced on the day following no moon, and two on days following the new moon, and the highest time spent in feeding was shown by the first observation, this did not apply to the bucks. Of the adult buck observations, two were conducted on days following no moon, and two when it was in its first quarter, that is, approaching full moon. Yet there was no marked difference between the time spent in feeding. Spot checks with the sniperscope showed that both bucks and does still grazed on the blackest of nights, always assuming that the arrival of the vehicle headlights had not roused them, and the observed grazing was not a displacement activity resulting from this. The possibility of lunar cycles influencing animal behaviour is, however, an interesting speculation, which might well warrant further investigation.

Movement

The movement of territorial bucks, bachelor bucks, and does, was similar, in that the animals did not restrict themselves to one place, but tended to make use of the whole of the territory or home range, working over different parts on different days. Mapping the daily positions of a number of marked does ($n = 10$) revealed exceptions to this, younger does being shown to range more widely than older ones. On the Peninsula, the does made equal use of the low and the high ground, using most of the buck's territories and extending into the Ogsa. But from about 9 years of age movement appeared to be more and more restricted, the older does spending almost all of their time either on the low or on the high ground of the Peninsula, and rarely venturing into the Ogsa. Does with young fawns also showed limited movements. The daily tracks of two does, one with a fawn, are shown in Fig. 48.

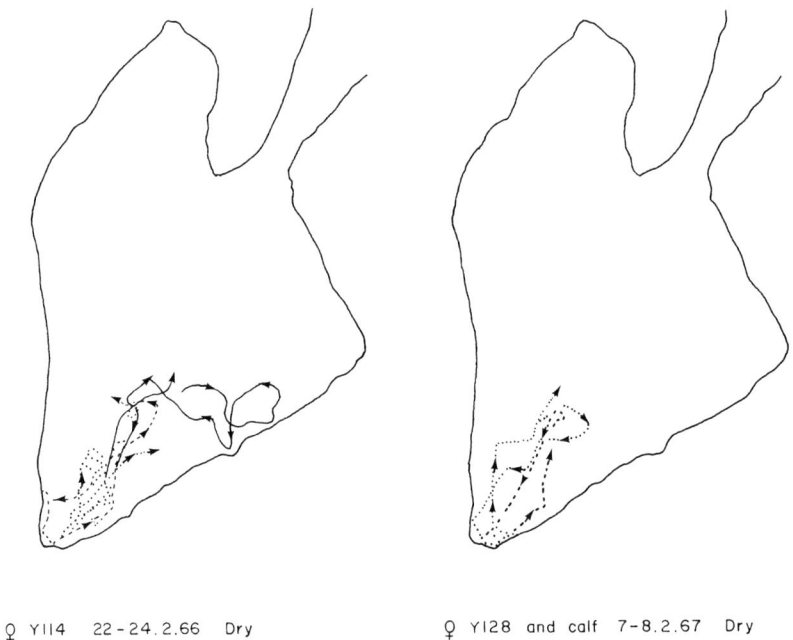

♀ YII4　22-24.2.66　Dry　　　　♀ YI28 and calf　7-8.2.67　Dry

Fig. 48. Diurnal movement of adult doe Y114 for three consecutive days, and adult doe Y128 with her 6-month-old fawn for two consecutive days. 1st day — broken line; 2nd day — dotted line; 3rd day — continuous line.

When a group of Peninsula does visited the Ogsa, their movement appeared spontaneous, and usually occurred on bright, sunny mornings at about 0730 to 0800 h. From time to time groups were observed to move in a determined manner from the eastern side of the Peninsula plateau. The directness of their walk, not stopping to graze, always told me that a group had made up its mind to leave. They followed a similar route on each occasion, down the escarpment and along the lakeshore to the Kanyeseswa ridge. When they first reached the edge of the escarpment, if the group was coming from the plateau, they would halt and stand motionless for several minutes, watching the ground below. I assumed that they were looking for predators or signs of suspicious activity. Satisfied that there was nothing untoward happening, they would continue on their way. Such movement was never seen to be a casual, drifting-feeding activity.

Feeding movement appeared to be partly imitative; if one moved, the rest did. Any doe, even a young one, might initiate this, except that the group had to be ready to move. There was a certain length of time spent in grazing one spot, before the group would walk forward to another. This had a certain cohesion about it, for if one doe went in the "wrong" direction it was simply left behind. There was no evidence of old does being group leaders, although when others were with the old grey doe, they would often follow her when she moved. But old animals were independent creatures, and tended not to follow groups themselves. The same pattern of movement, as regards the initiation and following, was shown by the bachelor groups.

Movement on the Peninsula showed a fairly set pattern both by day and by night, the diurnal movement being related to water. The length of the journey more or less governed the time at which they drank, this being at any time before about 1500 h, but most frequently around midday. Herbert (1972) found that they drank mostly in the morning and late afternoon, the peak being at about 1800 h, with minimal drinking at midday. In a journey down the Bamingui River in the heart of the Central African Republic, which I made in May 1977, I found that waterbuck were encountered on the river banks only between noon and 1530 h. Obviously therefore, the pattern differs from area to area, and it is probably related to the feeding grounds and the proximity of water to them; as well as perhaps to diurnal temperatures.

Doe groups preferred to pass the night on the plateau, this may have been to avoid tsetse flies.

Reactions to Predators

When waterbuck are grazing and moving there is no apparent orientation with respect to wind direction, unlike Altmann (1956) found for the elk in Canada, neither are there any protective arrangements for fawns. Fawns of only one or two months, which had passed the lying-out stage, were often found resting or feeding a good 100 m from the adult group. On a number of occasions, when I found an entire group lying down, which was usually the Peninsula bachelor group, I noted the orientation of individuals in the group, with respect to wind direction. Neither for the bachelor group, nor for doe groups, was orientation with respect to wind direction significantly different from random, at the 95% level of probability (Table 49). My observations were limited, partly due to the few occasions on which I encountered an entire group lying down together, and partly due to the pressure of other activities which did not allow me enough time to carry out such recording. But the relatively benign environment of the Peninsula, as far as predation was concerned, during the period of my study, may have contributed to negligence on the part of the animals with respect to anti-predator behaviour. But on the whole, I am inclined to think that waterbuck do not consciously orientate themselves with respect to wind direction; if for no other reason than that in hot environments it can often be very variable in direction.

Table 49 Mean orientation of individual waterbuck in lying groups, with respect to wind direction, expressed as mean % of total number in group.

Wind	Type of group	Number of groups	Leeward					Windward				
N — E			S	SW	W	Σ	NW	N	NE	E	SE	Σ
	Buck	9	3·9	18·2	15·6	37·7	24·7	0	7·8	16·9	13	62·3
	Doe	2	7·1	14·3	7·1	28·5	17·9	14·3	10·7	17·9	10·7	71·5
S — W			N	NE	E	Σ	SE	S	SW	W	NW	Σ
	Buck	1	0	22·2	0	22·2	11·1	11·1	11·1	3·3	11·1	77·8
	Doe	1	0	30	0	30	0	10	30	30	0	70
S — E			W	NW	N	Σ	NE	E	SE	S	SW	Σ
	Buck	1	28·6	14·3	0	42·9	0	14·3	0	0	42·9	57·1
	Mean	14				32·3						67·7

Although waterbuck are not infrequently taken by lions, the horns of the buck present a formidable weapon, and a buck will confront a lion from as little as 20 m without flinching, snorting an alarm signal and pointing at the enemy.

On one occasion I witnessed a Peninsula territorial buck suddenly find itself face to face with a crouching lion at just this distance. The buck's immediate reaction was not that of flight. Instead it carefully faced the lion and stood quite motionless, watching it, for a full 2 min. Suddenly, it leapt at least 15 m to one side, and turned to study the lion's reaction. The lion crouched a little lower, but, like Brer Fox, "he lay low". The buck then began to give warning snorts at intervals of 10 to 25 sec. Still there was no response from the crouching lion, so the waterbuck deliberately turned its back and slowly walked away for a short distance, then turned round again. After regarding the lion still playing Brer Fox, the buck then unhurriedly walked off altogether. Not until it was out of sight did the lion relax its crouching attitude.

This buck's behaviour was not atypical for a buck encountering a lion; not fleeing in panic but warning others by snorting and indicating the position of the predator. The lion, for its part, had sense enough not to impale itself on the waterbuck's horns. But if a lion under surveillance disappeared from view, then the buck instantly ran to a safe distance and tried to locate it again. There is a great advantage in having other pairs of eyes to help in this, and the lion often succeeded in outwitting its prey. One old buck, which lived all alone high up in the crater area, fell victim within a few days when a roaming lion pride paid it a visit.

One can always find exception to the expected behaviour pattern, and there is a well-authenticated observation of a young buck in the Park who stood confronting a lion, apparently so mesmerized by it that the lion simply walked up and seized the waterbuck by the nose!

The preponderance of bucks compared with does, which is killed by lions, has been confirmed in the Kafue National Park of Zambia (Hanks *et al.*, 1969). This sexual disparity may be partly because does are more alert, and try to get out of the way at the first sign of danger, although some will point and snort just as vigorously as a buck. But the does main advantage would seem to lie in their allomimetic behaviour. As they graze, the does, like other herbivores, repeatedly pause to look up and glance around. If one should notice anything unusual, or something which has intruded into the field of vision, then she stares at it to see if it will move. As soon as the rest of the group notice her staring, they gradually all do the same thing, focussing perhaps dozens of pairs of eyes from several angles on the possible source of danger. This enables a perception and recognition of form which cannot be met by one pair of eyes alone (Plate 42).

A warthog moving in head-height grass always elicited this behaviour; for the only difference in frontal silhouette between its head and that of a lion, is that the ears of the warthog are pointed, and those of the lion rounded. The waterbuck wanted to see a warthog move before they were satisfied as to its true identity. Older does possibly recognized an object for what it was, and either gave warning, or continued grazing, as the case warranted. Younger does learned from this, and were always the last to stop staring after the group had resumed grazing.

Plate 42. Does keeping an eye on a lion; Akagera Park, Rwanda.

Waterbuck Reactions

Does use the snort as an alarm signal more often than do bucks, but others do not react to it without they themselves see the cause of the alarm; unless the one giving warning flees. Hurried movement is always an alarm signal which all react to instantly. Some observers erroneously label the waterbuck as "stupid", because it is not easily put to flight, its movements being more deliberate than those of most antelopes. If a flushed bushbuck suddenly dashes through the undergrowth, this does not precipitate a headlong rush of waterbuck from the spot; they first satisfy themselves as to what the bushbuck is running from. It is perhaps this necessity to be convinced which leads the waterbuck to allow a closer approach by man, than does any other antelope in my experience, especially in areas where it is not harassed. But capture a waterbuck and you have a complete reversal to this apparent docility, waterbuck suffering greater stress in captivity than do most other antelopes. Animal trappers are well aware of this fact, and have substantiated that the waterbuck is one of the most difficult to keep in captivity. Usually, after capture, it will not eat or drink for 4 days; if it continues to refuse nourishment beyond this period, trappers are forced to release it knowing that otherwise it will die.

I found that sometimes sitting does would instinctively try to avoid detection by me, by adopting the fawn-like behaviour of lowering their heads to the ground, with neck extended before them, a behaviour obviously never used by bucks as the horns would give them away. I had already spotted such does long before, and when I "found" them they simply reverted to their upright posture, taking no further avoiding action.

An interesting behaviour was shown by a Peninsula buck who was proving difficult to capture for marking. In trying to get close enough to him for a shot with the capture crossbow, I was often confused by his simply disappearing. He would literally walk into a bush and vanish! I was always certain that he had not had time to pass through, and out of the other side, without being seen, and yet there was no sign of him. Then I discovered the secret. He simply walked into a bush and sat down, while I was left searching for him a metre above where he really was. This poses an interesting question as to whether the animal reasoned this behaviour, or whether it was simply an instinctive reaction which is not often brought into play, but is equally used to avoid detection by any pursuer. I have seen the same reaction with a juvenile hippopotamus on land, but that is an animal which is pretty close to the ground anyway.

Relationships With Other Species

The relationship of the waterbuck with other animals is generally one of indifference, although both curiosity and playfulness may be shown towards smaller species. I once came across a doe staring curiously at a metre-long monitor lizard which was making its zig-zag way towards her. She moved away at the last moment. A young doe, after watching two warthogs having a tussle, leapt at them to join in, but this was too much for the warthogs, who took flight. Fawns were often seen to chase warthogs, because they always ran away. Complete indifference was shown by a doe who walked up to a bush and sat down within inches of a young bushbuck. The bushbuck rose to its feet uncertainly, and sniffed the doe gingerly. When a buck came up, the bushbuck sniffed the doe once more and moved off.

With large animals, like the elephant and the buffalo, there is a minimum distance to which an approach is permitted; when this is exceeded the waterbuck move out of the way. Usually such a contact is accompanied by mutual indifference, but sometimes an elephant will make a rush at a waterbuck if it stays too long in its path. The waterbuck, for its part, simply skips nimbly aside.

No tendency was seen for waterbuck to associate with other animals. Warthogs were often seen feeding in their company, but this was simply because their ranges were co-incident (Plate 15). In other areas with a greater variety of grazers, they may be more often seen in the company of other species.

Waterbuck were seldom host to the yellow-billed oxpecker *Buphagus africanus,* which feeds upon the ticks carried by animals. Whenever waterbuck were in the vicinity of buffalo, a few birds would often fly across from the buffalo to investigate them, and, during her lifetime, invariably singled out the old grey waterbuck, ignoring the younger animals. But she would not tolerate them, shaking and galloping about until they were discouraged. But after her death I saw some on a young doe which was lying down, who let them climb all over her face, and other parts of her body, without making any attempt to dislodge them. Elliott (1976) also reports a dislike of oxpeckers by waterbuck.

Most of the time the does are serious creatures, ambling about with a restrained dignity and looking somewhat alarmed when the youngsters indulge in high spirits. Rarely, but perhaps most frequently after a shower of rain, they may briefly join in the youngsters' games. Gambolling about in an extraordinary exhibition of elephantine inelegance, snorting and blowing, confronting the fawns head to head, then frisking clumsily away. To see an 185 kg adult leap into the air and land with all four legs rigid, the stotting of the fawn, is a ridiculously unforgettable sight.

10. Social Organization

Introduction

A waterbuck's limited habitat preferences lead it to avoid some areas and to select others. When conditions are undisturbed, this results in the selected areas harbouring a full complement of waterbuck. How this complement might be arrived at, I can only speculate upon. It may be dependent upon food availability, the waterbuck–hippopotamus interaction suggesting this to be the most likely factor, or it may result from social interactions, with increasing irritability accompanying increase in density. But if the diverse accidents of life which befall a waterbuck fail to operate to the extent of limiting the density of waterbuck within a certain area to a level within the carrying capacity of that region, how then is population size controlled?

Observations suggested that the answer to this question was that control of numbers was effected by the waterbucks' behavioural interactions. Whether this was a conscious social reaction to density, as the theory proposed by Wynne Edwards (1962) postulates, or whether it was a random ejection of animals, perhaps those which had not the temperament to stand up to aggressive threats, could not be determined within the short span of my study. But those which were ejected from one society, had to seek a place elsewhere, and whether a doe was accepted or not in another area, probably depended upon her coming into oestrus and being retained there by the territorial buck. There seemed no reason why young does, ejected from one area, should be accepted in another, unless one supposed the latter to be an area which was not fully exploited. This, however, was the pattern which emerged during my study; does were ejected from one area and accepted in another; but unfortunately I was not able to determine if a two-way exchange of does ever took place.

For the buck, with its adult territorial system, the pattern was more defined. Young bucks associated in bachelor groups, but these groups did not appear to accept outsiders, while emigration from the group took place when it became too large. Such emigrants appeared to form new groups of their own in other areas, and in some cases joined other bachelor groups, but probably only if they were old enough to subdue opposition or too young to invite it. When old enough to leave the bachelor group the buck has to fight for a place in society, and unless a new habitat is being colonized, one of the contestants has to fail. It is a dead mens' shoes society.

Doe Group Structure

A doe likes the company of others. More than once I watched a small group run eagerly to join another, although no mutual sniffing or bodily contact took place when they came together. But doe aggregations were essentially dynamic, continually expanding and contracting, although these interactions were limited to the total population of a home range. This resulted in group size ranging from one to 70, and I attempted to define whether increase in mean group size was related to population density, as has been established for other species (Caughley, 1977), or whether there was indeed some structural arrangement which might be indicated by a tendency to form groups of a certain size.

For analysis, "groups" comprised adult and young does, but not single adults with fawns, single adults with bucks, or single adults with fawns and bucks, as these situations represented special behavioural interactions, unrelated to group aggregation. The results from each game count area, and from the study areas, were analysed.

Frequency distribution of groups against group size showed that with the exceptions of the low Ishasha river and Kikeri area densities in which the observations were too limited to draw any conclusion about size distribution, the distribution was skewed to the right (Fig. 49). Attempts were made to analyse this, by fitting distributions for discrete data usually employed to describe clustering. These included the positive binomial, Poisson and negative binomial distributions. Of these it was found that the truncated negative binomial, using the method of fitting described by Brass (1958), was the most satisfactory. This could only be fitted to results from six out of the 12 study areas, where the shape of the frequency distribution indicated a central tendency without undue fluctuations and irregularities.

The finding that the truncated negative binomial fitted the observed data somewhat satisfactorily in half of the areas studied (Table 50), suggested that groups were formed in accordance with the laws of

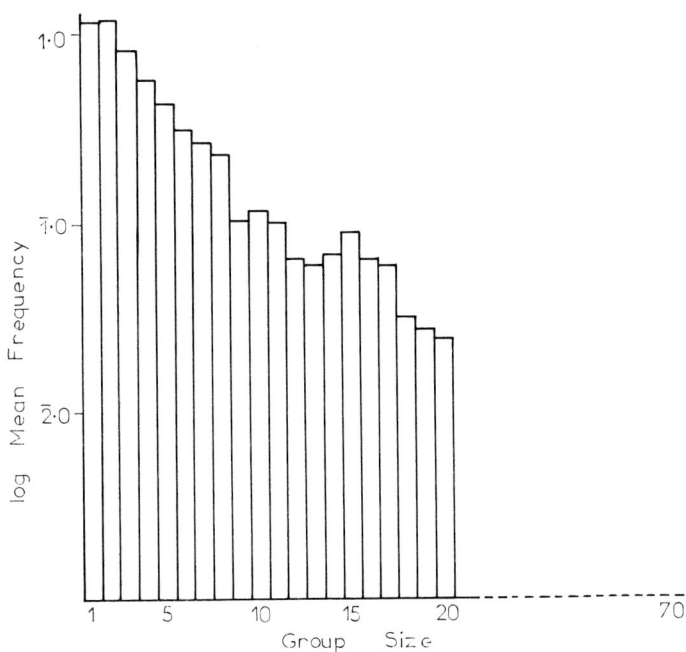

Fig. 49. Frequency distribution of mean doe group size.

Table 50 The goodness of fit of doe groups for different areas and seasons to the truncated negative binomial distribution.

Area	Season	Chi-sq	df	Probability associated with goodness of fit
Peninsula	Wet	55·8	21	< 0·01
	Dry	15·2	17	0·50 to 0·75
Nyamagasani	Wet	22·9	10	0·01 to 0·025
	Dry	8·69	8	0·25 to 0·50
Ogsa	Wet	8·58	8	0·25 to 0·50
	Dry	8·47	6	0·10 to 0·25
Lion Bay	Wet	7·68	10	0·50 to 0·75
	Dry	7·47	6	0·25 to 0·50
Katwe	Wet	11·1	7	0·10 to 0·25
	Dry	10·3	5	0·05 to 0·10
Kikorongo	Wet	8·82	6	0·10 to 0·25
	Dry	0·94	4	0·90 to 0·95

The Kayanja, Katunguru, Royal Circuit, Craters, Kikeri and Ishasha areas data were too variable for analysis.

clustering, relative to this discrete distribution. This might indicate a certain randomness of grouping pattern, although one must remember that the goodness-of-fit of a statistical distribution does not imply anything more than the assumption basic to the derivation of that mathematical function.

Mean doe group size was 5·3 (SD 1·77), but the group size in which the does were most commonly found, was given by the mode of the distribution as 1·7 (SD 0·33). In other words, does are most commonly encountered hovering between being solitary and in groups of two. Increase in mean group size was shown to be significantly related to increase in population density ($r = 0.877$, df $= 8$, $p = \ll 0.001$), which supports the hypothesis that groups are formed at random, and that there is not a tendency to form structural groups, or groups of a certain size. Thus the doe waterbuck's gregarious behaviour seems to be imposed by circumstance (i.e. density), and random interaction. But does of a home range do not repel one another, so that one could say that there was some exercise of choice in associating with one another.

At extremely low density, as found at Ishasha, and at high density found at Kayanja, there were groups, whose average size was larger than expected for differing reasons. In high density areas the animals may be forced by limitations of space to associate in larger groups, whereas in low density areas one may be either observing the occasional large group that has wandered from a high density area, or, because their numbers are so few, that have tended to aggregate rather than to be separated from one another in ones and twos over large distances. The view of others grazing is perhaps reassuring.

When the six analysed areas were classified by season, it was found that there was no apparent difference in the goodness-of-fit of the distribution between wet and dry seasons. In other words, group size did not change (Table 51). In view of the unsatisfactory fitting for all areas, the parameters of the negative binomial distribution could not be used to compare wet versus dry seasons for differences in group size. Thus a crude analysis of average group size and variance was used. This confirmed that season had no effect upon the manner of grouping, nor did it have any effect upon average group size (Table 51).

The frequency of occurrence of different sized groups on the Peninsula, was analysed to determine whether a difference existed between those occurring in the morning and those found in the evening. The result of 180 observations showed a significant difference between morning and evening group sizes at the 95% level of probability (chi-sq $= 16.06$, df $= 6$, $p = > 0.975 < 0.990$). This resulted from there being more groups of six and seven in the morning, and more groups of four and five in the evening; but there was no change in size of groups of less than four

Table 51 Analysis of doe grouping in the count areas.

Area:	Kayanja		Peninsula		Nyamagasani	
Season:	Wet	Dry	Wet	Dry	Wet	Dry
Density/km^2	17·8		10·5		7·0	
Observations	11	4	348	139	29	16
Mean group size	8·56	12·45	6·17	5·65	5·68	4·73
Range	1–65	1–70	1–33	1–25	1–49	1–30
Variance	92·25	208·8	29·34	24·08	35·98	20·57
t	1·5965		0·6716		2·3676	
df	176		2295		472	
p	> 0·1 < 0·2		> 0·5 < 0·6		> 0·01 < 0·02	

Area:	Ogsa		Lion Bay		Katwe	
Season:	Wet	Dry	Wet	Dry	Wet	Dry
Density/km^2	3·7		3·4		3·1	
Observations	25	17	19	10	25	16
Mean group size	3·87	3·71	3·97	4·38	3·7	3·63
Range	1–15	1–14	1–20	1–24	1–30	1–16
Variance	8·91	8·56	12·88	13·73	41·02	9·45
t	0·1438		0·9180			
df	300		339			
p	< 0·8 > 0·9		< 0·4 > 0·3			

Area:	Kikorongo		Katunguru		Royal Circuit	
Season:	Wet	Dry	Wet	Dry	Wet	Dry
Density/km^2	2·2		1·2		1·0	
Observations	29	14	24	13	28	17
Mean group size	5·23	5·39	5·61	4·39	4·21	5·52
Range	1–24	1–35	1–17	1–15	1–22	1–18
Variance	9·54	44·76	22·25	15·88	17·52	22·26
t			1·1444		1·1253	
df			67		59	
p			> 0·2 < 0·3		> 0·2 < 0·3	

Area:	Craters		Kikeri		Ishasha	
Season:	Wet	Dry	Wet	Dry	Wet	Dry
Density/km^2	0·5		0·3		0·15(3·1)[a]	
Observations	30	12	31	13	31	13
Mean group size	4·98	3·50	3·15	4·14	13·0	3·0
Range	1–21	1–10	1–11	1–9	5–18	—
Variance	22·33	8·40				

Notes: Density refers to mean waterbuck density and not to doe density.
[a] Refer to Table 28.
Katwe, Kikorongo and Craters variances for wet and dry seasons are not equal, thus the t test cannot be applied. The Kikeri and Ishasha data are insufficient for further analysis.
t = Student's t, df = degrees of freedom, p = probability that the two seasonal groups do not differ.

or greater than seven. The results suggested that there was some tendency to break up into smaller groups during the day, perhaps the aftermath of some concentration during the night as anti-predator reassurance.

Density Control

After weaning, agonistic behaviour is shown by adults to young does. I could not determine whether this was a parent-to-offspring relationship, or a feature common to all does. Juveniles making too close contact with the adults, or getting in their way, were butted vigorously but harmlessly. This caused no apparent reaction in the juveniles who merely moved aside and continued their activity. Sometimes a buck joined in. On one occasion when a doe threatened a young one, a nearby territorial buck witnessing the threat, rushed at the youngster and chased it vigorously, then suddenly stopped to graze. The young doe stopped also and watched him for a few moments, before abruptly trotting away. This was the signal for the buck to give chase again. The doe escaped, but the buck did not bother to return to the doe group from which he had driven her.

This antagonism displayed by the adults had the effect of separating a proportion of the young does from the adult group. Not all young does appeared to be separated in this manner, and it would clearly be of interest to learn what qualified them to remain with the adult groups. The occasional young doe, separated from its mother, led a bewildered, solitary existence, hovering on the outskirts of the group for some months. But most of the rejected young tended to group together, possibly forming spinster groups akin to the bachelor groups of the bucks.

Some antagonism existed among the young does themselves, of a rather more determined nature than that shown by adults to young. Two yearlings, while I watched, butted one another for about 2 min, during which a third joined in, attacking from the side. This may have been the makings of a hierarchy within the doe social structure, but it was so seldom encountered that I could learn nothing from it, and adult does were rarely seen to spar with one another.

The spinster groups ranged in age from 18 months to 3, and even 4 years of age. They tended to wander about, some eventually emigrating from the area.

Y111, aged 18 months in February 1965, left the Peninsula in September 1965 and was not located again until June 1966, at Nyamagasani, a journey of 30 km distance. Y133, aged 2 years, and Y110, aged $3\frac{1}{2}$ years, both left the Peninsula after capture and marking, swimming across the Kazinga Channel. The latter was found killed by lions in the Chambura Game Reserve in October 1966, 32 km distant from the Peninsula. Y115, aged 4 years, after ranging back and forth between the Peninsula and the Ogsa, left the Peninsula for good in December 1965, and remained in the Ogsa where she calved. It was unfortunate that I was unable to mark sufficient young animals to obtain a clear picture, but the evidence suggests that the young does who are ousted from the parental groups, undergo a phase of wandering and emigration until they have found a place in which to live. This is probably where they first calve, for as we have seen (Chapter 6), the doe apparently always returns to the same spot to calve. The map, Fig. 50, shows some of the longest recorded movements of both bucks and does.

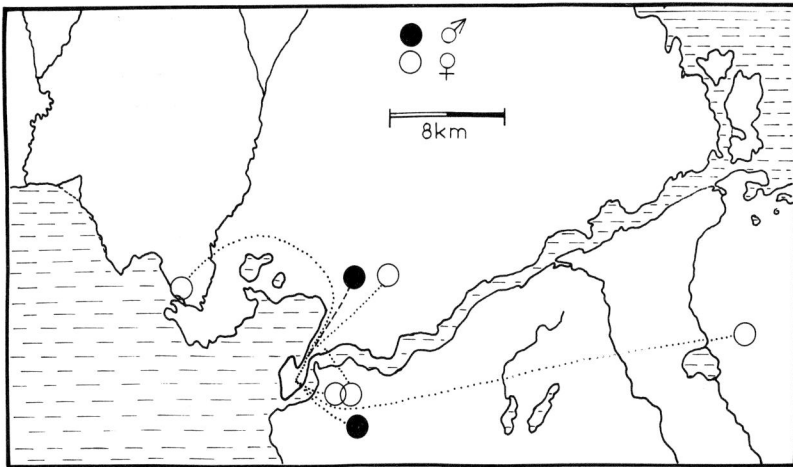

Fig. 50. Map of observed emigrations from the Peninsula.

Home Range

The place which a doe eventually finds to live in constitutes a home range, the doe limiting her movements within a certain area and sharing this area with other does of her choice; for, as we have seen in the preceding section, some selection process operates to determine which does will live together. I find Burt's definition (Burt, 1943) of a home range, adequately covers the concept for the doe waterbuck: "that area traversed by the individual in its normal activities of food gathering, mating and caring for young." The home range of the doe covers several buck territories; its size, as in the case of the territories, depending upon the suitability of the habitat for waterbuck occupation. In this respect, size of home range and number of territories included have some relationship to one another. The restriction of the doe to a home range, as opposed to a nomadic existence, is probably advantageous to her survival, as she knows where to find food, water and shelter, and the most favourable localities free from biting flies and other irritations. This concept seems to be borne out by dry season behaviour, when movement is usually restricted. Contrary to the popular belief of animals wandering far and wide in the dry season looking for food, the converse is the rule, the animals remaining where they know food to be. It is only with the onset of the rains that they wander far afield. The rains are also the time of emigration of the young animals.

Home ranges appeared to be the property of groups of does, but the does moved within them essentially as individuals, as the analysis of grouping has shown. There was a tendency for certain does to associate together, but these associations were very fluid. Those that I observed bore no relationship to age, but may have been familial.

The observed associations of 16 marked Peninsula does are shown in Table 52. Not all of these does were marked at the same time, but the total number of observations on each doe is shown in the table. The most frequent association demonstrated was between Y118 and Y116 (the prefix Y is not shown in the table), each aged 5 and 4 years respectively,

Table 52 Observed associations of marked Peninsula does.

Doe	Age	118	116	120	123	115	114	122	111	128	121	112	132	130	131	106	134
Doe	Age							Number of contacts									
118	5	327	120	107	78	66	93	59	30	78	7	20	44	26	50	14	0
116	4		302	90	78	66	103	67	32	59	17	34	32	31	44	16	1
120	10			260	53	40	69	48	23	50	7	12	21	24	39	14	0
123	?				256	40	68	60	32	75	40	59	69	36	58	17	2
115	4					132	44	43	30	50	16	9	18	9	26	5	2
114	7						321	59	16	45	24	65	54	54	52	39	12
122	2							155	46	57	25	58	42	27	53	14	5
111	1·5								118	31	12	26	19	9	12	14	11
128	3									276	68	34	48	23	39	8	0
121	9										215	55	37	24	22	3	3
112	8											292	65	72	56	48	33
132	3												242	60	71	20	15
130	7													175	66	45	28
131	7														159	30	15
106	7															206	27
134	18																88

suggesting that they may have been sisters. The next most common association was between Y118 and Y120, the latter aged about 10 years. In this case perhaps it was a mother–daughter relationship, as perhaps was also the case between Y116 and the 7-year-old Y114. That perhaps there is some familial relationship in the grouping is suggested by both Y118 and Y120 having least association with the next oldest doe to Y120, which was Y121, aged 9 years. The oldest doe of all, Y134, aged some 18 years, had the least recorded associations with other does, even when the much fewer number of observations on this animal is taken into account.

In the mountain sheep, Geist (1971) did not support the concept that ewe home range groups developed from a mother–daughter relationship, and suggested that the young animals probably inherited the home ranges by acquiring the movement habits of their elders. The same is true, in part, of the waterbuck, with does forming attachment to areas rather than affinities with particular bucks, wandering at will through the bucks' territories. Time spent in any one territory ranged from a few hours to perhaps as long as 3 months. Herding was sometimes attempted by territory owners in an endeavour to keep the does within a territory, but this was never seen to be successful. A buck intent on retaining a doe would run ahead and confront her, sometimes rushing at her with head lowered in threat, to try and drive her back. A typical instance was shown by buck Y7. When one of the does which was in his territory started to walk towards an adjacent rival buck, Y7 immediately ran ahead of her and halted at the boundary to his territory, some 30 m from the other buck, and confronted the doe. She simply ignored him, continuing her walk across the boundary. Y7 made a rush at her, head down, but did not follow her across. As soon as she had crossed the invisible (to me) landmark, the other buck, who had been quietly watching, immediately ran up to her. Y7 displayed no further interest, turning and walking away. Had it not been for the near presence of the rival buck, he probably would not have shown any interest in the doe in the first place, leaving her to wander as she pleased.

Bucks with the largest territories had the most doe visits (Fig. 60). This may have resulted from the larger areas having a greater chance of including those areas favoured by the does. Equally, it could have been the result of the bucks competing for the localities favoured by the does, either for the buck–doe contacts, or, because the areas had the same attraction for the bucks that they did for the does.

With increasing age the doe restricts herself to a smaller and smaller range of movement, and is thus frequently left alone by the more active, younger animals.

Does showed no signs of altruism within the group. Injured members were attacked and driven away from the others. This was witnessed on several occasions: one lame doe, who could only hobble short distances (Y132), was forced to live a more or less solitary life for 4 months until her injured foreleg healed, for whenever a group came her way the members would butt her vigorously, trying to drive her away from them. The object of this behaviour could be seen as having a survival value for the group, by not attracting the attention of predators to it, but one could equally argue that the presence of a lame animal would mean that it would be more likely to be taken by a predator than one of the other members of a group. Whatever its basis, if predation had been high in the area, there seems little chance that this doe would have recovered as she did. A dam will not abandon an injured fawn, unless it will not follow her where she wants to go.

Home Range Structure of the Peninsula and Ogsa Areas

Ranges of ten does, of all ages, were plotted over the period of study (Fig. 51). Generally the whole of the Peninsula was made use of by these does, and from time to time most of them ranged into the Ogsa as far as the boundary of the first buck's territory (Y33). Thus the total range of the average doe was about 600 ha, with a mean density of one doe to 23·5 ha. Examination of the maps, portraying the movements of these does between April 1965 and March 1967, shows how the younger ones ranged more widely, and how, with age, the range becomes more and more restricted. See, for example, Y118 aged 7 years, and Y120 aged 12 years (n = 327 and 260 observations respectively).

One doe, Y106, aged 7 years at the beginning of the study, was apparently anomalous in that she occupied two discrete home ranges, 5 km apart, commuting between them and not lingering in the intervening region (Fig. 51). Perhaps, if more does had been marked, this behaviour would have been found to be more common, but it was not the norm.

The maps illustrate some evidence of spatial separation among does in the home range, as shown by Y106, Y121 and Y120 (n = 206, 215 and 260 observations respectively). The map of Y120's range shows how the doe may return to the same locality each time to calve.

I had only six Ogsa does marked, who ranged in age from 3 to 12 years, but the limited number of observations which I was able to conduct on these animals (10 to 18), confirmed the Peninsula picture and showed that the resident does used about half of the Ogsa area, or 730 ha (Fig. 52).

In her Kenya study Elliott (1976) had a density of 36·6 to 53·9 does/km², or one doe from 1·9 to 2·7 ha, while the average home range size was only 33·6 ha, in the range 21 to 61 ha; one eighteenth of that found among Peninsula does at a density of 8·7 to 12 times less.

Such wide variation in home range size would seem to disprove McNab's (1963) optimum yield hypothesis, in which he related the size of an animal to the size of the home range that it occupies, by the formula:

$$A = 6 \cdot 7 W^{0 \cdot 63}$$

where *A* is equal to the area in acres and *W* is body weight in pounds. For a waterbuck doe this would imply a home range of 60 ha.

Kayanja

Lack of time prevented the determination of home range areas at Kayanja, but they were probably more extensive in this region. The does tended to move eastwards inland at night, and return to the lake shore in the morning, providing a regular east–west movement. Individual does were recorded moving distances of up to 5 km, but does marked in September 1965 were still found in the same localities one year later.

Bachelor Buck Group Structure

At 8 to 9 months of age, when the horns first appear, the adult territorial bucks separate the young bucks from the does. Antagonism may manifest itself earlier; I have, for example, seen a territorial buck chase a 6-month-old fawn from its dam. When the buck gave up the chase the fawn returned to its dam, who gave it a reassuring lick on the flank. Had it been 9 or 10 months of age, then she would probably have repulsed it. But whether this was a monosexual antagonism on the part of the adult buck could not be determined.

These initial aggressive harassments, in contrast to the gentle butting of the does, consisted of short, vigorous chases by the territorial bucks, lasting over about 30 m, the adult soon stopping and the youngster wandering back to the doe group with one cautious eye on the bully. This antagonism soon causes the young buck to seek more congenial company, and if there is a bachelor group in the vicinity he usually joins it. But separation may not be abrupt; at least two juveniles were seen which appeared reluctant to break the maternal bond. They repeatedly left their dams to join the bachelor group, only to rejoin their mothers again a few days later. This may, of course, have simply resulted from chance contacts between the bachelor group and the parents. When they return in this manner they meet antagonism from the dam also, so that sooner or later they are forced to remain apart.

Other youngsters, ousted from the maternal group by a territorial buck, may take up a virtually solitary existence, attaching themselves at a

CALVED c.110.65

Y116   6.

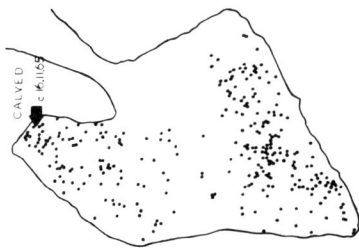

CALVED
c.16.11.65

Y128   5.

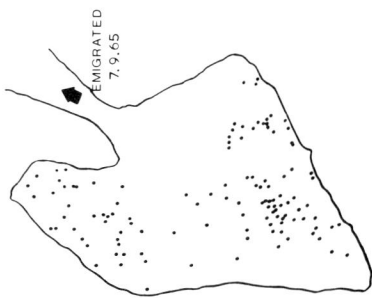

EMIGRATED
7.9.65

Y111   2.

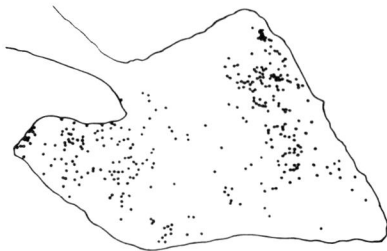

Y114   9.

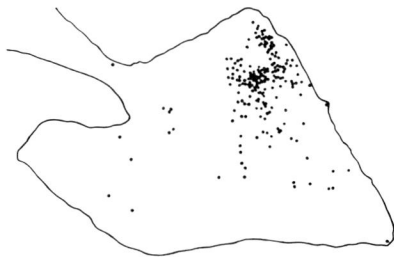

Y106   9.

Y118   7.

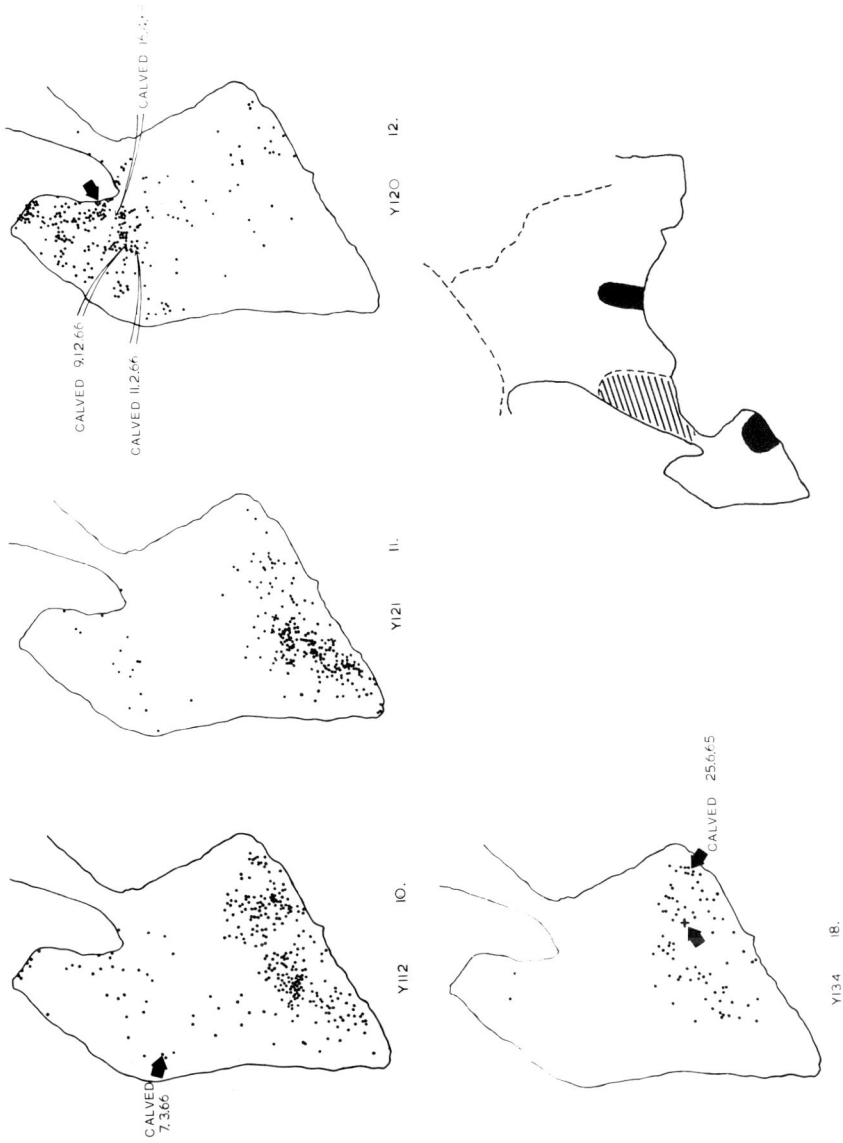

Fig. 51. Home ranges of Peninsula does. Inset shows the extent of some ranges into the Ogsa (hatched area), and the two discrete home ranges of doe Y106 (blocked areas); 13.4.65 to 7.3.67. Ages of the does are given.

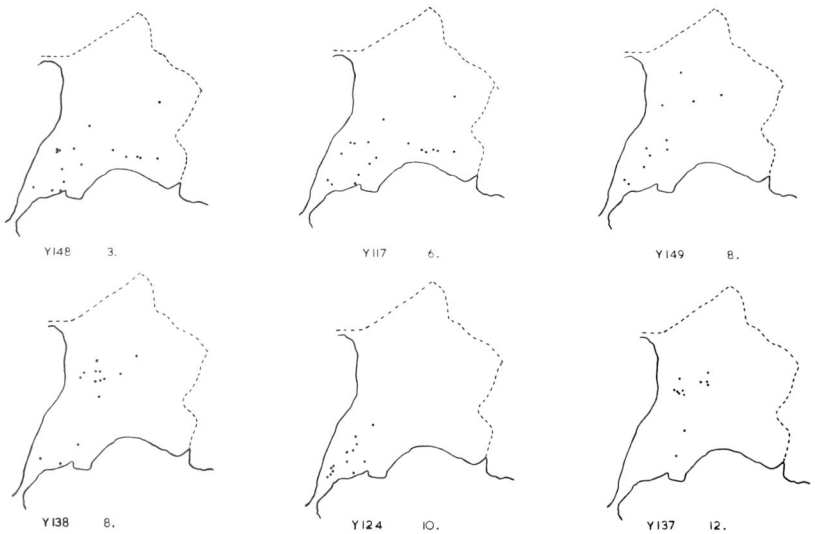

Fig. 52. The ranges of some does in the Ogsa, 7.8.65 to 11.3.67. Ages of the does are given.

respectable distance, to another territorial buck. One was observed to live in this manner for 10 months on the Peninsula. It associated with the bachelor group when the latter entered the area, but it never left with the group. Generally it appeared to be tolerated by the territory owner, but from time to time bursts of antagonism were vented upon it. On one such occasion I watched the youngster chased in circles for a full 5 minutes' duration, until the adult was exhausted, while all the time the youngster uttered plaintive, protesting bleats. The adults' behaviour led me to suppose that basically they were not inhibited from attacking juveniles, but they tolerated them if they were in groups.

Most youngsters joined the bachelor group quickly, and remained firmly integrated within it. The youngest that I recorded doing so was a Peninsula animal of about 8 months of age. By the age of one year a buck is usually completely integrated, and he then remains in the bachelor group until he is 6 years old.

Bachelor groups are thus a characteristic feature of waterbuck populations, but although they may be reported in the same area for many years, they are by no means permanent fixtures. In large areas their occurrence is dynamic, with groups developing and dying out in different parts of the area. If two or more young bucks leave their dams in an area not currently ranged over by a bachelor group, then they may form the nucleus of a new one, eventually perhaps being joined by other, younger members. This dynamism was well shown in the Ogsa (Fig. 53). From July 1965 to July 1966 there existed only two bachelor groups in this area: an eastern group consisting of a mean number of eight animals, of all age classes up to 5 years, and a western group of three yearlings. From December 1966 the eastern group was no longer seen in the area, but a new group of four animals up to 3 years of age had developed in the extreme west, while there was still a western group further north of three animals under 2 years of age. The large eastern herd, apart from the older members which may have left it, might still have been in the vicinity although it was not seen. An approximation of the age structure of the bachelor bucks in the Ogsa in July 1965 is shown in the histogram in Fig. 54.

Fig. 53. The Ogsa bachelor groups. Open circles up to 12.7.66, blocked circles from 5.12.66. Figures show mean group size.

In November 1964 the Peninsula bachelor group had 12 members. Two of these were in their first year, five in their second, one in each of the following 3-yearly age groups, and two in their sixth. The 2-year-old group was rather crowded, and during August of the following year three

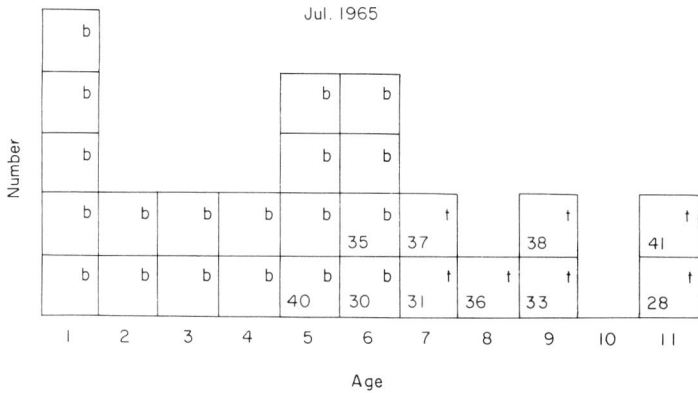

Fig. 54. Age structure of the Ogsa buck population, July 1965.

of them emigrated, swimming across the Kazinga Channel. One of these emigrants I found a year later with another bachelor group. The 4-year-old emigrated in March 1965 after being captured and marked, but I was not able to determine what precipitated the action of the others.

Five recruits were accepted into the group during the period of study, so that by November 1966 there were three one-year-olds, two in each of the next 3-year groups, and one 5-year-old (see Fig. 58). The group, however, operated a "closed shop", accepting into it only those animals born on the Peninsula, or in the Ogsa to a Peninsula doe. During my observations no outsider succeeded in becoming a permanent member, although there were several attempts lasting for no more than 2 or 3 days, and one exceptional one of 3 months. The latter occurred in March 1966 when a 3-year-old appeared with a large and obviously tender abdominal swelling, which looked like an abscessed horn wound. He avoided sparring with any buck when he approached, and his deferential manner enabled him to survive with the group for some time. But abruptly he disappeared, and I never saw him again.

Bachelor Group Size

Bachelor group size was firstly analysed by ignoring groups of "one", so that there was no confusion with the self-imposed solitariness of territorial bucks. As with the does, the low density Kikorongo and Ishasha areas were ignored in the calculations.

Like the does, mean group size frequencies in the buck showed a positively skewed distribution. This could not be fitted to the truncated negative binomial distribution, owing to the high number of groups of two, followed by relatively small deviations from a mean herd size of 5·5 animals (SD = 1·1). This suggested that there was a tendency to form groups of a certain size rather than randomly.

Although groups of a size larger than expected were found in high density areas, an analysis of group size on density showed that it was not significantly correlated at the 95% level of probability ($r = 0·382$, df 8, $p = \gg 0·1$), supporting the hypothesis of a tendency to form groups of a certain size. The mode of the distribution indicated this size to be 2·3 (SD = 0·15); thus one could say that this tendency was not strongly developed, as the groups were barely larger than one.

When mean group size was compared with mean doe group size, there was no significant difference between the means at the 95% level of probability ($t = 0·023$, df 18, $p = \ll 0·1$), suggesting that the same determinants of clustering were operating in both populations.

This apparent randomness in clustering could be a spurious effect relating to the randomness of encounters by observers, the distances between "groups" being arbitrarily assessed, so that social groups were fractionated by the observers. Or it could be that the smaller number of bucks encountered (1536 bucks compared with 3610 does), has truncated the frequency distribution to impart an apparent uniformity to mean group size, which, in reality, is as randomly arrived at as in the doe.

It was thus of interest to include groups of "one" in the buck analysis also, although a group of one in this sense did not imply a solitary animal, for many such single bucks were in the company of doe groups. The inclusion of groups of one reduced the mean group size to 2·5 (SD = 0·93), and the mode to 1·1 (SD = 0·11), thus emphasizing the preponderance of single bucks in the population. Mean group size is now obviously very different from that of the does, and the mode is also significantly different at the 95% level of probability ($t = 5·225$, df 18, $p = \gg 0·999$). The null hypothesis that group size is related to density is still rejected at the 95% level of probability, but is tenable at the 90% level ($r = 0·566$, df 8, $p = < 0·1 > 0·05$), indicating that it is the social structure of the adult buck sector of the population which marginally influences density. This suggests that, fundamentally, the tendencies relating to grouping in bucks and does are the same; but the territorial behaviour of the adult buck, in which those of like kind are repelled, influences this basic randomness of association. We might say that the tendency to form groups of a certain size among bachelor bucks is more apparent than real.

A more detailed analysis of the data has thus led me to propose a view contrary to that expressed some 10 years ago (Spinage, 1969c). Fundamentally, there appears to be no difference in the degree of motivation of grouping between bucks and does, except that the doe groups could be visualized as extended buck groups; whereas the latter have the appearance of cohesive units, the doe groups are more fluid by virtue of the greater area over which they spread. But the associations of the bucks within their much more limited range (100 ha for a Peninsula buck compared with 600 ha for a Peninsula doe), are probably just as random in their occurrence as are those of the does within their greatly extended range.

The bachelor group carries out all of its activities, feeding, ruminating and resting, in a gregarious manner without intraspecific strife. Due to the distinctive growth rate, neither too rapid nor too slow, sub-adult annual age classes are very defined, and an animal one year older than another is easily superior to it in size and weight alone. A younger animal will therefore almost always defer to an older one, so that no strong antagonism for position, within the group hierarchy, develops. The hierarchy becomes a linear one related to age. If, as is not infrequently the case, there are two bucks of approximately the same age and size, then presumably a dominance hierarchy is established between them as a match of their relative agression. When territorial age is approached, such contacts probably result in the emigration of the weaker buck. Where, at high densities bachelor groups are of the order of 40 to 60 animals, one may suppose that linear hierarchies develop between groups of approximately the same age. Thus in a group of 60, there might be six levels of linear hierarchies, each separated from the next by age. I had no opportunity to study large bachelor groups to determine this (the large group at Kayanja was broken up by poaching early on in the study), but it would seem logical to expect that such a complex structure would develop. Elliott (1976) claimed that a linear dominance hierarchy existed among the bachelor bucks in her study.

Any member of a group will approach and spar with another, whether it is older or younger. Sparring is spasmodic, but most frequent first thing in the morning from about 0730 to 0830 h, especially after light rain. A typical sequence of sparring contacts is shown in Fig. 55, which illustrates how age does not influence the contacts, although the oldest member of the group was only challenged once, and did not initiate any approaches. Should two contestants. become too determined, an older buck often comes up and takes on the bigger of the two. This is probably not through any altruistic motivation to aid the smaller animal, but more likely is to check that the particular aggressor is still in his right place. This has the incidental function of ensuring that no fight between two young bucks becomes too violent (Plates 43 and 44).

As in doe groups, the bachelor group occupies a home range, covering the territories of one or more territorial adults who tolerate its presence.

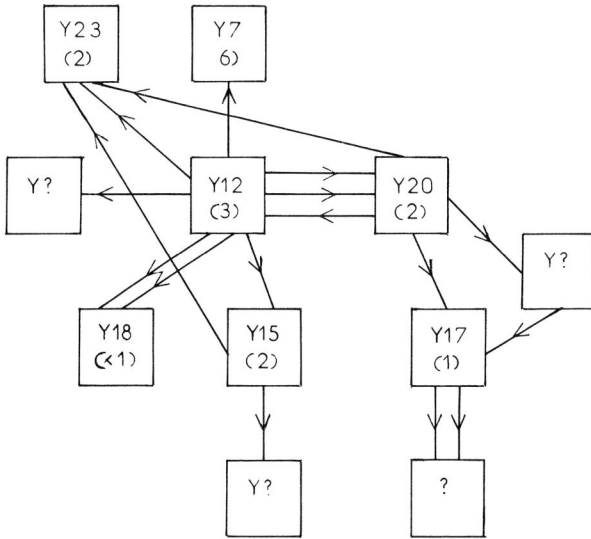

Fig. 55. Some sparring contacts amongst the Peninsula bachelor group.

Analysis of the Peninsula occupation (Chapter 8) showed that only one territory was consistently avoided, Y8's, while the most favoured buck was Y7. But I never saw any intolerance from territory owners which might have accounted for the bachelor group avoiding certain areas. Although tending to move within a circumscribed region for an extended period of time, as Fig. 56 shows, in time they covered most of the Peninsula (Fig. 40b), and, as I showed in Chapter 8, had access to 77% of the range covered by the does. Overall, the mean home range for eight bachelor bucks was 100 ha (Table 53); whereas in the Ogsa an average of 4·3 bucks occupied, very approximately, 340 ha (Fig. 56). This compares with a mean range of 29·7 ha (range 24 to 38 ha) in Elliott's study (Elliott, 1976).

Bachelor groups may be found in peripheral waterbuck habitats, but I never found them occupying areas where there were no territories. The exclusion of young males to unfavourable habitats may be possible in small species with a rapid turnover, which can quickly compensate for any loss which may be sustained, but it is manifestly illogical to suppose that in large, slowly maturing species, potential recruits to the male population sector should be forced to occupy fringe habitat; since this would be counter-productive to the survival of the species.

Exclusion can be permitted for short periods, such as during a defined breeding season, as Hanks *et al.* (1969) found in Zambia. These workers

Plates 43 and 44. Young bucks sparring.

Table 53 Changes in the Peninsula bachelor group home range area, in ha.

Period:	10/64–11/64	4/65–5/65	6/65–7/65	8/65–9/65	11/65–12/65
Area:	97	68	89	89	93

Period:	1/66–2/66	3/66–4/66	9/66–10/66	2/67–3/67
Area:	97	140	145	93

Mean size of area = 101·2, SD = 25.

showed the waterbuck here to have a seasonal breeding cycle, and the bachelor group to occupy a home range "the position of which depends upon season, fire and male territorial boundaries". During the breeding season the bachelors were excluded from the adult territories.

Some 10° further south from where Hanks and his co-workers found seasonal exclusion, Tomlinson (1979, 1980) also found that bachelor bucks were excluded from territories for a part of the year; possibly the breeding season, although this is not stated. Tomlinson analysed the crude protein and crude fibre content of the faeces of territorial bucks, bachelor bucks, and adult does, at different seasons, and claimed that, as the faeces of the territorial bucks and the adult does showed higher crude protein contents than did those of the bachelor bucks, territoriality therefore enhanced food supply and quality for the territorial bucks, and for the does and their young; the lower crude protein content of the bachelor bucks' faeces relating to the bachelor bucks exclusion from the territories.

According to Tomlinson's figures, in the rainy season both bachelor bucks and adult does took in slightly more bulk than did territorial bucks, with a crude protein content much lower than that of territorial bucks. In the hot dry season both bachelor bucks and adult does took in much less bulk than did territorial bucks, although the ratio of crude protein to crude fibre was approximately the same for all. This could simply mean that the young, growing bucks, and the physiologically more demanding does, were using their crude protein more efficiently, by more thorough digestion, which is what we would expect. Taylor and co-workers (Taylor *et al.*, 1969) have shown also that, when water intake was restricted, as it might well be in the dry season, the waterbuck ate less food and the apparent dry-matter digestibility increased. Tomlinson's assumptions are therefore inconclusive, and without further information on conversion rates we can only speculate as to the cause of the differences in the crude protein content of the faeces.

Although they become sexually mature at 3 years of age, bucks have little chance of reproduction until they become territorial, 3 to 4 years later in a healthy, competitive population.

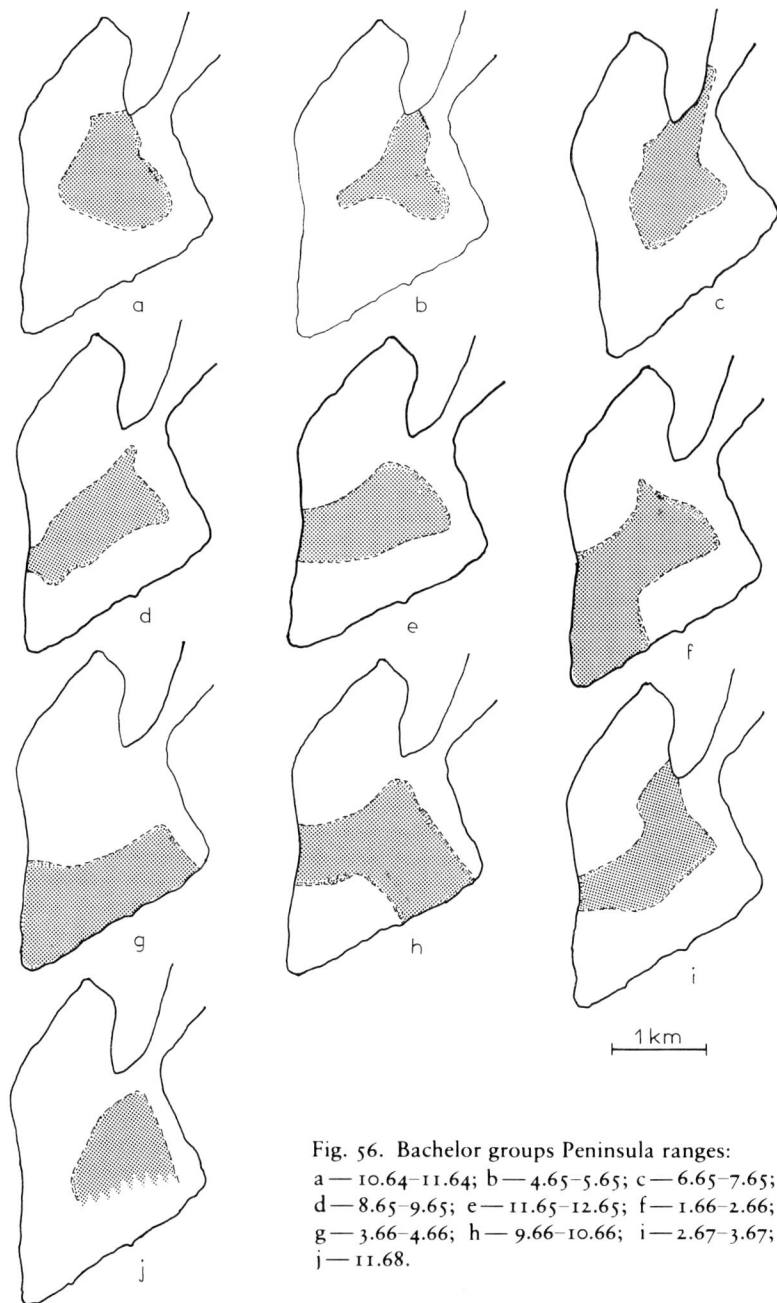

Fig. 56. Bachelor groups Peninsula ranges:
a — 10.64–11.64; b — 4.65–5.65; c — 6.65–7.65;
d — 8.65–9.65; e — 11.65–12.65; f — 1.66–2.66;
g — 3.66–4.66; h — 9.66–10.66; i — 2.67–3.67;
j — 11.68.

Herbert (1972) has summarized a few brief references to group size in waterbuck in other areas, but objective data are sadly lacking. Haagner (1920) referred to groups of five to 15 in South Africa; Dalquest (1965) in Mozambique reported groups of 30 or more does, and buck herds of four to 12 animals. Child and Von Richter (1969) reported a mean group size in Botswana of 8·1, not counting groups of one, with a maximum of 37. Herbert, in his study, observed a mean group size of 6·9, excluding groups of one, and a maximum of 23. When groups of one were included, mean group size was 5·7. These observations cannot be directly compared with mine, as all social groupings were considered, whereas I excluded those which were due to special circumstances, e.g. a dam with fawn, as these confuse the intra-group relationships of the average animal. Nevertheless we can see that mean group size could be said to be similar in these other regions.

Territorial Organization

The existence of year-round territorial ownership by the adult buck was established from plots of positions of marked and otherwise known bucks from the Peninsula, Ogsa and Kayanja regions, as well as from numerous isolated observations.

Bucks leave the bachelor group at from 5 to 6 years of age to try to establish a territory for themselves. This is co-incident with the age at which growth in body-weight, testes weight, and horn length, approaches its asymptotic level. The normally growing buck has no chance of accelerating this process of territory establishment, except perhaps in cases where adult territory owners have been removed by some other agency, although I had no opportunity to determine whether, under such circumstances, the gregarious instinct would not be greater than the solitary one. Elliott (1976) thought that some bucks in her study area became territorial at 4 years of age, but this seems unlikely to me.

The age at which a territory is successfully held will depend, within limits, upon the degree of competition for territorial space. But even if all available space is occupied, a healthy buck is usually able to obtain a territory of its own by at least the end of its seventh year, because bucks over 9 years of age are already in physical decline, and unable to compete successfully with the younger ones.

Peninsula observations showed that, in the absence of a suitably ageing territory owner who could be successfully challenged for his territory, a young buck could use either one of two strategies. One of these was to move into an occupied territory and share it against the owner's will, by restricting itself to one part of the edge of the area without trying to compete for the territory itself. The other strategy was to adopt the border area between two or more territories. In the former case the owner of the territory seemed to tolerate the intruder at first, but eventually attacked it and drove it out, or forced it to remain in a small part of the area. The second strategy appeared to be a fairly successful method of establishment, for, as in the case of Y21 who took up residence where the borders of three territories met, it was unlikely that the three owners would all be in the same area at once. If challenged by any one of the three, Y21 had a choice of two other areas to retreat into. In this way he seemed to be relying upon the inhibition against one owner crossing into the territory of another.

Examples of these strategies were provided by several animals. The 6-year-old Y6 quit the bachelor group about 9 February 1965, and shared the territory of Y22, an 8-year-old buck. He was apparently only accepted after a fight, for two days later he had a large horn wound on the left shoulder, and several smaller ones on the neck and flanks. As a result he was forced into a smaller part of the area, but by August had given up the unequal struggle and returned to the bachelor group. In January of the following year he adopted a small territory within that of a 10-year-old buck, Y9, who had himself been driven there towards the end of the previous year. Y6 remained there until August of 1966, when he returned to challenge Y22 again, 19 months after his first attempt. This time he successfully set up a territory for himself in a part of the latter's, taking roughly a third of the area away from him in the same area that he had originally competed for (Fig. 57).

This aptly demonstrates how a younger buck can impose upon an older one, once the latter begins to age, for when Y6 returned Y22 was probably by then just beginning to pass his prime of life. Y6, on the other hand, was probably approaching his eighth year, but although of large size he was an extremely timid animal, as far as my subjective appreciation of his behaviour could discern. Had he not been so, presumably he might have established a territory for himself at an earlier age, but on the other hand he always took it upon himself to try and compete with one of the bucks in the prime of life. What dictates such choice would be interesting to discover. It could be due to learning the area from bachelor group days, and a lack of exploratory courage in a timid animal to seek new ground, although Y22 had the largest area at the time and thus the most difficult one to defend.

Another timid buck was Y24, who could be driven from his first territory at the approach of a motor vehicle, which most other bucks ignored. By the end of my study, when he was in his eighth year, he was still occupying only a very small territory.

A territory was defended against other adult bucks, but not, as we have seen, against bachelor bucks of less than territorial age and size. That a contestant could eventually win a part of a buck's territory from him, suggests that a certain economy of defence was practised. A territory owner, possessed of a large enough territory, did not have the inducement to completely exclude rivals, but in the face of strong competition simply resorted to defending an area sufficient for its needs. Once a territory was obtained there was no evidence of attempts to expand the size of it, although, as I shall show, bucks in the prime of life hold the largest territories; this is because they start with small ones, graduate to larger ones, and then as they get older are forced to return to small ones again.

This lack of territorial aggrandizement was indicated by a failure to attempt to take over a neighbour's territory if the neighbour was killed in battle. On 26 January 1965 I found the cadaver of 9-year-old Y1, who had apparently been killed by a horn wound in the thorax. His territory, one of the four largest on the Peninsula at that time, remained vacant until it was taken up by Y9, another 9-year-old, some time before April of the same year. I think it highly probable that Y1 had been killed in a fight with the neighbouring 8-year-old Y8, his other neighbour, Y11, being only a 6-year-old. Of course Y8 may not have been aware that he had killed Y1, but whether he was or not, he made no attempt to invade the vacant territory. It was not until 8 September of that same year that Y9 was driven out by the 6-year-old Y11, who did not occupy all of the area. Most of it was then taken over by Y8, who, however, retreated to his original boundary again by December, leaving his territory to Y11.

Territorial mastership is brief, for after they have passed their ninth year, bucks are already in decline. After 10 years of age a buck has little chance of maintaining his position, and is then driven by competing younger bucks into less favourable areas. These areas are either, very small in size, sparse in vegetation, or far from water. Their average size on the Peninsula was 17·7 ha; this compares with the lowest density observed in the Park (Kikeri) of one animal per 30 ha; and the lowest recorded elsewhere (Kruger Park, South Africa, Pienaar, 1966) of one animal per 23 ha. Thus these territories appeared to be of a size which was less than adequate for a waterbuck's needs. That occupied originally by Y13, and later by Y9 (Fig. 57), on the Peninsula, was a sandy ridge with little food or cover.

I found no evidence of such old bucks attaching themselves to bachelor groups, although they tolerated young adult bucks in a manner which was not shown by vigorous territorial bucks. Bachelor groups were not forced to occupy marginal areas, and thus had no necessity to join up with old bucks. Several authors have stated the contrary, but I believe that they are confusing sub-adult bucks, whose size is to all intents and purposes adult, with mature adults. What these authors believe to be ousted adults, are probably in fact 5- to 6-year-olds who have not yet obtained a territory. Of course, the pattern may be different in latitudes where breeding is seasonal, and it warrants further investigation.

There was no indication that old bucks lost their territorial urge; ejected from their favoured territory they often tried to return to it. Such behaviour was shown by Y9 when he was in his tenth to eleventh year; he had been ousted from his original territory on the upper plateau of the Peninsula. Forced into a small area in September 1965, he returned to his former area at least three times, to be chased away by its new owner. On the third occasion that I witnessed this I was following Y22, who entered some thick *Capparis* bush on the extreme eastern side of his territory. The next moment, to my great surprise, the old buck, Y9, came streaking out of the other side of the bush and fled across the Peninsula, with the other in hot pursuit to the limit of his boundary.

In riverine populations territorial boundaries are apparently governed by water frontage, but in lacustrine regions, access to water was not always the dominating factor. This was particularly evident at Kayanja, where behaviour was consequently modified. In the Ogsa, however, territories which lacked water frontage were usually areas to which the old bucks were driven. On the Peninsula there was ample access to water, which may have been the reason for its popularity as a territorial ground, mean territory size being smaller in this area than that recorded elsewhere.

Territorial Structure of the Peninsula

To determine spatial distribution, after a preliminary survey I traversed the Peninsula daily, each morning and evening, for approximately one month, every other month. The majority of visits was made between 0730–0930 h, and 1700–1830 h, when the animals were found to be in the most used parts of their range. Counts conducted from noon to 1500 h were often low, due to animals being obscured in bush near to the water's edge. Between October 1965 and March 1967, I made a total of 487 visits, each of approximately 90 min duration. I thus had well over 3000 plots of individual buck positions.

There is some difficulty in presenting territorial data, as both temporal and spatial distribution are confused on composite plots. Thus either all juxtaposed bucks can be shown on one or more plots over a period of time, as I have done here, or a composite map can be prepared for each individual. In the former case two territorial bucks shown side by side at their common boundary may have been separated in time, and in the latter case spatial fluctuations between adjacent bucks are not shown. Of the two, the former method is considered to be preferable, the confusion of time–space relationships being borne in mind when interpreting the maps.

There was no evidence that waterbuck had a preferred spot within the territory, the daily plots showed a fairly random occupation, although I could usually find a territory owner within a general area. The twice daily plots revealed that during the period of study, from six to eight bucks maintained discrete areas ranging in size from 4 to 146 ha (Fig. 57, maps a to j). The boundaries shown drawn around these areas on the maps, can only be considered as a general guide to the positions of these boundaries, but the fact that their positions were confirmed on 11 occasions by observation, adds confidence to the approximations of the remainder of the boundaries.

In a competitive system, where the only means of obtaining a territory is either to take one from another buck, or to take a part of another buck's, the boundary would not be expected to change significantly in outline. This was found to be the case. One might thus say that territories were "inherited" as regards their boundaries. We have seen that territorial bucks appear to have been reduced to a low number again on the Peninsula, and it would be interesting to see what would happen if others were encouraged back to it once more. If the outlines of territories adopted were similar to those pertaining at the time of my study, this would indicate that the boundaries had some real substance in the form of the topography of the habitat.

The age structure of the Peninsula showed that all of those animals which possessed territories were either in, or had passed, their sixth to seventh year (Fig. 58). When mean territory size was plotted against age it was shown that the smallest territories were occupied by the youngest and the oldest animals, while the largest were occupied by bucks in their ninth year (Fig. 59). The difference in size between the territories of 6- and 7-year-old bucks, together with 10- to 11-year-olds, when compared with those of 8- and 9-year-olds, was highly significant at the 95% level of probability ($t = 3.73$, df 52, $p = \gg 0.999$ and $t = 4.17$, df 33, $p = \gg 0.999$ for 6–7 and 10–11 years respectively).

Although this relationship between age and territory size might suggest that the size of territory alone was the object of territorial competition, which might be inferred if the primary objective of the territory was to supply the owner's nutritional requirements, analysis of the number of doe visits enjoyed by the buck suggests that this may be the reason for territory size. The larger the territory the greater is the chance of does entering it at random. In this analysis a doe "visit" refers to one or more does found in the company of a territorial buck, the actual number

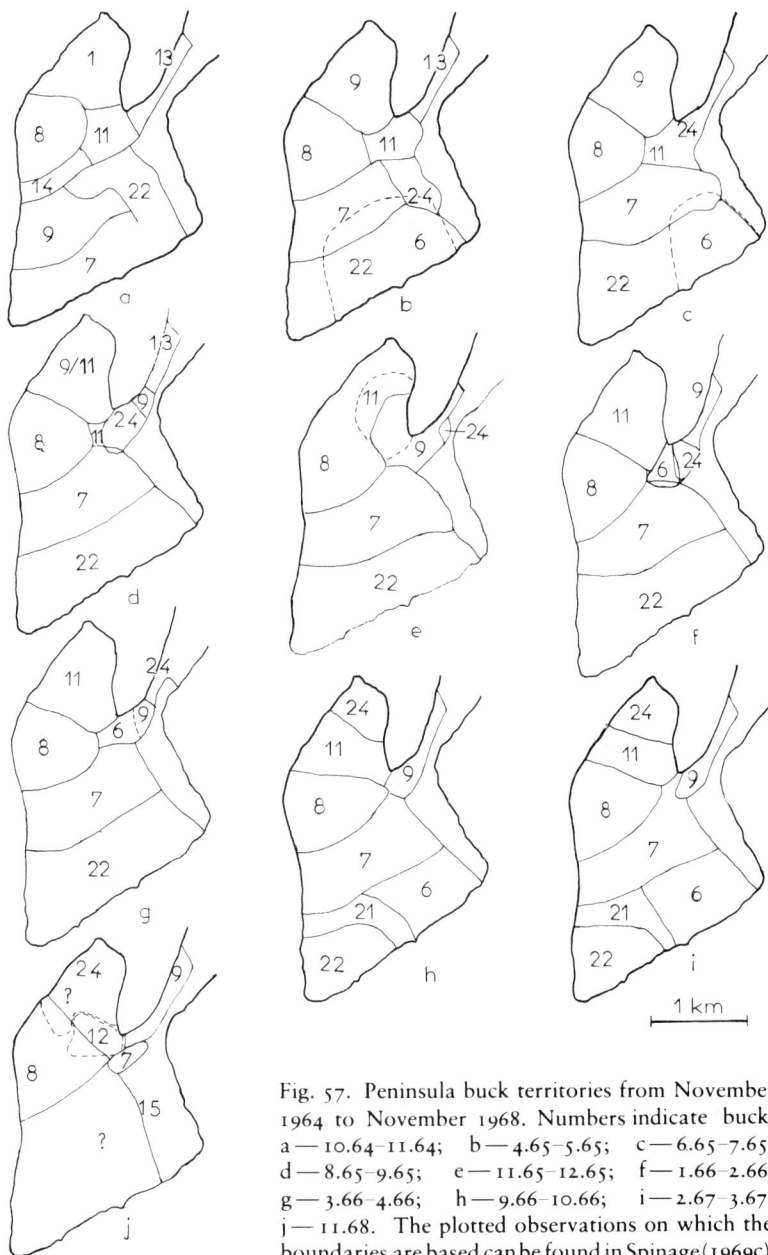

Fig. 57. Peninsula buck territories from November 1964 to November 1968. Numbers indicate buck. a — 10.64–11.64; b — 4.65–5.65; c — 6.65–7.65; d — 8.65–9.65; e — 11.65–12.65; f — 1.66–2.66; g — 3.66–4.66; h — 9.66–10.66; i — 2.67–3.67; j — 11.68. The plotted observations on which the boundaries are based can be found in Spinage (1969c).

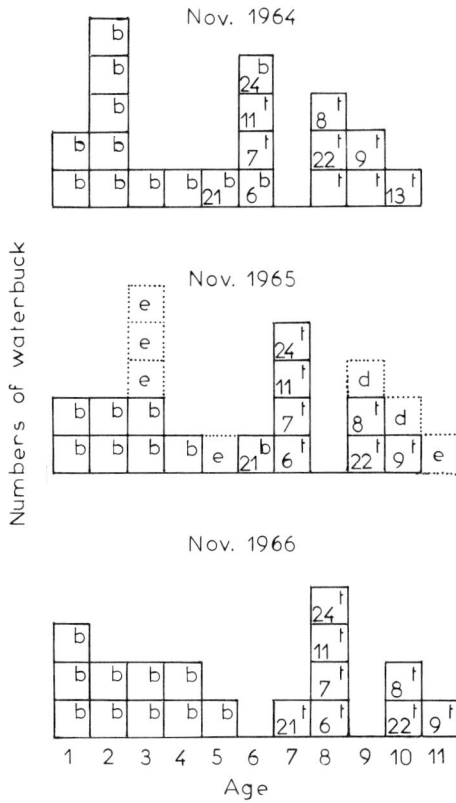

Fig. 58. Age structure of the Peninsula bucks 1964–1966. t = territorial, b = bachelor, e = emigrated, d = died.

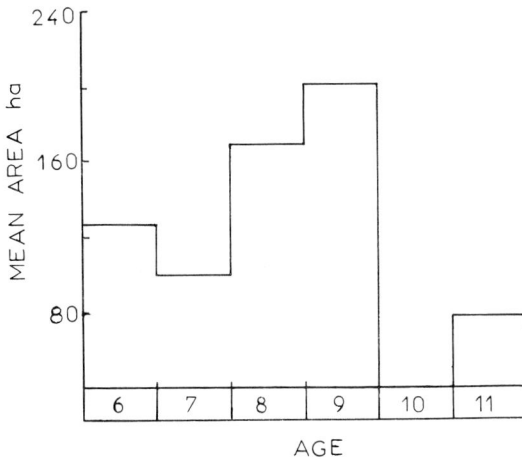

Fig. 59. Mean territory size in ha according to age of the territory holder for Peninsula bucks.

of does in a visit being considered irrelevant since, if the object was to enhance the chance of reproductive contacts, the territorial buck would not be able to accommodate more than one doe at a time. Hence one doe in a group of 30 is as good as one doe alone if she is in oestrus.

Analysis showed that the mean number of doe visits (expressed as a percentage of the total) and mean size of area (expressed as a percentage of the total) were not significantly correlated at the 95% level of probability (Wilcoxon matched-pairs signed-ranks test); although the shape of the frequency distribution was the same for both parameters (Fig. 60). This lack of agreement was because both young and old bucks had relatively fewer visits by does than would be expected from the size of their areas. Those bucks aged from 7 to 9 years had proportionately more visits than would be expected, with 9-year-olds, as expected, having most of all. This could be interpreted as selection for certain bucks on the part of the does, avoiding for example the aged and the inexperienced, but I think it more likely to be due to the fact that all of the area occupied by the bucks is not equally attractive, and that the small areas occupied by the youngest and the oldest have no attraction for the does. What emerges clearly, is that the prime of life for the buck is its ninth year.

On the Peninsula the largest territory averaged about 143 ha in extent, and in the neighbouring Ogsa about 226 ha. From this I assume that the

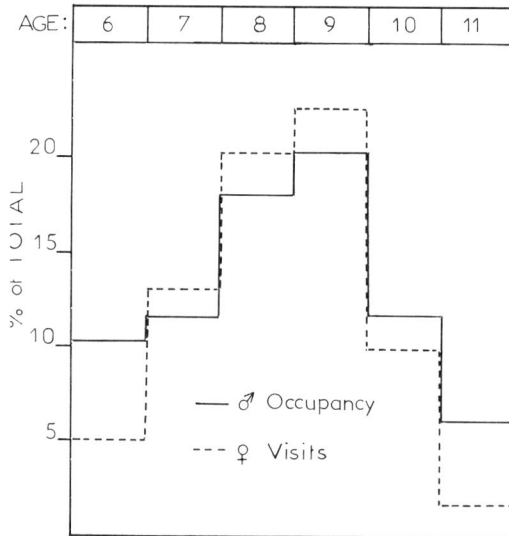

Fig. 60. The percentage of the total territorial area available expressed as buck occupation according to age, compared with the number of doe visits, for the Peninsula.

size of territory which a buck can defend is dependent, primarily, upon the pressure of competition, and not upon the surface area of the territory. The mean size of territory for a 9-year-old on the Peninsula was 81 ha, and in the Ogsa 203 ha. This difference is explained by the density of the territorial bucks in the two areas: one to 60·5 ha on the Peninsula, and one to 141 ha in the Ogsa. Thus a buck in the Ogsa had, on average, 2·3 times as much land as had one on the Peninsula.

The changes which took place among the Peninsula territories, during the period of study, are summarized in Table 54. Altogether a total of 20 changes took place, some of which, such as the history of Y6, have already been described.

Y24, who had previously been a member of the bachelor group, was seen to restrict his movements during April to May 1965 (Fig. 57, map b). In the period June to July 1965 he moved to another area which he shared with another young buck, Y11. Later he occupied a small part of Y9's area, leaving this and moving into his own, larger territory in September

Table 54 Summary of the territorial changes taking place on Mweya Peninsula between October 1964 and March 1967.

Animal:	Y13	Y9	Y1	Y22	Y14	Y8	Y11	Y7	Y24	Y6	Y21
Age:	10	9	9	8	8	8	6	6	6	6	5
Period											
10/64 — 11/64 Wet	17	71	63	61	16	59	23	81			
4/65 — 5/65 Wet	E	75	D	128	D	71	22	75	15		
6/65 — 7/65 Dry		65		146		85	26	68	34		
8/65 — 9/65 Wet	14	67		132		67	20	85	16		
		8[a]					67[a]				
Age:	10+	10		9		9	7	7	7	7	6
Period											
11/65 — 12/65 Wet		10		142		120	45	93	4		
1/66 — 2/66 Dry		12		144		75	73	97	6	8	
3/66 — 4/66 Wet		22		140		67	73	97	(24)	14	
9/66 — 10/66 Wet		22		51		67	43	95	26	50	29
Age:		11		10		10	8	8	8	8	8
Period											
2/67 — 3/67 Dry		25		54		69	38	87	32	71	30

[a] Indicates occupied two different sized territories during the period of observation.
Figures in parentheses indicate shared territory.
E = emigrated, D = died.
Areas of territories in ha, age in years.

1966 by taking over half of the territory that had been occupied by Y11. This was the only instance that I recorded of young bucks of similar age competing for the same area.

The buck Y21, who, as we have seen, established himself between the borders of three territories, appeared in March 1967 on a steep slope by the Kazinga Channel. I could only assume that he had been stampeded there by lions, which were active in the area the previous night. It is unlikely that a territorial buck would have pursued him to such a place, for disputes usually terminate at the boundary of a territory. When my fieldwork ceased in May, he had still not returned to his former territory.

Lions also appeared to be the cause of Y9 losing the area that he took over after the death of Y1. Three lions occupied the centre of Y11's territory in early September 1965, causing him to prudently withdraw. As a result he moved into Y9's territory, which he was successful in winning from the ageing buck.

On 28 March 1965, Y14 was accidentally killed during capture and marking. His area was taken over by Y7, who appeared to have been ousted by Y22 and, in turn, chased out Y9 (Fig. 57, maps a and b).

The oldest buck present at the commencement of the study, Y13, I judged to be about 10 years of age in 1964, but he may have been a little older than that. I well remember when I caught him, at the end of March 1965; his incisor teeth were very worn and he had bad breath. Other animals always had the sweetish, milky-smelling breath that is normal in herbivores.

Two days after capture and marking he left his territory, a small marginal one on the side of Kanyesewa ridge where all the old bucks seemed eventually to end up. Later I saw him in another fringe habitat, relatively far from water, about 6·5 km distant in the Royal Circuit area. He returned to the Peninsula on 26 July of the same year, during the dry season; but left again on 8 September, probably having been driven out by Y9. Eight days later I saw him again in the Royal Circuit area, but that was the last occasion.

I have already discussed the emigration from the Peninsula of the bachelors in 1965. A hypothetical age-structure histogram shows that the emigration of the young animals in that year appeared to indicate that this prevented the Peninsula from becoming overcrowded with territorial bucks four years later in 1968 (Fig. 61).

In fact, the picture turned out to be different to that which this extrapolation would suggest, for despite the initial stabilizing of the population density, the pattern of territoriality seemed to alter radically.

In November 1968 I was able to return to the Peninsula for a week to continue my twice-daily plots of the animals' positions. This already suggested evidence of a change taking place, there apparently being a predominantly senescent population of territory owners, with only two recruits expected in the next four years (Fig. 61), such that by 1973 there would have only been four to five territorial bucks, assuming no deaths and no immigration. In November 1968 I was able to identify with certainty: Y9, Y8, Y7, Y24, Y12 and Y15, bucks which were now aged 13, 12, 10, 10, 9, 7 and 6 years old respectively, out of a total of eight territorial bucks and eight bachelor bucks. Of the latter, five were in their first year, one in its second, and one each in their fifth and sixth years, suggesting marked losses from the Peninsula of those born in 1963 to 1966. The sex ratio had changed from 1 : 1·8 to 1 : 2·5 (Table 38).

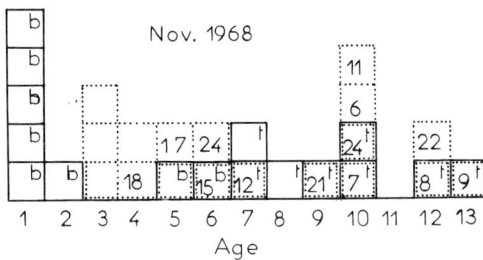

Fig. 61. Real and hypothetical age structure of the Peninsula bucks for November 1968. Continuous lines — observed presence, dotted lines — hypothetical presence.

Although of limited duration, the twice daily plots showed a considerable overlap of occupation between Y8, Y12, Y24, and an unmarked buck, but both Y8 and Y24 were by now ageing bucks whose territories were probably under pressure from the younger Y12 and the unmarked buck. Y15 appeared to be establishing himself in the territory of Y21.

By this time, 18 months from my last analysis, the Peninsula had lost Y6 (found dead), and apparently Y11 and Y22, from its territory holders, and Y17 and Y20 from its bachelor group.

Eltringham (personal communication) reported that in the first part of 1971 the adult bucks appeared to abandon their territories on the lower Peninsula flats, possibly in response to the presence of lions, but later in the year one buck returned. However, in 1972 it was reported that the waterbuck generally remained together in one herd on the high ground.

From 1968 to 1973 Eltringham (1979) conducted regular monthly counts of the Peninsula animals, in which he considered solitary waterbuck bucks as territory holders. These counts seemed to show a decline from a mean number of 5·3 territory holders in 1968 to 1·9 in 1973, for which he could offer no satisfactory explanation. It is difficult to interpret this data without knowledge of the composition of the "single large herd" which Eltringham reported on the high ground. Also a territorial buck is not necessarily solitary, and indeed may spend much of his time in the company of a doe group. But if the analysis does reflect the true state of affairs, then I suggest that it was attributable to the lack of territorial recruits. At the end of 1968 there was an ageing population of bucks with gaps in the bachelor group age structure. By the end of 1973 all of the 1968 territorial bucks would most probably have been dead from old age if nothing else, and the bachelors would have become one each of 11, 10 and 7, and five of 6 years. Thus we would expect only three territorial bucks at the most, assuming no immigration, if the five 6-year-olds had not yet sorted themselves out into territory owners. Eltringham's data does not give the number of bachelor bucks, but the expected territorial structure for 1973 seems to have been much as Eltringham found it.

Territorial Structure of the Ogsa

Due to the thicker cover, and the time taken to traverse the area, it was not possible to define the territories in this area with the same degree of accuracy as it was for the Peninsula. I found, nonetheless, that in this region of average waterbuck density, a clear system of territoriality still existed.

In July 1965 there were eight territories, ranging in size from 54 to 226 ha, with a mean size of 146 ha (Table 55). Their approximate dispositions between July 1965 and March 1967 are shown in Fig. 62. There was little change in the territories during this period. Those which did were as follows. The 13-year-old Y41 died from natural causes in about August 1965, his territory being taken over by a 7-year-old, Y31. There was an extension by Y38 into the northern part of Y33's territory, but then Y38, now about 11 years old, died in February 1967, and Y40, who had been

Table 55 Summary of the territorial changes in the Ogsa as at July 1965 and March 1967.

Animal:	Y41	Y28	Y33	Y38	Y36	Y37	Y31	Y35	Y30	Y40
Age:	11	11	9	9	8	7	7	6	6	5
Area:	54	107	226	183	172	118	86		129	
Age:	13	13	11	11	10	9	9	8	8	7
Area:	D	C	226	D	172	118	140		129	183

D = died, C = shot in July 1966.
Area of territories in ha, age in years.

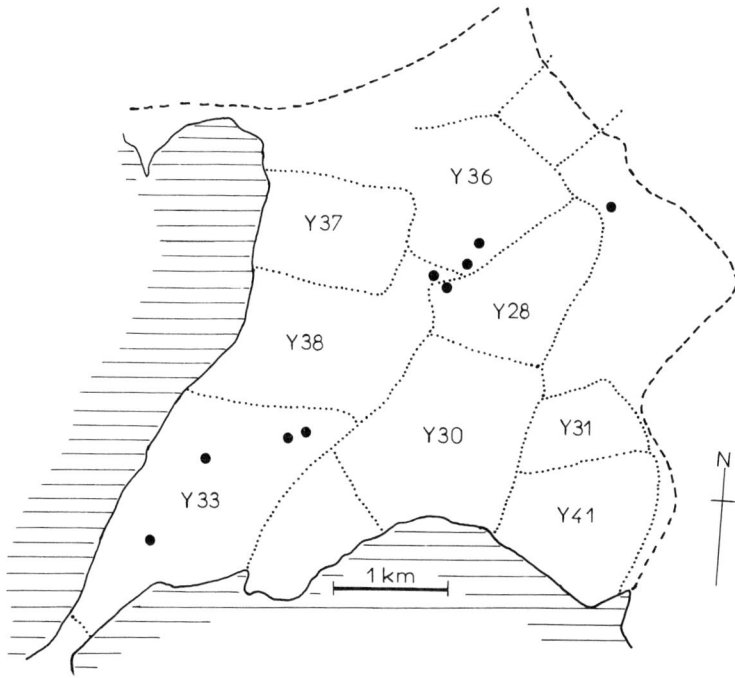

Fig. 62. Territories in the Ogsa, July 1965–March 1967. Blocked circles indicate observations of Y35.

sharing Y38's territory, took over the deceased buck's territory. He also extended it into the tip of Y33's. An unoccupied territory in the south was taken up by an unknown buck in March 1967.

A histogram of age against mean territory size, for July 1965, shows close agreement in its shape with that of the Peninsula, the largest territories being held by the 7–9-year-old group (Fig. 63). Disturbance resulting from capture and marking possibly created a rather artificial picture after this, so that further histograms have not been constructed. The fact that Y30 had a large area of territory at such an early age (6 years) may have been the result of the irresponsible shooting of two adult bucks in the area in October 1964, possibly making this territory and the unoccupied territory, vacant. Two bucks, one aged 5 and one aged 6 years, died from darting, and a further 6-year-old disappeared after capture and marking. The approximate age structure of the population in July 1965 and March 1967 is shown in the histograms, Fig. 55. The mean age of the population was not significantly different from that of the Peninsula buck population at the 95% level of probability, when

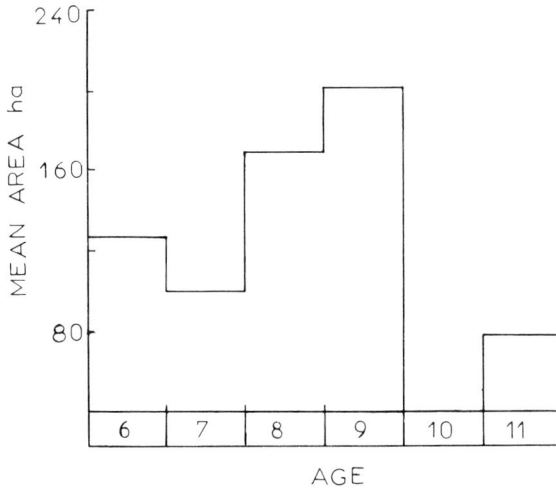

Fig. 63. Mean territory size in ha according to age of the territory holder for Ogsa bucks, July 1965.

tested for July 1965 and April 1967, against November 1964 and November 1966 ($t = 0.46$, df 19, $p = > 0.3 < 0.4$, and $t = -0.19$, df 22, $p = \ll 0.1$ respectively).

The Ogsa bucks had approximately 2·3 times as much area available to them as did the Peninsula bucks, and mean territory size was 2·5 times larger. But the 9-year-old bucks had territories only 1·6 times greater than their Peninsula counterparts. This leads me to suggest that the buck is only able to defend a territory of a certain maximum size, perhaps about 200 ha, as opposed to the 290 ha which we would expect on a *pro rata* basis, extrapolating from the Peninsula observations. As a result, the other age groups are able to possess proportionately larger territories, and the oldest animals, for example, occupied territories averaging 140 ha in size; more than adequate for a single waterbuck's needs. This might account for the fact that much older bucks were encountered here than on the Peninsula, their mortality not being accelerated in old age for want of sufficient feeding grounds.

The absolute size of the Ogsa territories, and the relatively thick cover in the area, may have meant that the territories could not be as well defended as on the Peninsula, and from time to time apparently unknown adult bucks were found wandering in the area, probably looking for territories. One buck, Y35, was a particular puzzle to me for I recorded him from time to time throughout the area (Fig. 62), and yet he appeared to be adult enough to be a territory holder. Somehow he always managed to evade capture, but at the end of my study I made a determined effort to catch him and to find out if my estimation of his age really was wrong. Unfortunately he died under the drug, and when I came to weigh the carcase both the cause of death, and his anomalous behaviour became obvious. He was exactly 45 kg under weight for his age, which I calculated to be 8 years, exactly as his appearance had told me. As a result of his light weight he had received a gross overdose of the drug. Why he was under weight was not apparent, but it was little wonder that he could not hold a territory when he was only the weight of a 4-year-old. This was a convincing proof of the hypothesis that it was size which ultimately determined when a buck could become territorial.

The most unfavourable territories in the Ogsa I postulated to be those inland, without water frontage. Examination of the age of the territory owners supported this hypothesis, for the territory furthest inland was held by Y28, who was about 11 years old.

Territory in the Kayanja Area

We have seen that this area carried the highest densities of waterbuck recorded in the Park, and at first a territorial system was not apparent. There appeared to be a rapid turnover of adult bucks, and of seven marked in August 1966, only three were located in January 1967. Poaching may also have been a significant factor affecting waterbuck here. This rapid disappearance of marked bucks created difficulties of identification.

Over much of the area the territories appeared to be of large size. Thus counts in area "A" (Fig. 16) revealed a mean occupation of four single adult bucks in 959 ha; implying a mean territory size of 240 ha, and of the order of magnitude that we would expect from the Ogsa studies. The mean number of adult bucks encountered was 9·5, suggesting that a number of 5- to 6-year-olds might be sharing territories. In a small area of approximately 66 ha in region "B" (Fig. 16), four bucks were identified whose estimated ages were two of 6 years old, and two of 7 years old. One of these, R33, was captured and his age determined; the age of another, "M", was estimated from the time that he left the bachelor group; and for the remaining two estimates of their ages were made from their horn length and general appearance. Thirteen plots of their positions, between November 1966 and February 1967, showed that each was restricted to an area of roughly 8·5, 8·5, 14·6 and 20·3 ha respectively (Fig. 64).

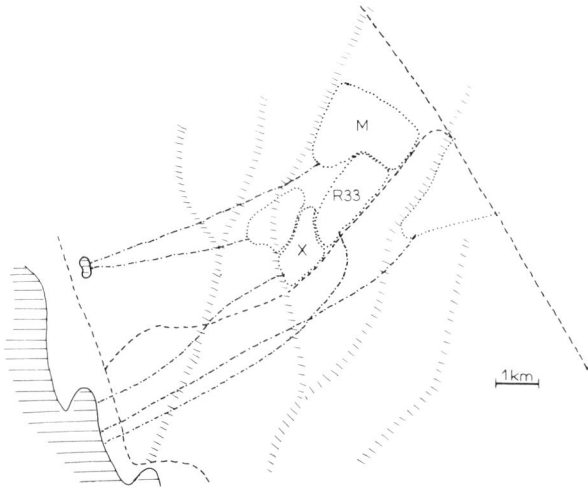

Fig. 64. Territories at Kayanja, November 1966–February 1967. Dot-dash lines indicate routes taken to water, dashed lines indicate park tracks.

This area of intensive occupation was approximately 1·6 km from the lake shore, and six periods of continuous observation during the dry season showed that there was a daily journey to the lake to drink. In the wet season this was not necessary as there was standing water inland. The indication was thus that there existed a very different pattern of territorial occupation here, for during the dry season the territories were only used for a part of the day, the remainder of the time being spent in a slow move to the water and back again, taking up to 4 h as the buck grazed by the lake shore and associated with does. The unoccupied area between the territories and the lake shore had the appearance of a "neutral" zone, for no agonistic displays were seen between adult bucks travelling through it.

In 15 observations in area "A", 18 pairs of adult bucks were seen, suggestive of animals of 5 to 6 years of age seeking territories together, or sharing the territory of an established buck.

It was unfortunate that more attention could not be given to this area, for it provided a contrasting type of territorial structure, at least in some aspects differing from the other areas which were studied. The preponderance of young *Acacia sieberiana* thorn trees suggests a possibly recent change in land use in the area, perhaps also indicated by the high waterbuck density, which might have been another instance of an exponential rise in numbers in response to a newly-exploited favourable environment, in which case, if the zone was newly occupied, a clearly demarcated system of territorial boundaries may not yet have emerged.

Territorial Organization at Extreme High Density

In the high density population studied by Dr Wirtz, at Lake Nakuru, Kenya, where the numbers of waterbuck in one study area of 4·5 km² were estimated to be of the order of 72 to 106/km², compared with an overall density of 9/km² for the 157 km² of the park's total land surface; there was no evidence of an intensive "lek-type" of territorial organization developing — the organization which is found in high density kob populations (Chapter 12). Although buck density is estimated at about 39 to 46/km² (all ages), as we shall see in Chapter 12, this is far greater than that at which territorial organization in the kob changes from a spaced-out "living area" type of territory to the intensive "lek-type" of breeding territory, the latter type being found in the Rwenzori Park at a density of 12·7 bucks/km².

Wirtz (1980) estimated territorial turnover time in his study area on average to be 1·5 years, which is probably similar to that which took place on the Mweya Peninsula for a prime territory. Nevertheless, it appears that the average territory size in his study area was about 32 ha, compared with 57 ha on the Mweya Peninsula. Thus, assuming the situation on the Peninsula to be more or less balanced with respect to territorial occupation as a percentage of the total buck population, that is, that the percentage of bucks recruited into the territorial population from the young bucks, is average, then the number of bucks of territorial age which are not recruited into the territorial population in the Lake Nakuru study area, is of the order of 59%, of the total buck population.

Wirtz identified a number of apparently adult bucks (which were probably in the 5–6 year age class) regularly encountered in a territory, which showed deference to the owner and which were not repelled by him, being permitted to co-exist there. Some of these bucks used several adjacent territories, and also joined in defending the territory against strange bucks. In all, 13 out of 25 territorial bucks had such other bucks present, which Wirtz has termed "satellite males". Unlike the bucks in my study which tried to obtain territories by "insinuation" at the edges of existing territories, these satellite bucks had the use of the whole territory. They were also able to copulate a number of times, certainly more than if they had been in the bachelor group, and also had a higher chance of becoming a territorial owner.

Although Wirtz has interpreted this as a form of co-operation between the territorial and the satellite bucks, from which both benefit, it seems to me to be indicative of a reduced tolerance by the territorial owner as a result of the high population density, for the bachelor bucks as a whole were excluded from the territories. The high population density meant a high number of bachelor bucks who wished to share the habitat, and thus a greater pressure on territorial bucks from the high number of bucks of territorial age seeking territories. The territorial buck, therefore, only allowed a limited number into his territory, which would presumably be the most persistent young bucks. The waterbuck is not a seasonal breeder at this latitude, and the fact that the bulk of the bachelor bucks were excluded from the territorial network suggests a form of population regulation coming into play, since their survival rate would be potentially lessened.

At a buck density of 19·3 to 28·5/km², Elliott (1976) found that mean territory size was only 13·3 ha, in the range 4·1 to 27·9 ha, thus this was much smaller even than the territories at Lake Nakuru; although in this case it was perhaps attributable to restriction by fencing. She found that some territories changed ownership frequently, as often as six times in 2 years, while others did not change ownership at all during this period. Despite the small size of the territories, there was no evidence here either of a "lek-type" organization, neither did she record satellite bucks or the exclusion of bachelor bucks; in fact she concluded that the territorial system "appeared to be particularly stable".

Wirtz (personal communication) claims, however, that the phenomenon of "satellite males" has been reported from Zimbabwe and from the Serengeti in Tanzania; although I know of no other workers who have reported this. It could be, nevertheless, that there is more than one form of social organization in the waterbuck, and that its form is not particularly dependent upon circumstance. This would accord with the theory of a geologically recent expansion of the species and variability of behaviour that might be supposed to accompany this.

The explanation for the lack of a "lek-type" of territorial organization developing at high density may be found in the fact that, whereas the kob, for example, has a tendency to exist at high densities where conditions are favourable, this is unusual in the waterbuck. Thus the latter species may not have evolved the plasticity of social organization shown by the former in this respect. Or, it may be that the waterbuck's specialized physiology does not permit it to indulge the same type of strenuous activity that the kob's lek-type of territorial organization demands. It might, for example, have to visit water so frequently under such conditions, that the system just could not operate. Another factor, of course, is that the waterbuck does not advertise its presence in the same manner as the kob, it does not have to crowd together to be seen and heard, since it can be smelt from a distance.

11. Buck Behaviour

Territorial Boundaries

Adult bucks have been shown to occupy distinct, discrete areas called territories. But, unlike several territorial antelopes, there was no evidence that the boundaries of waterbuck territories were actively marked by odours or by urine, dung deposits, or odoriferous gland deposits. As far as I could determine, the waterbuck lacks any conglobate odoriferous glands, but Hoffmann (quoted in Elliott, 1976) alleges that there is a concentration of apocrine glands in the centre of the forehead. Although there is a hair whorl at this point, I never observed any signs of an exudate there. But my attention was drawn by Dr T. Nay (personal communication, 1965) to the very large sebaceous glands present in the dermis of both sexes. These glands produce a dark-coloured exudate which covers the coat profusely at times, and comes off readily when the animal is handled. The characteristic sweet, cloying, musky odour of this secretion is well known to those who have been close to waterbuck. When climatic conditions are right, it is readily detectable in the air, and I have smelt it carried on the wind up to half a km distant from its source.

Size and frequency of these sebaceous glands in different parts of the skin is shown in Table 56. The results suggest that the glands are smaller under the tail, but similar in size on all other parts of the body. There is no significant difference in size between those of the buck and those of the doe, at the 95% level of probability, nor is there a significant difference in the number per unit area between the sexes; thus predisposing against the hypothesis that their purpose might be that of territorial marking. Nevertheless, the strong odour of the animals suggests that they are able to recognize one another by smell, and the marking of territory may be passive, occasioned simply by the buck's smelly perambulations within it. I have smelled places where bucks have been lying, but the odour was not particularly apparent at such sites. The horn bases, although apparently no more richly endowed with glands than most other sites, appear to have very active sebaceous glands, whose function is to lubricate the soft growing bases of the horns, preventing them from drying out and splitting.

Table 56 The mean diameter, and mean number per unit area, of sebaceous glands in the skin of the waterbuck ($n = 5$).

Area:	Inguinal	Anus	Flank	Neck	Horn base	Belly	Under tail	Mean	SD
Size in microns									
Buck	257	241	268	297	285	293	195	262	35·84
Doe	271	198	287	330	—	249	172	251	58·29
Number per unit area									
Buck	5·5	8·0	9·0	7·7	5·6	7·0	7·0	7·1	1·52
Doe	9·2	6·0	11·0	6·5	—	6·0	7·0	7·6	2·04

To determine whether the scent of another buck within his territory would perturb the owner, I conducted a small experiment. I took a clean, well-rinsed sponge, and when I captured a territorial buck in the Ogsa for marking, I sponged his coat vigorously. Immediately afterwards I went to the Peninsula and looked for a suitable buck to test. I saw Y7 walking along, so I drove across his path, dropping the sponge unobtrusively out of the offside door of my vehicle. Y7 continued his line of travel uninterruptedly and came upon the sponge lying on the ground, sniffed it gingerly, and shied away with a snort. He then continued walking as if nothing whatever had happened! It appeared that the fresh scent of a strange adult buck meant nothing to him.

I then repeated the experiment with fresh droppings. Inducing an Ogsa territorial buck to defaecate, in the manner explained in Chapter 7, I leapt from my Landrover, scooped up the fresh prize in a small shovel, and rushed to the Peninsula. Here I repeated the experiment in the same manner as I had done with the sponge, dropping the faeces out of the offside door in the path of an oncoming Y7. In this case the buck walked right past, and did not even deign to sniff the offering.

I never saw waterbuck deposit droppings or urine in the same spot as others, and neither was any interest shown in such deposits; but since they can adequately permeate their environment with their powerful sebaceous gland smell, there seems no reason for them to use accessory odoriferous deposits.

Schenkel (1966) has criticized the concept of odoriferous territorial marking, as there is no evidence, in his view, that conspecifics note the marks, or avoid an area because of them. Rather he sees odour as a form of indirect contact among individuals of a given population. Ewer (1968) has developed the view that odour marks the environment, leading an animal to know when it is "at home". As Mykytowycz (1973) has expressed it:

> "By setting down its own smell, the animal marks an area 'safe', indicating that it belongs there. This alone will restrict straying and either exclude strangers or inhibit their behaviour, placing them at a disadvantage with respect to the owners. It will increase the confidence and aggressiveness of the residents and assure their participation in reproduction."

Thus it is only necessary to show that a territorial owner marks what it considers to be its own; whether this is strictly observed by others or not, is another matter, for the owner succeeds in making itself feel "at home"; but territorial boundaries usually are observed by competitors.

Mykytowycz has pointed out that, not only do animals saturate their environment with their own odours, but they try to mask strange ones, thus explaining the apparent anomaly of a territorial owner adding to a neighbour's dung pile. In cases where masking is not possible, for example a large dung pile or a dead carcase, animals may attempt to become a part of the new smell, for example by rolling in it. However, in a relatively socially developed species such as the waterbuck, in which vision and visual interpretation have probably become more important than olfactory interpretation, I do not think that saturation of the environment with the owner's smell is as important for the development of various stages of reproduction as Mykytowycz suggests it to be in the case of smaller mammals. Indeed, we have seen that a waterbuck will enter the territory of another to try to mount an oestrus doe. Whether it is important in conditioning, or bringing about a sense of well-being, in the doe, is another matter, and, if so, may explain why at low densities does form larger groups than expected (Chapter 10).

We should not overlook the fact that what appears to be intentional, may not necessarily be so. Gosling (1974) records how the hartebeest, a territorial species, deposits its droppings at its territorial boundaries to mark them. But every criminologist is aware of the fact that burglars not infrequently defaecate at the scene of the crime. This is not a contempt for the aggrieved, as is so often supposed, but is a fear response deriving from high adrenalin levels which relax the bowels. The same effect can be seen in ungulates; when they are in a situation which results in an output of adrenalin, they defaecate. This could be considered as most likely to occur when confronting a rival, especially at a territorial boundary.

I conclude that, although scent plays some part in the recognition of conspecifics, as I shall show later, territorial boundaries in the waterbuck are probably based upon real topographical features which are apparent to the animals. In some cases boundaries appear to follow a natural feature, such as a ridge of high ground, but others run across open grassland, the positions of which are nevertheless clear to the owners.

During the periods of continuous observation I also plotted the positions of the animals which I was following. Most of the time Y7 kept well within the boundaries which I had allotted to him on my map, derived from the daily plots. His movements appeared to be rather erratic, circuitous meanderings, accompanied by a visit to drink at the lake, or an inland waterhole. When he did stray over his boundary, and was seen to do so, then he was instantly challenged. Figure 65, maps a to d, describes these movements.

It can be seen in map a of this figure that Y7 overstepped his boundary into what was Y21's territory. The arrow "A" indicates the following events which took place as recorded in my field notes:

"16.2.66, 1228 h. Y7 watching Y21 who has approached a doe near the boundary. He moves towards Y21 . . . comes closer and Y21 displays. Y21 walks round, past Y7, and horns a bush. Makes a short rush, horns the ground, and trots near. Y7 advances in 'proud posture" and stops. Y21 horns the ground again, displays, and walks away when Y7 approaches. He repeats the same activity. Y7 then grazes while Y21 continually displays, walking 'to and fro' in front of him. Suddenly Y21 rushes at Y7 when the latter turns away, but runs straight past when Y7 turns round. 1304 h. Y7 moves off, Y21 follows parallel to him to the edge of the scarp."

The interaction was concluded and Y7 went off to have a drink. At this time Y7 was 7 years old, and Y21 was only 6.

Three months later a confrontation again took place in exactly the same spot (map b). Of course, there may have been many more in the intervening period, but they were not observed by me. The arrow "B" in map b indicates the following events as recorded in my field diary:

"18.5.66, 1232 h. Y7 stands watching Y21 and Y22. Y21 walks forward, Y7 walks forward to meet him. Y21 displays at 50 m. Y7 stands, then walks forward again. Y21 displays then walks away a little. Y7 approaches, closes with Y21 then walks away and grazes."

On the next occasion that I watched Y7 there were several breaches of boundary etiquette. The first ("C" on map c), was when Y7 chased an 18-month-old bachelor buck into Y8's area, and was in turn instantly chased out by Y8. Much bigger breaches were shown in the east, probably due to

a

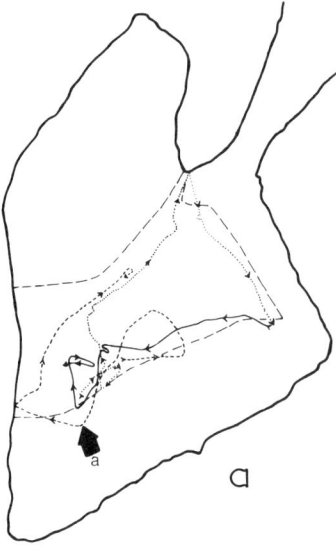

16 to 18.2.66   Dry

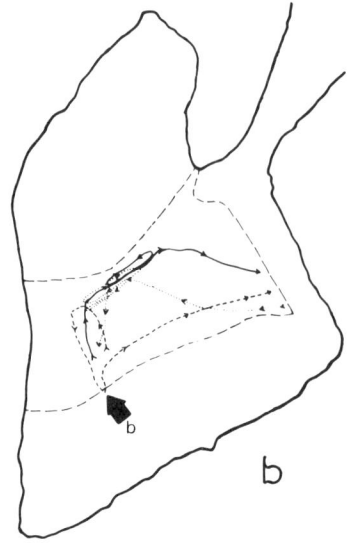

b

18 to 20.5.66   Dry

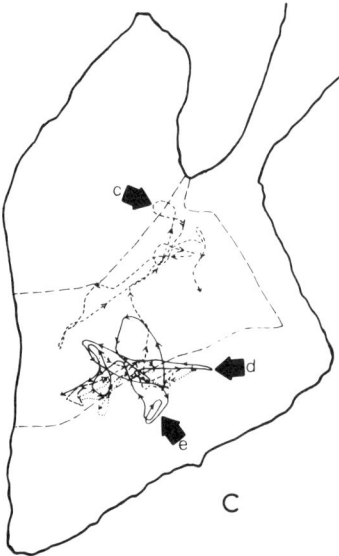

c

19 to 21.9.66   Wet

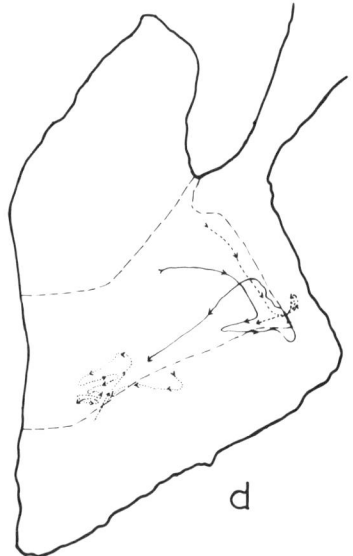

d

| 7-8.3.66 | 28-29.10.66 | 25-26.1.67 |
| Dry | Wet | Dry |

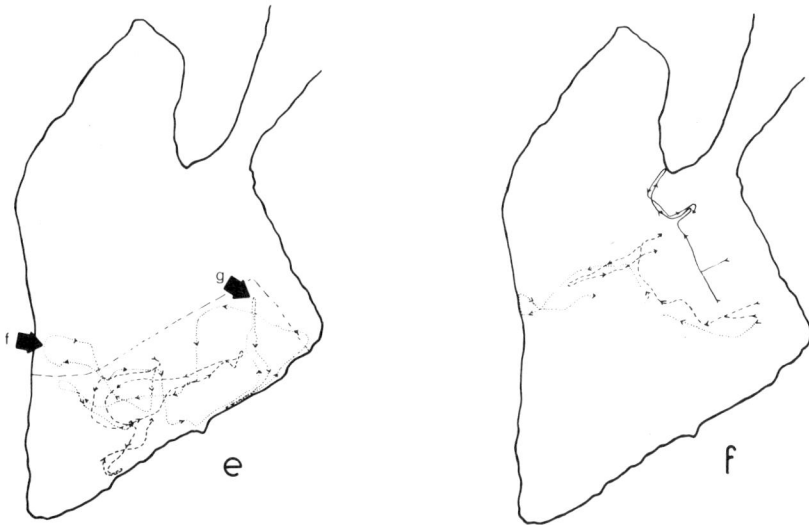

♂ Y22   28-29.6.66   Dry   Bachelor group   13-15.7.66   Dry

Fig. 65. Buck movements on the Peninsula. a–c — diurnal movements of Y7; d — nocturnal movements of Y7; e — diurnal movements of Y22; f — diurnal movements of the bachelor group. 1st day — dashed line; 2nd day — dotted line; 3rd day — continuous line. Long dashed line indicates the maximum plotted boundaries of the territorial bucks.

a doe coming into oestrus. "D" indicates from my field diary:
"21.9.66, 0840 h. Y7 sees activity about 350 m to the east. The bachelor group also sees it and gallops across. Y7 gallops after them. Y6 is chasing a doe at speed. He comes near to Y7 who immediately engages him and they fight vigorously until Y6 backs away. The doe has run to Y21. The bachelors and Y7 trot after her. When Y21 faces Y7 the latter walks away and slowly returns to his own area."

It is clear that the boundaries were fairly rigidly respected. As soon as the boundaries which I had allocated were crossed, the trespasser was challenged by the rightful owner of the territory. The nocturnal position was the same, although here there was some slight boundary overlapping which did not meet with a challenge, probably because it was not observed.

Observation of the buck Y22 indicated that he showed a similar respect for boundaries; although "F" (Fig. 65, map e) indicates where he descended the scarp into what I attributed to be Y7's territory. This was possibly because he had seen Y21 there, a buck whom he later chased up onto the plateau. The point "G" indicates where he stopped pursuing the old buck Y9, whom he had flushed out of some bushes near to the lake shore (see Chapter 10).

Some other incidents out of the 11 which I observed were:

"25.4.65: Y9 pursues Y8 to his boundary.

7.5.65: Y7 pushes Y24 backwards, nose to nose, until the latter is in his own territory.

18.5.65: Y9 walks an unidentified buck (Y11?) backwards, nose to nose, out of his territory.

19.9.66: Y7 walks Y21 backwards across his boundary, followed by much displaying. Y21 fights the earth vigorously in front of Y7, who horns a shrub a little and then walks away.

29.9.66: Y7 approaches Y6, both display, engage horns briefly, then part and graze."

Thus in almost every case where the allotted boundaries were overstepped, display or fighting between the adjacent bucks took place. On several occasions one buck in pursuit of another would stop for no apparent reason other than that he had reached his boundary line, which was often invisible to me in the open grassland, with no distinguishing feature.

Territorial bucks were seen to recognize, and tolerate neighbours. If one was on the boundary the adjoining owner might approach and display, but neither would normally cross to fight. Some of the incidents recounted above also illustrate this tolerance; the deliberate pushing, nose to nose, of a straying neighbour back into his territory, without fighting, is one example.

An example of the acknowledgement shown of an owner's rights is given by my field diary for 16 February 1966, on the Peninsula:

"Y7 lying ruminating. Y22 about 200 m distant with *post partum* doe, both lying. Doe and Y22 get up and graze in a north-east direction. Y7 suddenly rises and gallops up to them, stopping about 30 m away and displaying. Y22 approaches Y7 and displays, then walks away leaving Y7 with the doe."

My inference from this sequence is that the boundary into Y7's area had been crossed, and this fact was acknowledged by Y22. Several similar observations were made of bucks stopping at their boundaries, while the does continued into the next territory.

Fighting was reserved for territorial challenging, when there was an attempt by one buck to take over, or install himself in part of another's territory. Strange adult bucks were pursued instantly, and on the Peninsula this had the effect of preventing any immigration. Clear evidence that strangers could be recognized by sight was provided by my observations on the Peninsula on 20 April 1965.

I was on the high ground watching Y7 investigating some does, when he abruptly stopped and stared into the distance. I followed his gaze down the Peninsula with my binoculars, and, a good half-kilometre distant, saw a strange buck regarding us. He was easily recognized as a stranger by the shape of his horns, but Y7 had not needed binoculars to tell him that. The next moment Y7 began to walk quickly and deliberately towards the visitor. As soon as the latter saw Y7's studied approach he took fright and ran off out of sight, long before the two were within any distance of each other. Without stopping, Y7 continued straight to the spot where the stranger had been standing, and proceeded to sniff around. Picking up the scent almost immediately he followed it like a dog, nose held half a metre or so above the ground. Reaching his boundary, which was invisible to me, he abruptly lost interest and walked back the way he had come.

A little later the intruder came out of hiding. Making his way back he paused to sniff in turn the places where Y7 had passed. While occupied with this he suddenly came across Y24 with half a dozen does. But Y24 was a weak buck; instead of immediately showing aggressive intent he stared at the stranger, before cautiously advancing towards him. Thus the stranger did not run, but as Y24 stopped and displayed, he walked up to him. They touched noses and both engaged, testing each other's strength. Then they stood back and then came together again. At this moment Y22 appeared, following the stranger's scent. His attitude must have denoted aggression, although I did not detect it, for Y24 showed as much nervousness as the stranger. When Y22 rushed at the latter, Y24 made off also, as fast as he could. Y22 did not persist in his rush. The stranger mixed with Y24's doe group, walking back with them, and with Y24, towards Y22 again. The latter gave chase once more, but not very vigorously as they were outside his area. But by now the stranger was sufficiently agitated that when Y24 approached him, he took to his heels. Y24, emboldened by this, chased the trespasser at high speed, until they were both out of sight. I did not see the stranger on the Peninsula the next day; later he took up Y13's former territory in the Ogsa.

What was, to me, most interesting about this incident, was not only the demonstration of the acute eyesight of the buck, for I had to use binoculars to establish that the animal which attracted Y7's attention was a stranger, but also that waterbuck did use scent as a means of identifying where others had been.

On 11 September 1965 a young adult buck, who was in fact the oldest member of the Ogsa bachelor group, was seen on the Peninsula. He soon left, probably chased off by Y22; he was followed by Y21 and Y12 of the Peninsula bachelor group. The latter two returned to the Peninsula during the night. Young bucks which wandered on to the Peninsula from time to time were "worried" by the eldest members of the bachelor group. Such juvenile immigrants were not seen to arouse any antagonism from territorial bucks, but this was not true of the bachelors. These continually followed the immigrant, sniffing and pushing against its rump until it finally left.

From time to time a buck was found well inside of another's territory, but this was rare. When one did so trespass, it gave every appearance of knowing that it was doing wrong, moving cautiously, ready to run at the slightest alarm, and regarding every bush with suspicion. There seemed to be no reason for these visits, except perhaps to check up on an owner's presence.

On one occasion, on 6 November 1964, when Y8 had ventured far into Y11's territory, there was a vigorous fight between the two. Y11 eventually moved away, and Y8 walked slowly back to his own territory. The same two fought vigorously again on 22 November the following year, when there appeared to be a dispute over Y9's former area (Fig. 65, maps e and f).

At 1715 h I saw Y11 hurrying through Y8's territory, lowering his head from time to time as if following a trail. Soon Y8 spotted him and came running up. He broke into a walk when Y11 stopped. Y11 displayed, and Y8 defaecated nervously before advancing. The former walked away and entered a thicket. This was a cowardly move, and the signal for Y8 to break into a gallop. Uttering short bleats he chased Y11 back in the direction of the latter's territory.

Thinking that Y11 was now put to flight, Y8 abruptly swung round and ran back the way that he had come, without stopping. This apparently cowardly move was the signal for Y11 to turn and give chase. But Y8 soon put a stop to this by halting and displaying. Having checked Y11 he turned and walked off, the former standing watching him.

When a buck from a non-adjacent territory appeared in the territory of another, he was chased without ceremony. Often the mere threat-intent of the owner galloping towards the intruder was sufficient to put the latter to flight. If, however, the intruder was sufficiently bold as to stand his ground, then a different course of action had to be resorted to by the owner. The intruder meeting the threat-gallop stood still and usually displayed. This caused the owner to stop his gallop and to approach more cautiously. The two sometimes walked parallel to one another, possibly in small circles. This could be seen as what Geist (Geist, 1971) has termed an evasive technique; probably initiated by the weaker, or more nervous, of the two, to render it impossible for the horns to be used. In the cases which I witnessed, however, it was only a preliminary. Noses were touched, followed by foreheads, and accompanied by much licking of lips. I don't think that this tongue-flicking has any significance as a behaviourism; it merely indicates that the adrenal system works in the same way as it would do with us, and the animals' mouths become dry with fear or excitement. The horn bases were then engaged, and both contestants took the strain, with a vigorous wagging of tails.

At this point a younger buck who had merely strayed over the line, might simply be walked backwards by an older buck, forehead to forehead, back to his own area, as I have already recorded. In this case the older buck withdrew as soon as he had pushed the other over the boundary. In some cases the younger buck was walked backwards simply by confrontation, without touching heads. But usually the younger the buck, the more exaggerated the display in conflict situations. A pair of old bucks often hardly bothered with the preliminaries to battle.

Confronted by a dominant buck, a young sub-dominant often gored the earth, and fought with tufts of grass or small bushes. Serious in its intent, it was nevertheless amusing to watch. The older buck stood aloof and impassive, looking slightly away; while the younger buck vigorously vented his spleen on a small bush, rapidly working himself into a frenzy, until he was down on his knees, goring the ground and scattering leaves and branches into the air. Suddenly, he leaped to his feet, and rushed at the dominant buck, only to run straight past when the latter calmly turned to meet him. Such behaviour has been described in cervids, caprids and several bovids ranging from oxen to Uganda kob and impala (e.g. Graf, 1956; Burckhardt, 1958; Schloeth, 1961; Walther, 1964; Geist, 1965; Leuthold, 1966 and Schenkel, 1966). Generally considered as a displacement activity in a conflict situation, I believe that in the waterbuck at least, in view of the needle-sharp points of the horns, earth-goring is a demonstration of horn sharpening, rather than a displacement activity unconnected with meaning. As we have seen, there is considerable wear on the horn tips, which could only result from fairly frequent earth-goring.

The display behaviour to which I have referred, is typical of that of the Bovidae. What I have here termed "display" has been referred to variously as the "present-threat" (Geist, 1964), or the "proud-posture" (Schenkel, 1966). It consisted of the buck standing in lateral view to another, with the neck arched almost like a Viennese lippizano horse, and with the tail held out stiffly in line with the body. When displaying strongly the penis is extruded and the urethral process extended, indicating complete erection. If the opposing buck approaches, the horns are inclined stiffly towards him (Plates 45, 46, 47, 48). This attitude shows off the buck's fighting potential — his size, the thickness of his neck, and the presence of his horns. The penis display may indicate his maleness, or may be an incidental effect generated by excitement. The buck makes no attempt to make himself look taller by raising his hackles, as the doe does.

Plates 45 and 46. The approach of a territorial buck (Y7) by a bachelor buck; note how the adult's horns are inclined towards the bachelor.

Plates 47 and 48. Another approach to Y7 by a bachelor buck. Note the typical broadside display of the adult and the erect penis.

Unlike some animals with a gular patch, the white gorget of the waterbuck seems to have no part in display. It is no more evident than normal in the lateral-present, and indeed is obscured when the horns are tilted towards the opponent. Unlike in the mountain goat, for example (Geist, 1971), horn size is not a mark of rank, and the presentation of the horns is probably not intimidatory. In the waterbuck it is the ability to use them which counts, and all of those animals which I saw with very large horns were weak bucks, timid in conflict situations. Indeed, as I have explained, record length antelope horns often result from lack of aggression in the owner, who does not gore the earth and wear down the tips, and who avoids fighting. The same is true of the African buffalo, in which species the most aggressive bulls often have their horns worn down to stumps.

Whenever the older members of a bachelor group, that is those from about 3 to 4 years of age, saw a territorial buck nearby, they could seemingly not resist approaching him and investigating his reactions. To begin with, this was by no means bold. A young buck would approach cautiously, at first standing and sniffing from a distance, and then advance, with neck held low and outstretched in the submissive posture, champing in the same manner as a doe. The adult would adopt his display stance. This sequence was usual with any approaching buck, and if the latter was a young one, then this was usually as far as it went. The young buck, as if mesmerized, seemed compelled to approach until he could sniff the other at close quarters.

Elliott (1976) considered that the young buck tried to sniff the central area of the adult's forehead, and would leap away suddenly as soon as it scented the buck's odour at this spot. This may, of course, be the primitive behaviour lost in the defassa waterbuck, as may also be the alleged concentration of apocrine glands. In my observations the young buck merely sniffed noses and the side of the adult buck, as is clearly shown in Plate 46. He then sometimes engaged horns, leaping away as soon as he felt the strength of the adult. Then he sniffed the latter's flank, and was particularly attracted towards the region of the penis. The adult, for his part, remained motionless, or walked away. He gave every appearance of being at a loss as to what to do, apparently inhibited towards showing aggression. This passivity often emboldened the young buck to start horning him in the rump, and, as the young one got more excited, in the side. I have watched Y7 stand, looking completely discomfited, while a young buck with 45 cm horns jabbed away at his abdomen. But he needed only to turn his horns towards the youngster and the latter leaped away like lightning. Often such displays ended simply with the adult starting at a particularly painful prod, causing the youngster to jump away and run off.

This behaviour on the part of the young buck could be explained as "dominance testing"; an inherent urge to test the adult's superiority. For his part, the adult is inhibited against attacking the young buck, and after having given his display to warn him away, the adult lacks any further behaviour to cope with the situation if the younger buck is not warned off. Since adult bucks do not appear to be inhibited towards attacking juveniles or even does when out of this display context, it could be that we are witnessing here an evolutionary step in behavioural development; or conversely, a relict behaviour, in which the young buck's overtures suppress antagonism in the adult, who is thus unable to respond to the former's aggression. If this is an evolutionary step we might suppose that an inhibition on the part of adult bucks to attack young ones is in a stage of development in relation to behavioural responses. But if the converse, we could postulate that the adult's behaviour is the only remaining inhibited reaction.

Y7, for example, seemed to dislike a particular 18-month-old buck who was, at that time, the youngest member of the bachelor group on the Peninsula. All would be grazing peacefully, when suddenly Y7 would stop and fix an unfriendly eye on the youngster. The latter, aware of the studied gaze, would start to shift nervously. Suddenly, Y7 would charge. It was a determined charge too, but Y7 never succeeded in catching up with the youngster who was fleeing for his life. After a chase of perhaps 50 or 100 m Y7 would give up and wander back. Keeping a wary eye on the bully, the young buck would do likewise, drawn back to the bachelor group. How could the same animal stand meekly by and let himself be horned in the abdomen by a young buck? This selective lack of inhibition seemed to me rather inexplicable, unless one suggests that much more complicated patterns of behaviour exist than simple ritualistic interactions; patterns based on real likes and dislikes.

As far as I could determine, display behaviour was restricted to the adult, and did not characterize the sparring matches of the bachelors. These had no apparent preliminaries, one young buck would simply walk up to another and touch noses. This was followed by both bending their heads down so that first their foreheads, and then the horn bases, touched. After this it became a pushing contest, one merely trying to push the other backwards, and thereby show himself to be the stronger.

In their exertions they may fall to their "knees", but this seemed to be an incidental result of trying to maintain position, and not a predictable or "ritual" stance. During pauses, confronting one another head down, the contestants sometimes wagged their heads; an action that I never saw in the adult.

I seldom witnessed determined contests between adults. I saw only two real fights during my entire study. This is not because they do not often take place, but rather that an observer cannot be in all places at once to witness them. Just about every territorial buck has tears and nicks in his ears, as well as scars on his neck, testifying to the frequency of such combat.

The neck scars provide the clue to the stance in the lateral display. When the buck presents his lateral view before another buck, he is in fact presenting his most vulnerable aspect. But I believe that this is of no behavioural significance, and is incidental to the object of the display, which is to show the size of the neck; in other words, the animal's shield. The skin of the neck is up to 2 cm thick, and used to be prized by the Boers of South Africa for making boots; it is on this that the buck receives the rapier-thrusts of his opponent if he is not quick enough to parry them with his own horns.

As I have demonstrated (Chapter 4), unlike many antelopes in which chest girth and hind leg weight have been shown to vary in proportion to the animal's total body weight, this relationship does not hold good for the waterbuck. This is because, as the animal becomes adult, chest girth and hind leg weight do not keep pace with the weight that is being laid down in the neck. The latter is related to its defensive thickening, and the increasing size of the muscles required to support the head with its increasing horn weight. The horns and the neck are the waterbuck's sword and shield. I did not measure the thickness of the neck shield systematically, but this has been done for the impala, which has a similar protection (Jarman, 1972).

Would the waterbuck have such weapons of offence and defence if it did not use them? Of course not. Let there be no mistake about the violence of the adults' fighting when two opponents are well-matched. There is no covert ritualism about the thrust and parry. The techniques that they employ are the same as those of any boxer or fencer. With nearly 270 kg backing up the thrust behind the needle-sharp horns, the contestants are only unsuccessful in dealing a mortal blow because the other prevents it. There is no lack of intent. Geist (Geist, 1971) has explained the analogy with fencing and adds: "The reason why bloodshed is rare in severe intraspecific combat is not that the opponents disdain it, but that they usually protect themselves with a variety of defences."

Not much damage is done during fighting if one waterbuck horns his opponent in the shield, so both try to twist to one side in order to lunge at the opponent's vulnerable thorax or abdomen. When this succeeds the result is likely to be fatal.

Licking their lips vigorously and violently wagging their tails, penises erect in their excitement, the contestants strive head to head. Each tries to pull his head back and to lunge forward, at the same time preventing the other from doing likewise. The object is not, at least in this species, to lock horns in ritualistic combat as Walther (Walther, 1966) has suggested; horn locking is a consequence of the thrust and parry. Horn bases rubbed raw in the violence of their clashes, the contestants scrabble backwards and forwards. Now falling to their knees in a desperate effort to maintain positions, snouts covered in dust, their laboured bleats mingle with the clash of horns and the thud of hooves. Such violence is terminated only by the mutual tiring of the combatants; and there is no follow-up when both give in, each going his own way (Plates 49 and 50).

Geist observed that when two mountain rams were in combat another would sometimes join in against one of them. I observed the same thing in waterbuck. On one occasion, when Y8 and Y11 were fighting, Y7, who was at that time the oldest member of the bachelor group, stood watching (Plate 51). Waiting until the contestants were locked in combat, he suddenly charged Y8 in the side, knocking the wind out of him. Fortunately for Y8 he had used the flat of his horns. The attack was not repeated, and Y8 was far too occupied to do more than stagger from the blow. Eventually he won the bout, but Y7's action could well have been his death warrant.

Observations showed that aggression in the waterbuck was primarily directed at the defence of territory, with direct combat avoided, where territorial possession was not in dispute, by stereotyped behaviour patterns common to many ungulates of widely separated phylogeny. Geist (Geist, 1965) has listed five basic agonistic patterns which serve to avert direct combat, all of which can be attributable to waterbuck:

  (i)   the rush-threat, galloping at an intruder or opponent;
  (ii) weapon-threat, lowering or inclining the horns towards another;
  (iii) present-threat or proud-posture, which I have called "displaying", as I believe that this more accurately describes the activity;
  (iv) broadside-present, usually a preliminary to the weapon-threat, but which in the waterbuck seems to be indistinguishable from displaying;
  (v)   Scar-threat, described as the behaviour of a sub-dominant in an attempt to intimidate a dominant and which may comprise horning bushes, earth, etc.

Violent combats between neighbours, such as those witnessed between Y8 and Y11, are probably related to attempts by one of the combatants to move up in the territorial hierarchy; for, as I have shown, the largest territories are held by the bucks in their eighth to ninth year, when they are in the prime of life. Although territorial bucks defend their living areas against other bucks of territorial age, there exists, as I have indicated, a certain flexibility in the behaviour pattern, characterized by the represssion of aggression.

Plates 49 and 50. A fight between two territory owners.

Plate 51. A bachelor watching the fight, preparatory to attacking one of the contestants.

Repression of Aggression

Peninsula territories all had easy access to water, even those to which old bucks were forced to retire. In the Ogsa, by contrast, although one old buck held a waterside territory until his death, another was isolated in the centre of the area (Y28). To obtain his daily drink he had to pass through the territories of one or two other bucks. I found that he was permitted to do this, relatively unhindered, as long as he maintained a sub-dominant attitude towards the owners through whose territories he passed. This consisted of walking placidly with head held low, although not low in the extreme sub-dominant attitude, and taking no notice of the other bucks. At Kayanja, where extremely high densities of waterbuck occurred, this situation was developed further.

We have seen how in one small area, inland from the lake, there were four territories owned by young bucks of 6 to 7 years of age (Chapter 10), which were probably forced together by pressure from surrounding older territory owners. These surrounding territories were large, but in contrast to the Peninsula and Ogsa there was water on one side only. From the structure of territoriality in riverine waterbuck, we would have expected the lake shore to be divided up into territories, but this did not seem to be the case. Instead, there appeared to be a neutral zone adjacent to the water, where no agonistic displays, other than mild ones, took place. In the wet season, the four young bucks could drink water standing in hollows, but in the dry season they had to travel approximately 1·5 km to the lakeside each day; and probably other inland territorial bucks did likewise.

A typical day in the life of one of the 7-year-olds was recorded on 30 March 1967, in the dry season. In the morning the buck, R33, was found grazing and ruminating in his territory. At about 1030 h he moved to the boundary with his neighbour "X". Upon seeing him there, "X" moved up and R33 retreated a few paces, then both displayed. R33 then sat down and "X" walked away. R33 waited until the other buck was some distance away, and then got up and walked unchallenged through his territory. Further on he saw two more adult bucks, took a few paces back, made a little detour to avoid them, and continued towards the water, reaching a pool near to the lake edge at 1110 h. Here he drank, and then went on to graze at the lake edge on the *Panicum repens* and *Cynodon dactylon* which covered the sandy beach.

At 1515 h he returned, making a short detour again to avoid the two bucks, but otherwise following almost the same path. The buck "X" came to meet him when he reached the latter's territory, but this time R33 simply ignored him, and continued straight on to his own, reaching it at about 1600 h.

"X" probably used a seasonal waterhole close to Kabahango, which lay inland from the lake at the edge of a stand of *Acacia sieberiana* trees; this appeared to be a favourite drinking place among waterbuck because of the shade it offered. I spent several occasions watching this waterhole one year before it dried up. The visitors were all waterbuck from the adjoining regions, with the occasional warthog and bushbuck. During the morning, the waterbuck came to the pool to drink at staggered intervals, probably related to the distance that they had travelled, so that the place was never too crowded. An adult buck with a group of does would move into the trees, and 5 min later about 30 bachelors and does would appear from the trees to take their place, and so on. Minor displays took place among the bucks around the waterhole, but no fights occurred, and no buck was ever prevented by another from drinking. This produced the interesting situation of as many as eight adult bucks being present all at once within a few metres of each other, their aggression being temporarily repressed.

Typical sequences were recorded in my field diary for 1 and 2 January 1967:

"1.1.67.
1010 h: An adult buck sitting near the waterhole rises and moves into the trees, some does following. Another adult buck is sitting among the trees.

1015 h: About 30 bachelors and does appear. Two adult bucks are grazing.

1030 h: The buck that has been sitting in the trees gets up and slowly approaches. He watches from about 30 m and then grazes. The first buck rises when a young adult buck comes too close, but then walks away past the sitting buck, and grazes.

1125 h: Two adult bucks appear. One passes close by the first adult buck, now standing, but neither display.

1145 h: The buck amongst the trees approaches another who displays, but he continues past him. The latter then displays to another buck near some does.

1155 h: The buck in the trees comes over, a young adult buck moves away. The former investigates a doe, then grazes.

1230 h: A young adult buck arrives.

1300 h: There are six adult bucks in the area.

2.1.67.

0930 h: One adult buck, two young bucks, a few does and young, are present at the waterhole.

0945 h: A group of young bucks and does, with one young adult buck, come running in. The first buck approaches the young adult buck, who stops and then walks around the older one.

1000 h: An adult buck appears and sits among the trees. Groups wander away. Another adult buck seen to be sitting among the trees.

1015 h: A group of does and young appears.

1025 h: An adult buck appears.

1028 h: Another adult buck appears, and there are now five present, all spaced apart from one another.

1040 h: Three of the adult bucks are within 10 m of each other, one sitting. There is no displaying.

1050 h: A group of does with young, accompanied by two young adult bucks and three adult bucks, appears. One buck stands facing them while one of the approaching bucks walks straight past. Another champs and attempts to sniff the confronting buck, which displays. The former continues to water with the rest ignoring the standing buck.

1130 h: One young adult buck and several young bucks drinking. The rest are sitting among the trees. Eight adult bucks are now present."

This picture of mutual tolerance is very unlike the territorial system shown elsewhere in the Park, and exemplifies the flexibility of behaviour patterns in the higher mammals in response to changing circumstance. Whereas in the Uganda kob, as we shall see later, increased density leads to accentuated territorial behaviour which becomes isolated from routine behaviour; the waterbuck, in contrast, exhibits an attenuated territorial drive. It is not difficult to postulate why. At Kayanja, near to the lake, the density of waterbuck was 17·8/km², the highest recorded in the Park. Although this high density was mostly made up by does, the density of bucks was also the highest recorded in the Park with 3·79/km². This was only marginally different from the Peninsula density of 3·75/km², and probably not significantly different. But at Kayanja the habitable area was large in comparison with the water frontage, and any buck who tried to defend a waterside territory against incursions by other bucks, would find himself continually harassed. Death from such encounters would no doubt pose a threat to the continued existence of the population. I do not suggest that there is some innate behavioural response; I think that the bucks soon learn that it is simply not worthwhile trying to defend waterside territories. Thus under these conditions of high density the buck defends a place in which to live, but has learnt to share the essential commodity, water. In this way the pressure of competition is reduced. This is only different in degree from the territorial behaviour of the Uganda kob, which at high densities increases its territorial behaviour in small areas set aside for this purpose; but, like the waterbuck, suppresses its territorial drive when it leaves these areas to feed and to drink. The waterbuck, on the other hand, shows no apparent increase in territorial drive at high density when in its territory; its behaviour being marked only by the suppression of territorial drive when outside its territory.

12. Territorial Concepts and Function

Introduction

The problems of definition, associated with the concepts of territoriality and home range in mammals, have been discussed by a number of authors, among them Jewell (1966), and, more recently, Owen-Smith (1977); the latter author also providing a detailed review of the components of territorial behaviour. Based upon bird studies, Noble (1930) provided the first definition of a territory, considering it to be any defended area. This was enlarged upon by Nice (1941) who implied that attachment to the area was an essential component, as well as the defence of the area. Jewell circumvented the problem by considering a territory to be an area whose boundaries have some real significance for the owners; as probably also do the boundaries of a home range. But the latter is not defended, whereas I see the main distinction of a territory as being, first and foremost, that it is defended by its owner against conspecifics of competitive potential. A conspecific of competitive potential is one who is equally capable of maintaining a territory, and thus presents a threat to the established territory owner. Owen-Smith considers that it is the spatial location which primarily distinguishes territory, because if an intruder behaves submissively the owner's aggression is not released. But if an intruder behaves submissively then it puts itself in the role of a non-competitor; this submission is still subject to limitations, that is, it can usually only be of a temporary nature.

Any definition is likely to have many qualifications, and the one proposed above is no exception. In Zambia, for example, Hanks *et al.* (1969) recorded that during the breeding season the waterbuck territory owner defended his territory against bachelors, who could not be considered as competitors, except in the role of breeding. These variations, however, do not detract from the basic concept.

Territorial behaviour was first defined in birds and given significance by Howard (1920) in his book "Territory in Bird Life". But as Hutchinson (1978) has pointed out, there were several references and descriptions before this.

In the third century B.C., Zenodotus related a proverb that one bush did not shelter two robins, and thus, as Lack (1946) informs us, made the first unwitting comment on territorial behaviour. Other comments on birds, from Aristotle onwards, can also be seen as references to territorial behaviour; but as a phenomenon it had to wait until the middle of the last century to be recognized, when Altum (1868) gave an account of the role of defended territory in bird life.

Some time before territorial behaviour was thus described in birds, the first description of its occurrence in an African antelope appears to have been made by Burchell (1823), who described the solitary habit of the black wildebeest, and its territorial advertising, in South Africa. Later, Cumming (1850) described in the same species, a "cunning old bull wildebeest", which kept others out of the valley in which it was. But Cumming thought that the animal was warning others away from him, and not from itself. Neither of these writers were aware of the social significance of what they saw, and we had to wait almost a century for territorial behaviour to be related to a social context in a large mammal, when Fraser Darling described rutting in the Scottish red deer (Darling, 1937); although the red deer is not territorial as we understand the definition.

By 1958 it had come to be recognized in a range of mammals comprising rodents, seals, deer and primates (Carpenter, 1958). The first studies of ungulates, commencing in the early 1950s, began with roe deer and red deer, but were soon to include African ungulates, from whence the most informative information on mammalian territoriality has subsequently come.

Examples of Ungulate Territorial Organization

Uganda kob

Interest in the territorial behaviour of African ungulates seems to have been awakened by Buechner's discovery in 1957 of the "lek-type" of intensive territorial behaviour shown by the Uganda kob (Buechner, 1961); although territorial occupation had already been suggested to occur in the waterbuck (Bourlière, 1955).

The term "lek-type" describes the behaviour of the Scottish grouse, in which, prior to the breeding season, the cock birds gather together in one area and partition it among themselves. Each owner defends a patch of

Plate 52. Uganda kob fighting in the territorial arena. Note the proximity of the other territory holders. Kikorongo, Rwenzori Park.

this area, and by his posturing and display attracts the hens to himself for mating. But here the analogy ends, for once a cock is chosen by a visiting hen, pairing takes place and the hen remains and nests in the territory. Some cocks take two hens. Those birds which do not succeed in obtaining a place within the arena, and in breeding, do not survive (Watson, 1970). It is thus only the intensity of territorial behaviour in a small area, accompanied by the male's posturing, which can be equated with the behaviour of the Uganda kob.

Territorial grounds appear to occur wherever there are high densities of kob. In the Semliki Valley of northern Uganda, where Buechner first came across the phenomenon, there was an estimated population of 15 000 kob in 400 km², representing an overall density of 37·5/km², distributed among 13 well-defined territorial grounds. In the Rwenzori Park at least ten such arenas have been found where the average density of kob is 36·7/km², in the range of 20 to 52·1/km² (Modha and Eltringham, 1976). But whereas in the restricted habitat of the Semliki Valley, bounded by the Semliki river on one side and an escarpment on the other, there was one such territorial ground per 30·8 km², in the Rwenzori Park the density was only one per 146 km². Hence localized density seems to be the factor determining the establishment of such arenas. The phenomenon is of universal occurrence in the species; at least one territorial ground exists in the Saint Floris National Park in the north of the Central African Republic (latitude 9°40′N, longitude 21°20′E), and apparent territorial grounds have been reported for the Niokolo-Koba National Park of Senegal (Montfort, 1974) (Plate 52).

In studies of the Semliki focus (Leuthold, 1966), it was found that a territorial ground, or arena, comprised 10 to 20 central territories of 15 to 35 m diameter, with contiguous boundaries. These are nothing more than stamping grounds, or rutting grounds, to which the does come to breed. Although in this region the territories were considered to be roughly circular in shape, Floody and Arnold (1975), who conducted a study in the Katwe area of the Rwenzori Park, found the territories there to be extremely irregular in shape; but these authors did use a very precise method of determination of boundary limits. Adverse conditions, such as drought, can disrupt the territorial organization causing a temporary abandonment; but proximity to water is considered to be the least important requirement of a territorial ground site. Although some changes have been observed, with old areas being abandoned in favour of new ones, the fact that some have been known to exist for at least 30 years indicates their basic stability.

A doe enters the territorial ground solely for the purpose of mating, and consequently is usually in oestrus, and runs the gauntlet of the displaying bucks until one is permitted to copulate. Does appear to seek the most central territories for mating, which as a result are those most preferred by bucks, and have the greatest owner–occupant turnover. What motivates the doe in this manner is not known; it may simply be that rising excitement brings her to the peak of her desire approximately in the middle of such an area. As she passes each territorial owner the latter displays, prancing towards her in characteristic posture, stiff-legged and with nose pointed into the air; but he abruptly loses interest if the doe crosses his boundary into the next territory. After copulation the doe may leave the territory immediately; but if she remains the buck indulges in what has been termed a "postcoital display" (Buechner and Schloeth, 1965). Immediately after dismounting the buck remains standing with arched back and erect penis for 2–5 s. He may then whistle loudly several times, being answered in chorus by other bucks, or he may lick his penis. The doe, meanwhile, remains in the receptive attitude with arched back and hind legs straddled, and with the tail raised. The buck may now lick the vulva, udder, or inguinal area, and he may follow this up with the foreleg kicking movement, copulating again 10 to 15 min later. The sequence terminates when the buck retracts his penis, and one, or both of the pair, may then lie down or graze. If another doe enters the area the buck may turn his attention to her, so that this "postcoital display" seems to be nothing more than a continuous mating activity, brought about by the high pitch of sexual arousal resulting from the "lek-type" activity, and is not a specialized ritualistic display.

Breeding takes place throughout the year at the Uganda latitudes, so that the grounds are always active, even at night (Floody and Arnold, 1975). As one might expect of such an intensive activity, changes of territory ownership are very frequent. It was estimated by Leuthold (1966) that only one-third of the adults bucks used the territorial grounds at any one time, the remainder occupying more conventional territories of larger size. These, numbering ten to 20, lie adjacent to the arena and in the Semliki Valley were about 100 to 200 m in diameter, but were apparently not stable in numbers or in size. Some owners left these territories for several hours during the day, mostly to feed near the river when food was scarce. Frequently they were abandoned entirely in the dry season. No particular type of buck appeared to occupy them, and it was unusual to find a buck occupying both one of these and a place in the arena, so that they could not be considered as resting areas for arena bucks. The age distribution of the occupiers appeared to be the same as that of the arena owners, and young bucks might compete directly for a place in the arena without first occupying a peripheral territory. I suggest that they may, however, represent a step towards the occupation of an arena territory, harbouring bucks which have not yet attained the high level of sexual activity demanded of the arena.

A highly variable degree of tolerance was shown to sub-adult bucks by the territory owners, ranging from completely ignoring them to vigorous attacks and chasing. Single bucks, or small groups of bucks of adult age, were generally driven out; but if bigger herds invaded, territorial defence became impossible, and the owner then participated in the general activities of the herd until it left. Physical combat in defence of the territory was alleged to be seldom observed, being almost entirely confined to cases in which two bucks were actively competing for the same territory. But I found it frequent enough on the Kikorongo grounds in the Rwenzori Park to be able to obtain photographs with relative ease. Otherwise, mutual respect of territorial rights was generally observed.

No olfactory marking of territory was observed by Leuthold (1966), and defaecation and urination were at random. Despite Pocock's (Pocock, 1910) assertion to the contrary, the kob does not appear to have any facial odoriferous glands, and it also lacks interdigital glands. But it does possess inguinal pouches, although these are probably related to a sexual function unrelated to territorial marking.

Leuthold estimated that only approximately 5% of the adult buck population was engaged in territorial activity at any one time. The rest of the population was confined to bachelor groups which existed on the fringe of the territorial area. Unlike the waterbuck bachelor groups, these groups might contain young bucks, deposed peripheral territory bucks and deposed arena bucks. Doe herds in these areas numbered from two to three, to groups of more than 1000, but groups of 30 to 50 were the most common. Does used the peripheral territories, but rarely mated in them. This activity was reserved almost exclusively for the arena.

We can see that the kob has a very complex form of territorial behaviour, whose functions were summarized by Buechner (1963) as being fourfold: (i) stimulation to breed, (ii) maintenance of high natality to compensate for high mortality imposed by predation, (iii) ensurance of the best genetic combinations, and (iv) prevention of overuse of the forage supply by spacing out of the herds. But the type of intensive arena activity shown by the kob seems, on the contrary, to contradict all of Buechner's conclusions. Let us examine them in turn.

(i) *Stimulation to breed.* One might first ask why the kob should require more stimulation to breed than other antelopes. But, this question apart, I see the form of rutting ground activity, as shown by the kob, more likely to have a depressive effect upon reproduction, than a stimulatory one. By isolating the breeding bucks at one focus a complete reversal of the normal doe role must now take place. The doe must actively seek the buck for copulation, instead of vice versa.

As with most ungulates, oestrus is brief, lasting some 24 h (Morrison and Buechner, 1971). Thus, if a doe does not move to a territorial ground before coming into oestrus, the latter is likely to have waned before mating can be effected. It may therefore be that only after the doe has experienced one or more oestrus periods does she make an attempt to actively seek a buck and is drawn to the arena. Even if this only occurred in the virgin doe it would exert a significant effect upon the rate of population recruitment. This hypothesis seems to be substantiated by the findings of Morrison and Buechner (1971), who relate that the *post partum* doe commonly has up to four or five oestrus cycles before conception, with ovulation repeated at least one or more times.

Morrison (1971) stated that, in the kob, characteristics of reproduction were found "that are not commonly attributed to wild artiodactyls". But his assertion that this includes lactating does, "often nursing extremely young fawns", being pregnant or in oestrus, is no different to that found in the waterbuck; nor indeed in the Grant's gazelle (Spinage, unpublished), and is probably the norm among tropical antelopes. However, the kob does apparently differ from those other ungulates studied so far, in that oestrus intervals were as little as 6 to 13 days, compared with an expected 21 days. Also non-pregnant does in general contained several *corpora lutea* in their ovaries at once, in various stages of development. This is suggestive of several unsuccessful cycles, either with, or without, ovulation.

In the kob ovulation can apparently take place at 10 days *post partum*, compared with about 21 days in the waterbuck. At this time, as in the waterbuck, the average doe will be nursing a calf (of 10 to 21 days of age), which she must abandon in order to seek the territorial ground. Comparing this with the behaviour of the waterbuck doe, who remains tied to the vicinity of her calf at this time, it would appear that the kob's behaviour could put the calf at greater risk, and thus pose another contrary indication to the hypothesis that the system furthers reproductive success.

Kob does were observed to visit the arenas at as little as 5-day intervals, but without copulating, returning again on the sixth day to copulate. Thus they would seem in general to be motivated to arrive at the arena before oestrus takes place; but this was not observed in all cases, and intervals between visits ranged up to 29 days.

The apparently short di-oestrus period could serve to over-ride to some extent the negative effect of the lengthening of the calving interval, which inability to arrive on the territorial ground in time could entail. Although some does were considered to be capable of conception as early as 10 days *post partum*, seven were observed to conceive at "less than 65 days" *post partum*, which indicates a fairly long interval compared with the waterbuck's average of 45 days, for example. The evidence produced by Morrison and co-workers suggests that the doe's reproductive physiology is linked in some manner to the territorial ground system. This may serve to counteract, to a limited extent, the inherent disadvantages to reproductive efficiency which the system of buck isolation would seem to impose, and not, to my mind, to enhance reproductive efficiency in a general manner as other workers have supposed.

(ii) *Maintenance of high natality to compensate for high mortality imposed by predation.* Buechner (1963) observed large numbers of skulls on the territorial grounds, and thus concluded that the bucks were subject to high predation by lions while occupied with their territorial activities. No evidence has been brought forward to substantiate this, and indeed Leuthold (1966) states that the bucks move off their territories when lions are present. I believe it more reasonable to assume that the remains are the result of intraspecific fighting. I know of at least one case in which a buck was killed on a territorial ground, and do not doubt that its occurrence is frequent.

There is no evidence to suggest that natality is higher in the kob than it is for any other continuously breeding tropical ungulate of similar gestation length.

(iii) *Ensurance of the best genetic combinations.* It seems likely that the territorial ground bucks are representative of the most vigorous sector of the buck population, although even this is in doubt until we can determine what motivates the difference between peripheral and arena territory bucks. But this being so, it is doubtful whether the restriction of breeding to less than 2% of the male population is genetically advantageous. Obviously a buck carrying some genetic abnormality not affecting the temporary vigour required to occupy a territorial ground, could readily pass it into the population. More recently, Buechner and Roth (1974) have offered evidence to show that the territorial grounds represent relatively closed breeding units, the majority of bucks and does returning always to the same arena to breed. Wright (1931, 1960) hypothesized that closed breeding units were a feature of natural populations which provided a structure favourable to rapid evolution when occasional outbreeding occurred. Buechner and Roth considered that the territorial ground system might be important in this respect by maintaining heterogeneity in the kob, which lives in geographically isolated populations. The latter is a feature, however, of many African antelope species, which do not show such intensive territorial organization. It seems doubtful to me whether there is more inbreeding in a population which clumps in one part of its range, than in one which is more evenly distributed. The kob's system may well promote inbreeding, but not more so than some other systems.

(iv) *Prevention of overuse of the food supply by spacing out of the herds.* I do not find this a tenable postulate. The carrying capacity of an area may be affected in the long term by the populations concentrating on small parts of the area at a time; but this rarely occurs among ungulate populations under natural conditions. Centralizing the buck activity would, in fact, appear to contradict Buechner's hypothesis, for it is said that the bucks have to seek forage elsewhere in the dry season (Leuthold, 1966). This is not the case, for example, in the type of dispersed territoriality of the waterbuck.

I have dealt with these hypotheses at some length because kob territoriality has attracted considerable interest, and obviously repres-ents an important variant which is central to the understanding of territorial behaviour. But having dismissed Buechner's postulates one must ask to what, then, can we attribute the kob's extreme form of territorial behaviour?

I believe it to be simply an inherited response to density, of which the effects are incidental, whether they serve to enhance or to impair the health of the population. That they do, on balance, enhance the well-being of the population, can be inferred from the persistence of the response. If they did not, then the population would decline to a lower density and the response would change to a less intensive type of territorial behaviour. Its universal occurrence, however, suggests its genetic basis.

How does this response come about? I suggest that it results from the fact that an animal cannot maintain a territory without advertising it. If it has no scent glands with which to mark it (whether the scent is deposited on faeces and in urine or as an odoriferous secretion), then how can it be done? The kob does it by the relatively extraordinary method among ungulates (although also used by puku and reedbuck) of whistling.

But sound is a very imprecise manner of demarcating a boundary. According to Morrison and Buechner (1971) the "chorus" from an arena may be heard up to a kilometre distant, but the isolated whistle of a kob neither carries very far, nor does it tell where the territorial boundary is. It can only serve to announce that another kob is in the vicinity. Such a method can thus only lead to continuous boundary infringement, and Leuthold (1966) has pointed out that the boundaries of peripheral territories are not stable. Indeed how could they be unless they were marked in some way?

At high density, therefore, there must be continuous invasion of larger-sized territories, which as we have seen takes place with peripheral territories. Thus, it would seem, the answer at high density, to maintain negative association of breeding bucks, is to crowd together so that each territorial buck can be seen and heard in his area; and to inhibit the territorial response when feeding and drinking.

This crowding together, and its resultant negative association between the bucks, has the effect of lessening interference during mating which would otherwise occur at such high densities. We have seen how the peripheral territory owner does not attempt to defend his territory against a large group of invading bucks, groups which are large by virtue of the high population density. So how could he be expected to entertain an oestrus doe in the face of such competition? Clearly also the does would be unbearably harassed if there was not some sort of order imposed upon the bucks' behaviour. Being a continuous breeder, only a limited number of does come into oestrus at any one time. In a seasonal breeder the same situation does not arise, for there are many oestrus does available at the same time, dispersing the bucks' attentions. The voluntary restriction of the kob buck to an area where he can be both seen and heard by the other bucks, and of which the limits of the area can be readily seen by each other buck, ensures that order is imposed upon the bucks' behaviour.

That this crowding together may tend to have a depressive effect upon reproductive efficiency I see as an incidental effect, and not one that needs to be explained in terms of Professor Wynne-Edwards' theory of epideictic display (Wynne-Edwards, 1962), as selected to limit population growth. It appears to be partly counteracted by a form of hypersexuality in the doe, which, as Morrison and Buechner (1971) suggest, may result from increased sexual stimulation related to the intense sexual activity of the bucks on territorial grounds; but which I believe is nonetheless also an incidental effect, which would not be found among does at relatively low densities.

Thus the territorial ground system of the kob is seen as simply resulting from the method of territorial advertising, whose imprecise nature results, at high density, in the crowding together of the bucks in order that they may be both seen and heard by one another.

Lechwe

The waterbuck's other congeneric species, the lechwes and the puku, have yet to benefit from such behaviour studies, with the exception of Schuster's studies on the Kafue Flats lechwe (Schuster, 1976, 1976a). There are two species of lechwe. Firstly there is the Nile lechwe, also known as Mrs Gray's lechwe, *Kobus megaceros* (Fitzinger, 1855), which occurs in the vicinity of the Nile and the Sobat rivers in the southern Sudan, and in parts of western Ethiopia. Little is known of its biology, except that it occurs in large numbers. Secondly, there is the lechwe, *Kobus leche* Gray, 1850, of which four races are recognized: the red lechwe *Kobus leche leche* Gray, 1850; the black lechwe *K. l. smithemani* Lydekker, 1900; Robert's lechwe *K. l. robertsi* Rothschild, 1907; and the Kafue Flats lechwe *K. l. kafuensis* Haltenorth, 1963 (Ansell and Banfield, 1979) All of these races occur in Zambia; while the red lechwe extends into Zaïre in the north, and Botswana and South Africa in the south. Unfortunately, in the past 40 to 50 years all but the Kafue Flats lechwe have been drastically reduced to a fraction of their original numbers, and the latter is now threatened by a hydro-electric scheme.

An interesting species, the lechwe is characterized by extreme gregariousness, resulting in herds of over 3000 animals, confined to seasonally-inundated floodplain areas. Its physiology has not been studied, so that we know nothing of its water requirements, but in its behaviour it is much more adapted to water than is the waterbuck. An adept swimmer, it spends a great part of its time wading, although not usually in levels above the belly. It derives most of its food supply from floodplain grasses such as *Acroceras, Leersia, Echinochloa, Paspalum, Sacciolepis* and *Oryza* (Grimsdell and Bell, 1975). It appears to be unable to survive on savanna grasslands, which may be attributable to a high protein demand. Unlike its congener, the waterbuck, it does not meet the protein deficiency in savanna grasslands by browsing in the dry season.

In Zambia, De Vos and Dowsett (1966) presented a brief comparative study of some aspects of the behaviour of the three congeners: waterbuck, lechwe and puku; but their observations were too meagre to substantiate any conclusions concerning similarities or disparities in behaviour. These authors appear to have been the first to report a relatively weak form of territorial behaviour among the Kafue lechwe during the rut. Schuster (1976) found, however, contrary to De Vos and Dowsett, that territorial behaviour was very pronounced at the height of the breeding season, in December to January, some of the adult bucks forming a "lek-type" of territorial organization, similar to that of the kob. Schuster confirmed Robinette and Child's (1964) finding that breeding takes place throughout the year, but with the addition of a peak in activity from mid-November to mid-February. This is at the height of the annual flood regime, when the lechwes are forced into a much closer association by the restriction of their habitat. At this time, relatively small proportions of the adult bucks organize themselves into small concentrations in conspicuous areas. Between 50 and 100 such bucks are found within nearly circular areas of about half a kilometre in diameter, bucks being spaced about 15 m apart at the centre of the arena, with a greater spacing towards the periphery. Within this arena there is an almost continual occurrence of intra-male interactions, the bucks displaying with head held high, prancing, tail-wagging and head-shaking, forming a highly ritualized interaction consisting mostly of threats and chasing with little physical contact taking place. Although fights, when they do occur, are followed by an elaborate "victory" ceremony on the part of the victor. Such territorial behaviour was not observed outside of the main breeding season, and Schuster concludes that a large number of bucks rutting at the same time seems to be the inductive factor.

The does participate in this by coming to the rutting grounds to breed, where they seek out the bucks in the centre of the arena, clustering around them in tight groups of 10 to 20 animals. Although pairing thus has the appearance of being very selective, the central territories probably have the highest turnover rate of occupation, so that many more bucks take part in pairing than would appear to be the case at first sight. Little pairing takes place, however, away from the arenas.

Lechwe arenas are unstable in position due to the flood regime, appearing to consist of a series of temporary locations which are shifted in the face of the rising water level. The territories thus bear no distinctive marks, being indicated only by the behavioural displays of their occupants. Lechwes lack conglobate scent glands, except for rudimentary inguinal pouches, and they do not vocalize. We can see that the "lek-type" of organization meets the explanation advanced for the kob — the closer together that the defenders are, the easier it is for them to see what the others are defending. Their naturally massed concentrations would mean that whistling, as in the kob, would come from so many quarters at once that its audiolocation would be impossible; for here we are dealing with 50 to 100 bucks on one arena, compared with 10 to 20 in the kob.

In its undisturbed state this species must clearly achieve densities at the limits of social compression; but this seems to result in a reduction rather than in an intensification of territorial behaviour, the latter being limited to a short part of the season. Of course, we are considering the behaviour of almost relic populations in most cases, and were their numbers much closer to those of their original undisturbed state, territorial behaviour might be seen to be more common. But a relative lack of territorial behaviour in the species is probably attributable, firstly, to its ecology. Living in a continually changing habitat the bucks are forced to be continually moving and in close contact with one another. This leads to a reduction in territorial initiative as the resource is ephemeral in space. Also at the high densities experienced by this species, defensive encounters would be so frequent as to soon lead to the exhaustion of the defenders. Thus this activity is restricted to only a short part of the year.

It is unfortunate that lechwe populations have been so severely altered by destruction. Although Grimsdell and Bell (1975) found densities of only $7 \cdot 9/km^2$ and $13 \cdot 8/km^2$ for the black lechwe, in wet and dry seasons respectively, it seems likely that these densities should be many times greater in the undisturbed population. At a place called Lochinvar Ranch, the Kafue Flats lechwe reaches a density of well over $500/km^2$ in the flood season (Sayer and van Lavieren, 1975). The sex ratios for these two populations are estimated at $1 : 1 \cdot 2$ and $1 : 1$, bucks to does, for the black lechwe and the Kafue Flats lechwe respectively. This gives us densities of bucks during the rut of $6 \cdot 3/km^2$ and $250/km^2$ (all ages). Compare these densities with those of the waterbuck and the kob in the Rwenzori Park; $3 \cdot 3/km^2$ and $12 \cdot 7/km^2$ respectively for bucks of all ages based on Modha and Eltringham's estimates (Modha and Eltringham, 1976). Although the black lechwe shows a lower density than that of the kob in the Rwenzori Park, historical records suggest that its true density should be some seven times greater, that is, about $197/km^2$. Or, if the Lochinvar density is a true representation of normal lechwe density, then the density of the black lechwe should be some 40 times greater. A population of red lechwe in the Kafue Park was reported to have increased under protection from about 2/km² in 1948, to 32·3/km² in 1971, and to be still increasing (Grimsdell and Bell, 1972). Their social density was far greater than this, for they were observed to occupy much smaller areas at any one time, than the total area used for calculating density.

Puku

Another of the waterbuck's relatives, but one of which we have but scant knowledge, is the puku *Kobus vardoni* (Livingstone, 1857). This animal appears in most respects to be nothing more than the southern variant of the kob, but it is given specific status due to the nature of its glands. It has facial glands — although seemingly not well developed, pedal glands of variable occurrence, and the form of the well-developed inguinal pouches differ from those of the kob in that they are directed forwardly instead of backwardly (Ansell, 1960). Puku occur from southern Zaïre down to South Africa, but have a localized occurrence restricted to flood plain areas.

From observations in Zambia, de Vos (1965) claimed that it exhibits a similar type of territorial behaviour to that found in the kob; although its territorial grounds are bigger, of 5 to 21 ha in extent, and territorial activity is restricted to a seasonal occurrence when the plains which it inhabits are not flooded. It is said by de Vos and Dowsett (1966) to have a peak breeding period from November to February in Zambia; but Child and von Richter (1969) claim that it breeds throughout the year. These authors noted that the territories were well-defined but had overlapping boundaries. Since this species whistles like the kob, this may be the reason why the boundaries cannot be better defined; although de Vos and Dowsett (1966) report it only using the whistle as an alarm signal. During the floods they are forced to abandon their territories, but similar areas are occupied again the next dry season; although not by the same animals, and not in the same locations.

The mean group size, excluding groups of "one", was reported as 5·5 (de Vos and Dowsett, 1966), but Child and von Richter (1969) noted that the most common association was one adult buck with one doe, accompanied by a young one. A bachelor group of six young bucks was observed to move through territories unchallenged.

This meagre information allows us to make no constructive comparisons with other species, and we have no suitable information on densities, except for one small area of the Kafue Park, in Zambia, where Grimsdell and Bell (1972) recorded a density of approximately 2·8/km².

Reedbuck

The next closest relative of the waterbuck, the reedbuck, differs from the species so far considered in that it is an antelope sufficiently small in stature that it cannot see above medium grass height. There is a well-defined rule among African antelopes, that those which can see one another above the top of the grass, form herds, and those which cannot live solitarily, or in small family groups. Thus all of the smaller antelopes show a type of territoriality which differs radically from that of the larger antelopes, often having the doe assisting in the defence of the territory. Thus in these cryptic antelopes we cannot compare territorial behaviour in the same way, although there are many similarities.

The southern reedbuck *Redunca arundinum* (Boddaert, 1785) has been the subject of a general study in the Kruger National Park of South Africa by Jungius (1971), who contributed some brief observations on territorial behaviour. Essentially an inhabitant of long grass, avoiding bushed and wooded areas, this species lives in pairs or in small family groups, consisting of the buck, a doe, and her last offspring. These groups are said to inhabit a home range, within which is a defended territory. The size of this territory ranges from 35 ha in the dry season, to 60 ha in the wet season, and it is mainly found along rivers, the home range having a 300 to 500 m water frontage. Within the defended area itself there is also usually a core area, where the buck spends most of his time. This favoured spot is characterized by permanent water, good cover and a good food supply, but it does not have a fixed position. The territory boundaries also fluctuate in position, and adjacent areas overlap; while neighbours might enter 100 to 200 m inside one another's areas.

Those animals which occupy territories around permanent water have to reduce the sizes of their areas in the dry season, as others migrate in and exert pressure on them. It is normal practice for those who lack water in their territories to pass through the territories of others each day to drink; an analogous behaviour to that seen in the waterbuck.

Although living in pairs, the buck and doe do not always move about together, and does are not herded to maintain them in a territory, remaining of their own volition. Some bucks seeking a territory are accompanied by a doe, and does have been seen to take part in the defence of a territory. Although there is a general tolerance towards neighbours, territorial rights being maintained by display rather than by fighting, vigorous fights take place between contenders and defenders of territories. Indeed, it is well known that captive bucks soon become a nuisance as they mature because of their very aggressive nature, regarding human beings as intruders into their territory.

Young bucks appear to start off by occupying territories in

unfavourable regions, such as areas far from water or heavily bushed. When an owner is killed his territory may either be included within those of his neighbours, who extend their boundaries, or it may be taken over by a new buck. This observation of Jungius seems at variance with his assertion that the territories are each within a home range.

Like the kob and the puku, the reedbuck is a whistler. According to Jungius it marks its territory by a form of display in front of rivals, consisting of whistling and pronking. This may also be performed in the absence of any onlookers. The replies to its whistle may result in a chorus from other bucks, as is reported for the kob. Since defaecation and urination are at random, Jungius' assertion that this is territorial marking is unconvincing, as is also his supposition that olfactory marking takes place by means of the subauricular patch, which is characteristic of this species; although it appears to be variable in occurrence. A buck was never observed to rub the ear patch on objects, but Jungius suggests that particles of exudate would drop from the gland as the animal moves about.

But it would seem that a conglobate gland in this position, on the head, should be actively used in marking if indeed that is its function. I doubt whether the observation of reedbuck scratching the earpatch, is designed to transfer scent to the hoof for marking the ground, as suggested. In the oribi, which also has ear patches, I found, in the single specimen of an adult which I examined, that the underlying glandular tissue of the ear patch was heavily infected with microfilariae. If the concentration of microfilariae in this thin-skinned area is of common occurrence, then this would certainly explain the scratching. Although the reedbuck possesses inguinal glands, these are probably sexual in function.

The indeterminate territorial boundaries would seem to suggest, as with other "whistlers", that the reedbuck's main form of advertising is its whistle. Jungius found some evidence of a "stamping ground", perhaps used only during the peak of the mating season. It consisted of a trampled area about 2 by 3 m in size, with faeces and urine deposited in the centre; but it was only seen to be used at irregular intervals for a short period. The same author offers some possible evidence of territorial marking by horning vegetation, a buck often repeatedly horning the same sites; but more observation is needed on this point.

In the Rwenzori Park the highest recorded reedbuck density was 1·08/km² (Field and Laws, 1970), but in the Lake Nakuru Park, Kutilek (1974) reported a density of 8·76/km², while Holsworth (1972) reports a density of about 225/km² in the Dinder National Park, Sudan. Thus it is a species which, like its congeners, can achieve a relatively high density. What we do not know is what behaviour it shows in reaction to high density.

Reedbuck are sometimes found to abandon their territories, for example after they have been burnt, or when the vegetation has become sparse in the dry season, forming temporary concentrations in open areas, or on good pasture. Jungius supports the view that animals which can see each other in the open, band together as an anti-predator defence. But this is a behaviour which is more likely to benefit the individual than the group as a whole. Rather than its being designed to produce a "confusion effect", as Eibl-Eibesfeldt (1966) has postulated, such clumping appears more likely to be a question of spreading the chance of individual capture more widely. The reedbuck's temporary adoption of the herding habit, which is not shown by other cryptic antelopes, suggests the reedbuck to be on the borderline between the herding and the non-herding antelopes. This could be seen as a reflection of its size, which at about 75 kg in weight, and 0·9 m at the shoulder, far exceeds that of most other cryptic antelopes. The exceptions are the largest of the forest duikers, the yellow-backed, Jentink's and Abbott's, of whose social behaviour we are ignorant.

Oribi

A species with a similar social structure to the reedbuck is the oribi *Ourebia ourebi* (Zimmerman, 1783). Although classified in the far-removed *Neotragini*, it possesses several features resembling those of the reedbuck, and, as Ansell (1971) suggests, may indeed be an aberrant Reduncine.

A small antelope, weighing only some 22 kg and standing about 65 cm high at the shoulder, the oribi, like the southern reedbuck, occurs as single adult bucks, single does (perhaps comprising only those who have lost their partners), or in groups of buck, doe and offspring. In observations conducted in the Akagera Park, Rwanda, where the density ranged from 0·03 to 12·8/km², Montfort and Montfort (1974) found that out of 604 sightings, the most common social grouping was that of one buck with one doe (35%). But groups of up to six, consisting of a buck with two to five does and young, were seen. The buck and doe association appears to be a stable one, and is maintained all year round. Unlike the reedbuck, oribi were never observed to be gregarious.

Well-defined territories were maintained, which, although figures are not given, appeared to be about 25 ha in extent, and therefore quite large for such a small species. Some areas were discrete, while others had one or more adjacent boundaries. Related to the territory there was usually a neutral area, perhaps containing water or a salt-lick, which was visited by different groups. These groups, however, never mixed, and showed threat postures if others approached them too closely.

Both buck and doe defend the territory, which is extensively marked with an antorbital gland secretion. The buck has a very positive method of marking, first clearly identified by Gosling (1971). This consists of biting off the upper part of an upright grass stem, and inserting the prepared end into a very conspicuous, pouched antorbital gland which leaves a black, sticky exudate on the top of the stem. At regular intervals the buck revisits the sites, bites off the old marked piece, and anoints the stem afresh. Montfort and Montfort (1974) found that, although this activity was particularly conducted along the boundaries, it was also carried out randomly throughout the territory, when feeding, after mating, or during or after conflict situations. They found, however, that it was not frequent in short grass areas, where it is, of course, mechanically difficult to execute.

Alcelaphines

Let us now leave the waterbuck's relatives, close or distant, and consider some other territorial species, firstly, the Alcelaphine antelopes. The subfamily Alcelaphinae has two tribes; the Connochaetini, which contains the blue and the black wildebeest, and the Alcelaphini. The latter tribe contains the hartebeests and the bastard hartebeests, as well as the little-known Hunter's hartebeest *Beatragus hunteri*. With the exception of the latter, which still awaits study, the Alcelaphines have all been shown to be territorial antelopes, some of which favour extreme gregariousness of habit.

Perhaps the most gregarious of all is the western white-bearded wildebeest *Connochaetes taurinus mearnsi* Heller, 1913, whose concentrations may reach well over 500 animals/km² on open grassland plains under undisturbed conditions. An estimate of the size of the population inhabiting the Serengeti ecosystem of western Tanzania gave 721 000 head in 1971, having increased from about 439 000 in 1965 (Sinclair, 1973). Although this gives an overall density for the 30 000 km² of the ecosystem, of 24·3/km², the wildebeest's highly gregarious nature results in herds of several thousand head. Sinclair (1974) recorded a mean density of 79/km² in one study area, but densities of well over 500/km² have been recorded, especially during the rut. The former migrations of the South African springbok apart (and their numbers may have been exaggerated, but were in any case more of the nature of irruptions responding to adverse conditions, rather than social organizations), the western white-bearded gnu is the most densely occurring African antelope. Numerically a possible exception might be made for the lechwe, but when size is taken into account the biomass of wildebeest easily exceeds that of lechwe. At these high densities it must of course be nomadic, and the Serengeti population is noted for its migratory nature, as it is continually in search of fresh pasture. The gnu moves to areas where the rain falls.

This interesting species has been the subject of a detailed behaviour study by Dr Dick Estes. Estes (1969) conducted most of his studies in the 265 km² Ngorongoro Crater, adjoining the Serengeti plains. Here there was a sedentary population, estimated at that time (1964–1965) to number 14 222; an overall density of 53·7/km². Territorial behaviour had been first noted in the Serengeti population by Talbot and Talbot (1963), but their account was confused, resulting from their studying only the migratory population. By concentrating on the sedentary population of the Ngorongoro Crater, Estes found that there were two types of territorial behaviour, matching the two types of life style, sedentary and nomadic. These two types of territorial behaviour were fundamentally the same in pattern.

Wildebeest occur in one of three social arrangements: territorial bulls, cows with young, and groups of bachelor bulls. The latter may contain old and dispossessed bulls and "resting" territorial bulls, as well as young bulls. The sedentary populations comprise permanent territorial networks, with separate cow herds with young, and segregated bachelor groups. The mobile nomadic populations consist of aggregations of mixed bulls and cows with their young, with attached bachelor groups. Young bulls are separated from the dam as yearlings by the territorial bulls which do not tolerate them with the cows when the latter move into their territory. These young bulls then form bachelor groups on the fringes of the best wildebeest habitat. In the sedentary populations territorial spacing is wider and harassment is reduced. Sometimes young bulls insinuate themselves into cow and young groups to avoid identification and harassment. Estes estimated that up to 50% of the male sector of the population was confined to bachelor groups at any one time.

According to Watson (1969) the bulls of the migratory Serengeti population do not become territorial until at least 5 years of age, but in the Ngorongoro population they apparently become territorial at about 40 months of age, in the peak of that year's rut when they have just reached adult weight. This difference would seem to be attributable to a difference in population density. New territories in the sedentary population are won by persistence, the contender for a territory repeatedly returning until he eventually becomes accepted as a neighbour. This is somewhat similar to one method of obtaining a territory seen in the waterbuck.

According to Estes, serious fights are rare, as are serious injuries, but old bulls can often be encountered with their horns battered to stumps, suggesting that some fights might be more violent than Estes supposes.

Von Richter (1972), however, supports Estes' view, and it can be seen that as a result of the dense population compression, if aggression was carried to the point of mortal combat there would always be another to immediately fill the vacancy that might result, so that fighting to the death would become a continuous process. The wildebeest thus seems to have adopted ritual display, rather than overt aggression, in its intraspecific encounters.

In the sedentary territorial system we find a network of territorial bulls, each bull occupying a "stamping ground" consisting of a small, trampled bare patch. On this it defaecates, urinates, rubs its face, rolls, and horns the ground. The surrounding territory has no visible boundary, a fact confirmed by von Richter (1972) on studies of the black wildebeest in South Africa. Spacing appears to be maintained by visual estimation, the territorial owners spending a great deal of time displaying to, and interacting with, each other, grunting vocalization playing a dominant part. Estes estimated that up to 45 min each day is spent in this way. The spacing depends as much on the habitat as on the population density, but the most desirable territories are those where the population density is highest and the competition most severe. Estes estimated that one month before the annual rut there was a density of 47·5 to 70·7 territorial bulls/km² on the central plain of Ngorongoro Crater, with an average spacing of 130 to 160 m between bulls. During the rut this rose to a density of about 56·8 to 85·3/km², while spacing declined to 120 to 147 m; but the latter could drop to as little as 30 m between bulls when they were actively herding cows.

In the migratory populations the bulls mix with the cows and their young until the assemblage drifts to a halt. This is the signal for the bulls to immediately rush about forming a temporary territory round themselves. When the herd moves on again the territorial network breaks up, and only a few, widely scattered individuals may be left, the rest moving with the herd.

Despite its year-round territorial behaviour the wildebeest has a sharply defined breeding season, 80% of the cows being mated within one month. As a result of this, during the annual rut in May to June, territorial behaviour attains phrenetic proportions in both systems. In the nomadic herds the bulls divide up the cow aggregations into groups of about 20 cows with young; whereas the same effect is obtained in the sedentary populations when the cow herds wander into the territorial network. As with the kob, it seems that the bull concentrations, and their activity, have an attraction for the oestrus cows, for little, if any, mating takes place with non-territorial bulls. The bulls try to herd the oestrus cows within their territories, but there is considerable competition, and territorial rights are generally ignored by neighbouring contenders so that there is much interaction between the bulls, and territories may be lost through fatigue. At this time the nomadic bulls probably get a greater respite from continuous territorial activity than do the sedentary ones, for as soon as the herd moves the territorial activity stops.

After the rut territorial behaviour declines in intensity, becoming attenuated in the long dry season. But the sedentary bull still defends his patch; now also against large cow herd invasions. The cows, however, generally filter through the network, rather than being segregated. This is not difficult, as at this time bulls frequently leave their territories to feed and drink elsewhere, returning again in the evening. When absent from their own territories at this time they react with deference to those other bulls whose territory they might cross, as we have seen to be the case with other species. Some bulls associate together in temporary bachelor groups, losing all sense of territorial identity until they return to their patch again. Some bulls absent themselves for as much as 6 months, but shorter periods are more usual.

Territorial behaviour of the black wildebeest *Connochaetes gnou* (Zimmerman, 1780) of South Africa, the mammal in which territorial behaviour was first unwittingly noticed, has been described by von Richter (1972) from observations on a herd re-introduced into its former range in the Orange Free State; and from observations on populations in several other areas. The populations were all sedentary, restricted within small reserves, although historically the black wildebeest underwent migrations like its East African congeners. Unfortunately, von Richter's populations were much disturbed by catching operations, but nevertheless showed an identical system of territorial behaviour to that which was found by Estes.

In this species the cows were seen to occupy home ranges, whose owners did not tolerate strangers within their groups. The average size of these groups was 29·7, consisting of cows, yearlings and calves. Bachelor groups of non-territorial bulls, young bulls and yearlings were present, while 14 to 15% of the adult bulls, or 6 to 9% of the total population, was territorial, occupying stamping grounds within a territorial network. Spacing between bulls was greater than that found in Ngorongoro Crater, averaging 180 to 450 m. Some bulls occupied territories outside the network, but these were not seen to be visited by cows. As in East Africa, territorial activity was less pronounced after the annual rut, with bulls joining in bachelor groups for short periods.

Von Richter disagrees with Estes' suggestions that the bull marks his territory with the antorbital and pedal glands, and with faeces and urine; pointing out that the position of the antorbital gland does not permit the animal to rub it on the ground, and that bulls do not apparently take any notice of one another's scent. Von Richter suggests that the bull marks himself with his odours, by rolling in his dung etc., and that territorial position is advertised by display behaviour; a behaviour which is almost identical to that found in the East African wildebeest.

Estes concluded, that although there were several incidental effects, the primary function of territorial behaviour in the wildebeest was to impose a degree of order upon the breeding regime. During the rut, the intense competition between the bulls which fragments the aggregations into small groups, divides the oestrus cows among the breeding bulls more evenly, and also limits harassment of the oestrus cows in that the territorial bull defends the possession of them against others. This organization is maintained within the nomadic populations just as well as it is within the sedentary ones.

Von Richter suggests that the wildebeests are the most advanced of all the Alcelaphine antelopes, with their evolution of the nomadic habit to maintain continual contact with the best environmental conditions, and the lack of olfactory territorial demarcation. In contrast, the hartebeest and the bastard hartebeests, show a less pronounced nomadism, and all perform antorbital gland demarcation of their territories (although David (1973) questions whether this is an effective marking); while intra-specific encounters are less ritualized, and can lead to serious injury.

Topi

The topi *Damaliscus lunatus* (Burchell, 1823), one of the bastard hartebeests, is a species which also tends to be highly gregarious, but has gone a step further than the wildebeest in that it appears to have three types of territorial social organization. Low densities are occasioned by a lack of its favoured habitat of open grass plains, restricting it to spending much of its time in light woodland habitat. Studies by Montfort (1974) in the southern part of the Akagera Park of Rwanda, showed that at such densities, which ranged from $1.57/km^2$ to $7/km^2$, the bulls defend a territory of several hectares within which apparently live a group of cows and young numbering from 10 to 20. It is not clear whether these are permanent associations (Montfort seems to imply that they are), or whether the cows use a home range covering several bull territories, as in the waterbuck and other species. In marginal areas there are bachelor bull groups of 10 to 20 animals. At the end of the dry season, when burning of grasslands may bring on fresh flushes of grass, the territorial bulls leave their territories and join one another on the common ground. When the rains bring on widespread fresh growth the territories are reconstituted. This organization seems to be identical with that of the wildebeest at low density as described by von Richter (1972).

In the more favoured open grass plains, however, where densities reached 35·84/km², Montfort identified a territorial ground system which appears to be similar to that of the sedentary wildebeest as described by Estes (1969). Here the bulls defend small territories of 50 to 100 m in diameter, which are grouped together into territorial networks. The sites of these networks are relatively permanent, but the individual territory owners not infrequently change. The cow and the bachelor herds remain on the outside of the networks, but the former sometimes wander through, disrupting them. During the breeding season the cows deliberately visit the network for mating, as in the kob territorial ground system. When no cows are present in the area, the bulls leave their territories and graze together.

In the southern part of the Rwenzori Park, at Ishasha, we find one of the extremes of topi densities with some 3800 head restricted to within an open plains area of 80 km², giving an overall density of 47·4/km². With herds of 2000 forming, localized densities of up to twice this figure are created; densities which equate with those of the wildebeest in the Serengeti. Jewell (1972) studied this population during the three months' rutting period of January to March (there is some suggestion that it may shift from time to time); and although he chooses a different nomenclature to that of Estes (1969), his description of the social organization during the rut appears to be essentially the same as that for the migratory wildebeest.

During the rut, unfortunately, Jewell's study did not extend outside this period, a number of bulls form temporary territories within the moving population. He does not call them territories because they are not defined in any way, the bull appearing to defend a group of cows rather than an area. The situation seems to differ slightly from that found in the migratory wildebeest in that the topi tend to keep moving slowly as they graze, so that the bulls attempt to herd members of this passing crowd within small territories of 80 to 100 m in diameter, but the cows relentlessly move on without stopping, so that the bulls have to continually move their positions.

On the periphery are bachelor groups of non-territorial bulls, but some of these bulls continually change status, taking over territories and losing them again. As the rut wears on and the older bulls become exhausted, younger and younger bulls manage to take part. After the crowd has passed, lone bulls remain in its wake for a few days and then disappear; perhaps catching up with the crowd and establishing territories again. Although the most common group size which the bulls herd is seven to eight cows with young, at the leading edge of the phalanx the bulls herd groups of 30 to 80 individuals.

The bushed and wooded confines of the Ishasha plains mean that the population does not migrate in the true sense of the word, but rather drifts about the area in a random manner, the grass being at the same seasonal stage of growth throughout. For this reason Jewell has called it an extensive group home range. Perhaps this random movement in aggregated bands represents an evolutionary step towards the migratory habit of the Alcelaphines. Under suitable conditions numbers of topi build up to the extent where they form dense concentrations which, given an extended open habitat, would migrate; the numbers consequently expanding still more. In eastern Tchad large herds of topi are reported to undergo seasonal migrations which appear to be forced upon them by the topography of the habitat. Inhabiting the extensive open floodplains of the upper tributaries of the Chari River during the dry season, when the rains come they are forced to travel to drier ground, which by virtue of the extent of the plains is a considerable distance away. Unfortunately these populations have been sadly depleted in recent years.

The topi thus seems to parallel the wildebeest in its territorial behaviour, the type adopted depending upon density. Montfort sees the first two types as: dispersing the population among the feeding grounds and preventing overuse, co-ordinating sexual activity which results in a surfeit of young for predators, natural selection of the most vigorous bulls, etc. But its universal nature among Alcelaphines suggests a single selective factor, which is to be found in the lowest density populations, the organization at higher densities probably being directed to maintaining that originally selected advantage.

Bontebok

Another Alcelaphine which it is instructive to look at, the bastard hartebeest bontebok *Damaliscus dorcas* (Pallas, 1766), has been the subject of a study by David (1973). This is one of the most southerly of African antelopes, occurring formerly at the very tip of southern Africa. It is a particularly interesting species in that it demonstrates the deep-rooted inherited nature of territorial behaviour, for it is a species which was reduced almost to extinction as long ago as 1830; while the first Bontebok National Park, formed in 1931, contained only 17 animals. At the end of 1969 the total number from all localities was put at 800; while in 1970, the second Bontebok National Park, which was stocked with the population from the first, had a total of 265 animals in 28·1 km².

Despite being a strictly seasonal breeder, which is what we would expect at this latitude, the bull displays a well-defined year-round territorial behaviour, even at low densities, where there is no territorial competition. There may, however, have been some loss of learned, or inherited, information associated with this basic behaviour, for David found that there was a marked lack of aggression in the bulls, and that territorial rights were maintained in the most gentle of manners. Bulls chased away trespassers simply by walking or running up to them and informing them of the owner's presence, never challenging them to fight. This lack of aggression might, however, be attributable to a low density of population, for during the rut the bulls are capable of aggression.

The social organization consists of adult territorial bulls, a bachelor group which seems to contain bulls of all ages, as well as yearling cows, and which roams throughout the area, and cow herds with young. David's estimate of $1 \cdot 5$ territorial bulls/km² is comparable to that found in the waterbuck in the Rwenzori Park, but since the potential density of bontebok in the area seems to be low, density figures are meaningless for comparison.

David was unable to determine precise boundaries for the territories and considered that the boundaries were not fixed but were diffuse regions of lessened territorial drive. Although he does not refer to the existence of a "stamping ground", territorial bulls, in contrast to bachelor bulls, have several dung sites within their territories which they use exclusively for dunging, and which they also lie on. But no behaviour patterns have been observed connected with the recognition or avoidance of these spots by other bulls, and indeed, if a trespassing bull comes across a neighbour's dung pile he adds his own faeces to it.

The bull has a well-developed antorbital gland, present also to a lesser extent in the cow, which produces a black, sticky exudate, Like the oribi, although in this species stem biting was not observed, the bull inserts the end of a grass stem into the gland, but having anointed the stem, waves his head from side to side over the stem as if spreading the exudate, repeating this sequence several times. Very little, if any, trace of the exudate has been found on the grass afterwards, but this does not imply, as David seems to suggest, that it is no longer there! Although this activity is mainly conducted by territorial bulls, and is intensified during the rut, it is also sometimes seen to be performed by cows. David considered it perhaps to be a relic behaviour pattern, as no reactions to it were observed. I have already discussed this point, and consider that it is only necessary for a territory owner to mark his territory for his own guidance.

Territorial bulls were seen to take part in display activities with their neighbours, which appeared to be closely similar to those described for the wildebeest by Estes (1969). Bontebok bulls spent an average of 35 min each day in this activity, and would enter one another's territories to indulge it.

The cow herds move freely over a limited home range covering several territories, but some cows appear to form more or less permanent attachments to certain localities. Unlike the bachelors, which tended to move together in one group of 75 head, the cows showed little disposition towards gregariousness, the largest groups observed comprising only nine animals.

Hartebeest

The hartebeest *Alcelaphus buselaphus* (Pallas, 1776) is regarded by Gentry (1978) as of very recent descent. It is not markedly gregarious like its other Alcelaphine relatives, but has a territorial system with similarities to that found in the waterbuck. But the cow organization differs considerably. Gosling (1969, 1974), who studied this species in Kenya's Nairobi National Park, found that the cows were accompanied by up to four consecutive offspring. These cow groups occupied home ranges of 3·7 to 5·5 km², covering an area of 12 to 18 bull territories, through which they wandered more or less at will; although herding attempts were made by territory owners. At an age of 10 months the first reaction between juvenile and adult bulls was seen; the young bull reacting when approached by a territorial bull by calling in a juvenile manner, and adopting a head-in posture. The territorial owners did not usually chase the youngsters away at this stage, but if they did, then the dam followed the young bull, unlike in the waterbuck where the dam would take little notice of such a chase. The young bulls separated from their dams at from 10 to 30 months, this time span also indicating a much greater tolerance on the part of the territory owner than is shown by the waterbuck. When separated, the young bulls would either try to join bachelor groups, or to form groups of their own, or similar, age. These groups ranged in size from two to 100, but were of flexible composition, breaking up and then reassembling again. They used home ranges of from 6·7 to 10·3 km².

Maximum size is reached early in this species, at 2·5 to 3 years of age, before which bachelor dominance is based upon size, as it is in the waterbuck; but after this a linear hierarchy is established prior to attempting to take up territories at 3–4 years of age. Gosling (1974) suggests that this linear hierarchy may be an important pre-requisite to obtaining the best territories; but it seems unlikely that the status among the bachelor bulls could influence the challenging of the already established bulls.

Gosling identified 73 territories in his study area, of a mean size of 31 ha, but size varied in relation to habitat type; the tussock grass plains and woodland areas being less favoured than the short grass plains. The main observed method of territorial marking was, according to Gosling, by faeces, dung piles being most common along the boundary zones; although this tendency to defaecate at the boundary, could have been a nervous reaction resulting from boundary interactions, without marking significance, as I have discussed in Chapter 11. But some significance may be attached to the fact that these boundary dung piles are pawed by their owners, whereas those in the interior of the territories are left untouched. Although the hartebeest possesses both a large antorbital gland and interdigital glands, these are present in both sexes, and both cows and young were seen by Gosling to mark with the antorbital gland at least as frequently as did territorial bulls.

Unlike waterbuck, but similar to wildebeest, bulls were not continually resident in their territories but left them from time to time for anything from a few hours to several days, the mean percentage of occupation being established by Gosling at 76%, its highest level being in the most favoured habitats. During absences a territory might be taken over by another, and on the return of the former owner the territory was usually contested for, the strongest bull winning. It was estimated that 44% of territories was held for less than 3 months, and only two were recorded as being held by the same owners for the 3 years of the study. Bulls holding territories near to water were often swamped by bachelor bulls during the dry season, which resulted in a diminished aggressive response, the territorial owners often joining with the bachelor groups. Gosling identified several types of territory according to the degree of competition for them; one type was similar to that found in the waterbuck when a young buck attempts to set up his first territory, occupying a zone on the boundary between two adjacent territories. But in the hartebeest this appeared to be a temporary type formed mainly during the late dry season in areas near to water. Unlike in the waterbuck, territories were not uniformly defended, but owners had an "aggressive response" area which was more vigorously defended than were other parts. Gosling records that territory ownership did not confer freedom from interference during mating, and adjacent territory owners might challenge one another for oestrus cows. Although the territory owner was always deferred to, he might lose the cow while chasing away his competitors.

How representative Gosling's study is of Hartebeest territoriality in general cannot be assessed without complementary studies on other populations, for he was witnessing a dynamic situation in which the numbers of hartebeest had been reduced to some $5 \cdot 7/km^2$ in 1962 by drought, thence increasing to some $12 \cdot 3/km^2$ by 1967, with a predominantly young bull composition; and a correspondingly decreasing mean territory size during the 5-year period. Between 1972 and 1975 their numbers dropped again by 50% due to drought. Thus the possible destruction of an established territorial network, and the preponderance of young bulls, may have been the reason for the rapid turnover of territorial occupation. On the other hand, since the hartebeest bull becomes territorial 2 to 3 years earlier than does the waterbuck, we might expect the tempo of events to be accelerated.

Theories of Territorial Behaviour

Having considered the nearest relatives to the waterbuck, and the contrasting and important Alcelaphines, enough has been said to provide some insight into territorial organization, its density-dependent plasticity, and its fundamental similarity. It would be tedious to describe the findings in all other species which have been studied; suffice to say that territorial behaviour has been elucidated in a number of other animals, particularly the gazelles, such as the Grant's, Thomson's, and the closely related impala, and in such widely separated species as the roe deer of northern Europe and the pronghorn antelope of North America; testifying to its essential role in social organization.

Whereas Darling (1952) contents himself with the observation that "territorial behaviour as a whole is a social phenomenon, and it has survival value", others have speculated as to what exactly its role might be, and have tended to lump together every conceivable benefit to justify it.

Some years ago I attempted to analyse the role of territorial behaviour in the waterbuck in the light of the several hypotheses which had been put forth (Spinage, 1974), but especially in respect to whether it played a role in the regulation of population size, for Wynne-Edwards (1952) had stated that its dispersal role provided the "simplest and most direct kind of limiting convention that it is possible to have".

Firstly I considered the question of food supply, and whether the spatial distribution imposed by the adult bucks' mutual intolerance used the food supply more effectively. Were the bucks limiting the size of the population by hoarding food, or by rationing food to allow the population to exist at a level at which it could not otherwise do so, by ensuring fair shares for all? In the first place, the bucks would be denying to others food which was surplus to their own needs; in the second case they would be aiding its full utilization. But the answer to both questions must be in the negative, as the territorial bucks allow the bulk of the non-territorial sector, the bachelor groups and the does, to use the food resource without hindrance. Effective denial of food could only be to a minor part of the population. We have seen also that the difference in territorial density between the Peninsula and the Ogsa appeared to be attributable not to a difference in vegetation, since this was the same in the two areas, but to a difference in topography.

With increasing population density there was a corresponding decrease in the number of bucks relative to the number of does. The Peninsula observations showed that this could be attributable to the fact that territories were capable of being compressed only within certain limits. If the buck sector continued to increase beyond these limits, and emigration could not take place, then bucks would have to be eliminated from the population. All other things being equal, had the bucks survived at the same rate as the does, we could postulate that the density of waterbuck at Kayanja would have been 66/km², instead of 17·8/km². On the Peninsula it would have been 20·5/km² instead of 10·9/km², and in the Ogsa 4·6/km² instead of 3·2/km³. In the total park population there might have been 1800 more waterbuck.

But this sexual imbalance seems unlikely to have had any effect upon reproductive efficiency, as there were always enough bucks to accommodate the does, and there was not an unused surplus of breeding bucks. Those bucks which were excluded from breeding were the young and the old, which would later take, or would have taken, their turn in the territorial system. Some people tend to think in terms of the benefits of animal husbandry, and regard sex ratios, which do not depart widely from 1 : 1 in favour of does, as wasteful in biological terms, considering a buck to be taking the place of a breeding doe. In fact, the population with the greatest chance of continued survival in the face of adverse conditions, such as drought, starvation or epidemics, is that in which the sex ratio most closely approximates to unity.

Reduced numbers of bucks leading to a more effective breeding stock could, nevertheless, be seen as a short-term benefit, accelerating population increase. But let us recall the changes which took place in the Peninsula and Ogsa populations. At the end of 1956, Drs Petrides and Swank (personal communication) recorded an average of four territorial bucks and 16 adult does on the Peninsula, a sex ratio of 1 : 4 and a density of 4·6 animals/km². In 1964 the sex ratio was 1 : 1·8, and the density 10·8/km². In the adjacent Ogsa in 1956, there was an average of 5·3 territorial bucks and 28·4 adult does, with an overall sex ratio of 1 : 1·7, and a density of 3·1/km². In 1965 the density was unchanged and the sex ratio was 1 : 1·5. The number of bucks on the Peninsula increased four-fold during the 10 years 1956–1965, but the number of does only increased by a factor of 1·75. These changes thus suggest the antithesis of the postulate that territoriality may provide a means of suppressing the number of bucks. The territorially organized sector of the population had increased at the expense of the sector which was not thus organized.

If we regard the Peninsula and the Ogsa as one, thus taking into account possible movement of animals between the two areas, then the number of bucks increased by 1·5 times, while the does decreased by approximately one-seventh. But on the whole it looks as if the Ogsa maintained its *status quo* throughout the 10 years, and changes were confined to the Peninsula.

An increase of bucks at the expense of does could, in itself, provide a brake to population increase, as bucks do not contribute as much to rate of increase as do does. The absolute increase on the Peninsula has been greater, and one could suggest that it took fewer does to reach saturation point, so that further increases were restricted to the bucks. In the Ogsa, saturation level we may suppose was already attained, so the complement of bucks and does remained stationary.

The hypothesis that territorial organization allows mating to take place undisturbed was first advanced with respect to birds. Although some of the high density populations of antelopes that we have considered would seem to derive such benefit, this does not seem to be particularly true either for the waterbuck, or for the wildebeest. But spatial separation does curb sexual strife, and the fewer bucks that there are in proportion to does, the less competition there is for mating.

In territorial studies we have been rather carried away by the role of the male, which is more spectacular, and seems to be a more active one, than that of the female. But it is clearly the female which must supply the clue to a limiting factor to population increase, if there really exists such a factor, for she is biologically more valuable. Malthus (1796) was of the opinion that such a factor did not exist,

> "They are all impelled by a powerful instinct to the increase of their species; and this instinct is interrupted by no doubts about providing for their offspring. Wherever therefore there is liberty, the power of increase is exerted; and the superabundant effects are repressed afterwards by want of room and nourishment."

The question immediately arises that, if territorial behaviour does limit the numbers of males, and the same system does not pertain in the female, why do female numbers not increase exponentially, unless the females possess some independent curbing mechanism? If we postulate that the chance mortality factors which operate against the male are not in themselves sufficient to regulate population numbers, then likewise they cannot be sufficient to regulate the number of females. We have seen, however, that the waterbuck doe does appear to exert some sort of regulatory mechanism, resulting from the agonistic behaviour shown by adult does to young ones. Perhaps they are not as improvident as Malthus would have us believe.

Wynne-Edwards (1962) laid great stress upon territorial organization acting as a dispersal mechanism. In 1931 the Worthington's recorded that there was only one adult waterbuck buck on the Peninsula (Worthington and Worthington, 1933). This solitary buck, and the gradual increase in bucks thereafter (an identical situation was reported by Elliott (1976) for

Lake Naivasha's Crescent Island), could suggest that the buck is responsible for the dispersal of the species. There is a good chance that, wherever there are territorial bucks, then there is also that essential requirement, water. A doe therefore needs only to find a territorial buck to know that water is probably at hand. But the territorial system, with its competition for the best areas, suggests that normally the colonizers are more likely to be in the fringe areas to which they have been driven by competition, and thus are not likely to find the best environments for continued successful survival.

It is more logical to assume that it is the does which play the active role in dispersal, as they must seek the best conditions for the rearing and survival of their young. This would imply that, at Kayanja, for example, they reacted to some favourable change in the environment; perhaps nothing more than an increase in rainfall providing a better food supply. The does then moved in amongst the territorial bucks, as opposed to the concept that the population was a unit which developed a wide sex imbalance within itself, the result of male strife consequent upon the increase in population numbers. This did not happen on the Peninsula, nor did it happen on Crescent Island.

Thus, we may postulate, a favourable area attracts the does. Eventually the population increases to a level at which its density means that the area is no longer better than others. But before this level is reached, increasing agonistic behaviour, of adult to young does, means that the latter have already begun leaving the unit to seek other pastures. Thus, as not only the original complement of bucks, but also those produced by the does to fill all available territorial places, remain behind, the sex ratio begins to adjust itself.

A function of territorial behaviour in the waterbuck could thus be seen as that of anchoring the male sector of the population, so that the population unit as a whole does not shift from one area to another. This would ensure maximum dispersal of the species, a dispersal in which the role of the buck is essentially a static one, young does moving to new areas. The changes in sex ratio which took place on the Peninsula suggest this to be a more plausible explanation of sex imbalance than that of intra-specific fighting reducing the male sector. This confers no obvious advantages, other than that of enlarging the biologically most important sector of the population. As such it would not serve to regulate the population numbers, but to ensure their increase. The evidence does not support this.

Wynne-Edwards (1962), in favouring the concept of dispersal as the function of territorial organization, considered the critical resource to be food, as far as population density was concerned. It was highly advantageous to the species' survival, and thus strongly favoured by selection, for a species to control its own population density, and to keep it as near as possible to the optimum level for the habitat which was occupied. Food cannot be the proximate factor in this, only the ultimate; and a proximate factor, limiting the numbers of a species within the resources of its food supply, is provided by territorial behaviour.

Among birds, Wynne-Edwards considered that minimum territory size was inversely related to productivity of the habitat, and to food supply, but the contest was for the possession of territory and not for the food itself. This provides a completely density-dependent mechanism which allows the habitat to fill, amid mounting rivalry, to a maximum density. This spreads the population, avoiding clumping.

There are many similarities in the waterbuck's behaviour with this concept of dispersion, except that the territorial behaviour, confined to the buck, is seen only as a link with the behaviour of the doe. It is the doe which really determines population density within a given area.

Some authors have advocated the Neo-Darwinian view that territorial competition maintains a prime breeding stock. Territoriality ensures that breeding is denied to the young, and therefore untried, and to the old and the weak. But we are left with the old tautology, that the selection for bucks most capable of holding a territory leads only to those bucks most capable of holding one. We then have to explain what advantages the doe must derive from this singular attribute of the bucks. For one thing it does lessen the chances of inbreeding, a son with his mother, or a brother with his sisters. Weak strains that might result from this inbreeding are thus kept to a minumum; although all inbreeding is not necessarily deleterious. But assuming such a function for territorial behaviour, this would tend to maximize fertility rather than to act as a limiting factor to population increase.

The ultimate question appears to be, not whether territorial behaviour does, or does not, have an effect upon the numbers of a population, but whether, if it does, that effect is of greater significance, directly or indirectly, than the many other diverse mortality factors which militate against the animal's survival. In the waterbuck, territorial behaviour cannot be the ultimate behavioural factor in any populations' regulatory mechanism, should such a mechanism exist, as this must rest with the doe, which is not territorial.

In his book, "Sociobiology", Wilson (1975) dismisses the regulatory postulate with the words: "The literature is filled — indeed, it overflows — with tortuous discussions about the role of territory in the regulation of populations", and goes on to state that: "its function is the resource defended." As an example of territorial behaviour in which the primary function has been well-established, he quotes Kiley-Worthington's (1965) observations on the waterbuck, as establishing that the resource protected is the food supply.

Kiley-Worthington did not specifically advance such a viewpoint herself, and an examination of her findings suggests that, in the riverine environment, the further that the territorial bucks move inland, and thus the further into their feeding areas, the less defensive they become. The fact that their defensive behaviour increases, the closer that they are to the river, could suggest that, either they are defending the water resource, or the place in which they rest in cover. By no stretch of oblique reasoning can one deduce that their territorial aggression is proximally related to defence of the food resource. Owen-Smith (1977) has summarized the evidence contra-indicating food as the defended resource in territorial behaviour, and points out Estes' (1969) observation that territorial wildebeest bulls lost condition faster than did bachelors, even though the latter were supposedly relegated to less favourable parts of the habitat.

Jarman (1974) has proposed that the food habits of an antelope species determine that species' form, and this in turn determines its social behaviour. Thus, whereas Wynne-Edwards proposes the food resource as the ultimate factor explaining territorial behaviour — resulting from dispersion of the species as the proximal factor — Jarman sees the proximal factor as having reproductive significance. Of course, if Wilson is arguing in ultimate terms then he is right that the food resource determines animal behaviour and thus must account for territorial behaviour. But what we are interested in are the proximate causes which determine behavioural specializations; to think only in ultimate terms does not provide us with much insight into these specializations. Owen-Smith (1977) considers that territoriality based exclusively on food is the exception, and points out that the situation in small passerine birds with altricial young, on which many territorial hypotheses have been erected, seems to be an exception.

Although Owen-Smith has attempted to unify the concept of territoriality, pointing to exceptions, I feel that despite the catholic occurrence of territorial behaviour in the animal world, we should not be tempted into an oecumenical way of thought. What holds for the *Odonata* (Johnson, 1964) does not necessarily hold for African antelopes. On the other hand, within the antelopes we should look for a single, unifying, fundamental role. I believe that such a fundamental role, eschewing eclecticism, is not hard to find.

The Role of Territorial Behaviour

In the light of the studies which have been accomplished on African antelopes, a number of which has been outlined in this chapter, I believe that a unifying theme can be established. This supports Lack's hypothesis (Lack, 1966), although attributable to birds, that the role of territorial behaviour is simply that of ensuring a mate. Although it can go some way towards alleviating sexual strife, I reject Jarman's suggestion (Jarman, 1974) that its function is to confer exclusive mating rights; there are too many exceptions to this postulate. It can only serve as an aid to this end, at least in the majority of African antelopes. Owen-Smith (1977) also endorses a reproductive function, and presents an evolutionary model, indicating that selection could favour territoriality, if the adoption of an alternative mating strategy resulted in an excess of male mortality of greater than 10% per annum. But this is to suppose that other strategies do have such mortalities in excess of 10% per annum, and since none apparently do (see, for example, Sinclair's (1977) life tables for the African buffalo), the postulate becomes merely theoretical.

Like other authors, Owen-Smith proposes "several benefits", and refers, for example, to Geist's (1974) suggestion that a vast seasonal superabundance of forage produced during temperate summers, allows males to store extensive energy reserves in the form of fat, which are later used to sustain them in a brief, but intense, rut. A more important role for such reserves, however, is to see the animal through the winter, and those which use up too much during the rut, are unlikely to survive. Owen-Smith considers that tropical savanna populations remain closer to carrying capacity all the year round than do temperate species, so that breeding seasons are extended and intensive rutting eliminated. As a result, he considers, the potential for fat build-up is more limited, and a male engaging in very vigorous interactions would quickly bankrupt himself energetically. "This difference accounts for the general prevalence of territoriality among tropical ungulates, but does not explain the non-territorial exceptions" (Owen-Smith, 1977). The argument, however, is fallacious, for tropical savanna ungulates do lay down quite extensive fat reserves, and if they were used up in interactions, then they could be more readily replaced in the more benign tropical environment than could those of a temperate zone species, whose rut takes place at the beginning of the winter. There is, furthermore, no evidence to suggest that tropical savanna ungulates exist closer to the carrying capacity than do temperate zone species, when considered on a year-round basis.

Owen-Smith points out that, "in most ungulates territoriality is exclusively a male mating strategy". The importance of obtaining a mate needs no emphasis, but at low densities a male faces the risk of not obtaining a mate by virtue of the relative scarcity of females; especially when the few that there are may be clumped together, as we have seen happens at low density in the waterbuck, thus benefiting fewer males. At high densities a male also runs the risk of not obtaining a mate by virtue of the competitive interference which is occasioned by high male density. Territorial organization limits this risk of failing to obtain a mate, whether at low, normal, or at high density, by ensuring in all cases a partitioning of females among the males. The examples of territorial behaviour have shown how this partitioning is achieved.

But this does not explain why territorial organization should provide a better system than does a rank hierarchy; although Owen-Smith suggests that it is a "low benefit low cost" system, in that rank hierarchical males may have more matings, but at greater competitive cost. He poses the question: "what would happen, in evolutionary terms, were a [territorial] male to adopt an alternative mating strategy under the same ecological conditions?" He considers that the question is not why does territoriality exist, but "why has territorial dominance been favoured, rather than some alternative system of dominance organisation?"

I do not believe that the answer is as complex as this author suggests it to be, balancing costs and benefits relating to different alternatives under different conditions. Although I too believe that the answer is to be found in a consideration of the evolutionary consequences of alternative social organizations, which are limited to: rank hierarchy, as in the African buffalo; temporary dominance establishment in rutting species such as the red deer; possession of mobile harem groups, which is restricted to zebras in Africa; and the indeterminate home range system of the lesser kudu. Estes (1974) noted the difference in the physiology of growth between ungulates organized into rank hierarchy societies and those organized into territorial societies, males of the former continuing to grow throughout active life, in contrast to the latter in which growth ceases at maturity. He failed to realize the significance of this in explaining the territorial system.

A social system based on rank hierarchy among the male sector has the evolutionary drawback that there is selection for exaggerated male development. It tends to result in an evolutionary dead-end, selecting males solely for their ability to obtain mates. Thus, it is a system which is found only among the largest of the African ungulates, for example, the eland, buffalo, giraffe and elephant, all animals whose bulk alone offers some protection against predation. Smaller animals cannot afford to become bulky and cumbersome in the face of a high potential threat from predation. The territorial organization, in contrast to the rank hierarchy, operates to select for the male which can hold his place in the environment. He defends, first and foremost, a territory — a place in which to live. His mating activities become subordinate to this, and thus do not lead to exaggerated development. In evolutionary terms this makes it the most efficient method, and we find it to be the most common.

An exception to the large size hypothesis is provided by sheep and goats, which are significantly absent from Africa south of the Sahara as a wild species. In their natural range these are species whose method of escape from predation is by occupying inaccessible rocky habitats, where large size would become a disadvantage. This could be seen as exerting a counter-selective pressure to that which might result from the linear hierarchy system which these species possess. That they have not become territorial may be a reflection of their habitat, there being insufficient rocky areas to accommodate large, dispersed populations. Perhaps this is the reason why these species have never colonized Africa south of the Sahara, where there are few equivalents to the mountains of Europe and North America.

That sub-adult territorial males may be organized into a rank hierarchy, which could select for rapid growth, in that the quicker growing male would become the most dominant and take precedence in territorial occupation, is immaterial in respect of selection among the adults; for when rank hierarchy ceases, growth ceases. There is no selection for continued growth.

The larger cervids have adopted an intermediate system in which the stags associate together for the greater part of the year, competing only during a brief period of rut for groups of hinds. It can be seen that this competition for hinds must have some selective force related to a rank hierarchy, selecting for those stags most capable of defeating others. We find, indeed, that the larger male cervids do have continuous growth (although this may not continue into old age) and exaggerated developments connected with reproduction, but seemingly not to the extent that is shown by males in a rank hierarchy system, being offset by the fact that for the greater part of the year they must exist as ordinary animals, competing with the environment rather than with one another. But we find that the smaller cervids are territorial, for even the intermediate system presents some evolutionary risk for smaller animals.

It behoves me to mention the zebras because they exhibit a social organization not found among other African ungulates. Both the plains and the mountain zebras live in mobile, family groups, of mares and their young, headed by a stallion. Other stallions live in bachelor groups. These groups are non-territorial and move freely over a large home range, of from 80 to 200 km² in the plains zebra, but only 3 to 5 km² in the more limited habitat of the mountain zebra. Essentially this arrangement seems to be related to the mobility of these animals, but the equally mobile Grévy's zebra and wild ass have no family associations. Stallions appear to occupy large territories, from 2·7 to 10·5 km² in the Grévy's zebra, where exclusive mating rights are accorded to the territory owner by other stallions (Klingel, 1974). This difference in social organization is probably attributable to the semi-arid habitat in which the latter species lives, a habitat where food resources are thin on the ground, necessitating greater spatial separation between individuals. Hence the stallion cannot herd a close group and adopts the only alternative available to him.

Why not a system where the males and females live together and take mates equally, without strife? The answer seems to be that in order to ensure the continuation of a species, selection has produced a sufficiently strong sexual drive that male competition must always result, and this leads to evolutionary consequences. But such an equable system appears to be almost attained in the lesser kudu, which Leuthold (1974) studied in the Tsavo National Park of Kenya. He found that the adult bulls apparently occupied overlapping home ranges, in which they showed no aggression towards neighbours or intruders. He admits, however, that more observation is necessary to elucidate their exact relationships; but such a system can be seen as a result of the lesser kudu's ecology — a browser in a semi-arid environment. Where food is widely scattered, and may be hard to find at times, it does not pay to defend an area which may come to have nothing in it to eat. It is better for each animal to be able to search independently and unhindered, if the food items do not exist in sufficient concentration for the animals to compete for them. The lesser kudu's system, therefore, seems to be an adaptation to a harsh environment, and it is not unlike the "breakdown" of the system that one sees in some territorial species during unfavourable seasons.

There has been much speculation about the causes of giantism among mammals during the Pleistocene period. Undoubtedly climatic conditions favouring a continuing abundance of primary production were a prerequisite, but a hitherto unconsidered speculation is that such giantism may have resulted from the evolution of social behaviour; these giants being the grotesque end-products of selection resulting from the rank hierarchy system of social organization. None of the giant ungulates of the Pleistocene has any modern territorial representatives. There were no giant reduncines or alcelaphines; but there were giant bovids, cervids, pigs, giraffes and elephants, representative of species all having rank hierarchy systems today. There were, of course, giant carnivores, the modern representatives of which are territorial (although others, such as the hyaena and hunting dog, have hierarchical systems); but you cannot have giant carnivores without giant prey, for superabundant small prey cannot be easily captured by large and cumbersome predators. Hence these giants must have evolved in response to giantism in herbivores.

In conclusion, I find that selection for continued growth in the hierarchical system of social organization presents problems of balance between a species and its environment. A very fine distinction must be struck, producing males big enough to dominate, yet not continually selecting to this end so that inefficient giants result. The same problem, although perhaps to a lesser extent, is found in the intermediate system of the larger cervids. I suggest that it is the absence of these problems, in the balanced selection of the territorial system, which explains the latter's widespread occurrence among ungulates, serving to make it one of the most successful of systems of social organization in the context of long-term evolution.

References

Altmann, M. (1956). Patterns of herd behaviour in free-ranging elk of Wyoming, *Cervus canadensis nelsoni. Zoologica* **41**: 65–71.

Altmann, M. (1963). Naturalistic studies of maternal care in moose and elk. *In* "Maternal Behaviour of Mammals" (ed. L.H. Rheingold), pp. 233–254. New York: Wiley.

Altum, B. (1868). "Der Vogel und sein Leben." Münster: W. Niemann.

Amoroso, E.C., Kellas, L.M. and Harrison, L. (1954). The foetal membranes of an African waterbuck *Kobus defassa. Proc. zool. Soc. Lond.* **123**: 477.

Amoroso, E.C. and Marshall, F.H.A. (1960). External factors in sexual periodicity. *In* "Marshall's Physiology of Reproduction" (ed. A.S. Parkes), pp. 707–831, 3rd edn, Vol. 1, Part 2. London: Longmans Green.

Anderson, J. (1964). Reproduction of imported British breeds of sheep on a tropical plateau. *Proceedings Vth. Int. Congr. Animal Reproduction (Trento)* **3**: 465–469.

Anon. (1925). Report of the Acting D.C., Kigezi, 17.3.25. Uganda Government Medical Department Archives, Entebbe.

Ansell, W.F.H. (1960). Contributions to the mammalogy of Northern Rhodesia. *Occ. Papers. Nat. Mus. Southern Rhodesia* No. **24B**: 384–389.

Ansell, W.F.H. (1969). Addenda and corrigenda to "Mammals of Northern Rhodesia". No. 3. *Puku* **5**: 1–48.

Ansell, W.F.H. (1971). Artiodactyla. *In* "The Mammals of Africa: An Identification Manual" (eds J. Meester and H.W. Setzer), 1977. Washington: Smithsonian Institution.

Ansell, W.F.H. (1978). "The Mammals of Zambia." Lusaka: The Government Printer.

Ansell, W.F.H. and Banfield, C.F. (1979). The subspecies of *Kobus leche* Gray, 1850 (Bovidae). *Sängetierk. Mitt* **3**: 168–176.

Ardrey, R. (1967). "The Territorial Imperative." London: Collins.

Barnett, S.A. (1967). "Instinct and Intelligence." London: MacGibbon and Kee.

Beadle, L.C. (1965). *In* "Uganda National Parks Handbook". Kampala: Trustees of the Uganda National Parks.

Bere, R.M. (1959). The hippopotamus problem and experiment. *Oryx* **5**: 116–124.

Best, A.A. and Best, T.G.W. (1977). "Rowland Ward's Records of Big Game." 17th edn (Africa). London: Rowland Ward.

Best, G.A., Edmond-Blanc, F. and Witting, R.C. (1962). "Rowland Ward's Records of Big Game." 11th edn (Africa). London: Rowland Ward.

Blaxter, K.L. (1964). Protein metabolism and requirements in pregnancy and lactation. *In* "Mammalian Protein Metabolism" (eds H.N. Munro and J.B. Allison), pp. 173–223, Vol. 2. London and New York: Academic Press.

Bourlière, F. (1955). "The Natural History of Mammals." London: Harrap.

Bourlière, F. (1959). Lifespans of mammalian and bird populations in nature. *In* "Colloquia on Ageing, 5" (eds G.E.W. Wolstenholme and M. O'Connor). London: CIBA.

Bourlière, F. and Verschuren, J. (1960). "Introduction à l'Écologie des Ongulés du Parc National Albert." Brussels: IPNCB.

Brass, W. (1958). Simplified methods of fitting the truncated negative binomial distribution. *Biometrika* **45**: 59–68.

Brocklesby, D.W. and Vidler, B.O. (1965). Some parasites of East African wild animals. *E. Afr. Wildl. J.* **3**: 120–122.

Brody, S. (1964). "Bioenergetics and Growth." New York: Hafner.

Bubenik, A.B. (1965). Beitrag zur Geburtskunde und den Mutter-Kind-Beziehungen des Reh-und Rotwildes. *Zeit. Saugetierk* **30**: 65–228.

Buechner, H.K. (1961). Territorial behaviour in Uganda kob. *Science* **133**: 698–699.

Buechner, H.K. (1961a). Unilateral implantation in the Uganda kob. *Nature* **190**: 738–739.

Buechner, H.K. and Schloeth, R. (1965). Ceremonial mating behaviour in the Uganda kob. *Z. Tierpsychol.* **22**: 209–225.

Buechner, H.K. and Roth, H.D. (1974). The Lek System in Uganda Kob Antelope. *Amer. Zool.* **14**: 145–162.

Burchell, W.J. (1823). "Travels in the Interior of Southern Africa." London: Longman.

Burckhardt, D. (1958). Kindliches Verhalten als Ausdrucksbewegung im Fortpflanzungszeremoniell einiger Wiederkäuer. *Rev. Suisse de Zool.* **65**: 311–316.

Burt, W.H. (1943). Territory and home range concepts as applied to mammals. *J. Mammal.* **24**: 346–352.

Carpenter, C.R. (1958). Territoriality: a review of concepts and problems. *In* "Behaviour and Evolution" (eds A. Ŕoe and G.G. Simpson), pp. 224–250. New Haven: Yale University Press.

Caughley, G. (1966). Mortality patterns in mammals. *Ecology* **47**: 906–918.

Caughley, G. (1977). "Analysis of Vertebrate Populations." New York: Wiley.

Child, G. (1964). Growth and ageing criteria of impala, *Aepyceros melampus*. *Occ. Papers. Nat. Mus. Southern Rhodesia* **27B**: 128–135.

Child, G. and von Richter, W. (1969). Observations on the ecology and behaviour of lechwe, puku and waterbuck along the Chobe River, Botswana. *Z. f. Säugetierkunde* **34**: 275–295.

Clegg, M.T. and Ganong, W.F. (1969). Environmental factors affecting reproduction. *In* "Reproduction in Domestic Animals" (eds H.H. Cole and P.T. Cupps), 2nd edn, pp. 473–488. London and New York: Academic Press.

Cloudsley-Thompson, J.L. (1961). "Rhythmic Activity in Animal Physiology and Behaviour." London and New York: Academic Press.

Clough, G. (1969). Some preliminary observations on reproduction in the warthog *Phacochoerus aethiopicus* Pallas. *J. Reprod. Fert., Suppl.* **6**: 323–337.

Cott, H.B. (1961). Scientific results of an inquiry into the ecology and economic status of the Nile crocodile (*Crocodilus niloticus*) in Uganda and Northern Rhodesia. *Trans. zool. Soc. Lond.* **29**: 211–356.

Cowie, M.H. (1956). "Annual Report of the Royal National Parks of Kenya." Nairobi: Trustees of the Kenya National Parks.

Crandall, L.S. (1965). Record of African antelopes in the New York Zoological Park. *Int. Zoo. Yrbk.* **5**: 52–55.

Crile, G. and Quiring, D.P. (1940). A record of the body weight and certain organ and gland weights of 3,690 animals. *Ohio J. Sci.* **40**: 219–259.

Cumming, R.G. (1850). "The Lion Hunter in South Africa." London: John Murray.

Dale, I.R. and Osmaston, H. (1954). Flora of the Queen Elizabeth Park. *In* "Uganda National Parks". Kampala: The Trustees of the Uganda National Parks.

Dalquest, W.W. (1965). Mammals from the Save River, Mozambique, with descriptions of two new bats. *J. Mammal.* 46: 255–264.

Daniell, J.P.S. (1951). Hybridism in captive animals. *Zoo Life*: 47–49.

Darling, F.F. (1937). "A Herd of Red Deer." Oxford: OUP.

Darling, F.F. (1952). Social behaviour and survival. *Auk* 69: 183–191.

Darling, F.F. (1960). An ecological reconnaissance of the Mara plains in Kenya Colony. *Wildlife Monograph* 5. The Wildlife Society.

Dasmann, R.F. and Mossman, A.S. (1961). Commercial utilisation of game animals on a Rhodesian ranch. Unpublished, pp. 11, mimeo.

David, J.H.M. (1973). The behaviour of the bontebok, *Damaliscus dorcas dorcas* (Pallas 1766), with special reference to territorial behaviour. *Z. Tierpsychol.* 33: 38–107.

Dean, F.C. and Gallaway, G. A. (1965). A Fortan program for population study with minimal computer training. *J. Wildl. Mgmt* 29: 892–894.

Deevey, E.S. (1947). Life tables for natural populations of animals. *Quart. Rev. Biol.* 22: 283–314.

Dekeyser, P.L. (1955). "Les Mammifères de l'Afrique Noire Française." 2nd edn. Init. africaines. 1. Dakar: IFAN.

DeVos, A. (1965). Territorial behaviour among puku in Zambia. *Science* 148: 1752–1753.

DeVos, A. and Dowsett, R.J. (1966). The behaviour and population structure of three species of the Genus *Kobus*. *Mammalia* 30: 30–55.

Dittrich, L. (1972). Gestation periods and age of sexual maturity of some African Antelopes. *Int. Zoo. Yrbk* 12: 184–187.

Dollman, J.G. (1931). Development of auricular "glandular" patches in the Waterbucks. *Proc. Linn. Soc. Lond.* 144: 86–87.

Doss, Mildred A., Farr, Marion M., Roach, Katherine F.M. and Anastos, G. (1974). Index-Catalogue of Medical and Veterinary Zoology, Special Publication No. 3. "Ticks and tickborne diseases II. Hosts. Part 2." G-P. Washington: US Dept. Agric.

Ducker, M.J., Bowman, J.C. and Temple, A. (1973). The effect of constant photoperiod on the expression of oestrus in the ewe. *J. Reprod. Fert., Suppl.* 19: 143–150.

Eibl-Eibesfeld, J. (1963). Angeborenes und Erworbenes im Verhalten einiger Säuger. *Z. f. Tierpsychol.* 20: 705–754.

Eibl-Eibesfeld, J. (1966). 'Grundrisse der vergleichenden Verhaltensforschung." Munich: R. Pieper and Co.

Elliott, E.M.N. (1976). A study of waterbuck in a newly enclosed area. Ph.D. Thesis, Cambridge.

Elliot, G.F. (1896). "A Naturalist in mid-Africa." London: A.D. Innes.

Eltringham, S.K. (1980). A quantitative assessment of range usage by large African mammals with particular reference to the effects of elephants on trees. *Afr. J. Ecol.* 18, 53–71.

Eltringham, S.K. and Din, N.A. (1977). Estimates of the population size of some ungulate species in the Rwenzori National Park, Uganda. *E. Afr. Wildl. J.* 15: 305–316.

Eltringham, S.K. and Flux, J.E.C. (1971). Night counts of hares and other animals in East Africa. *E. Afr. Wildl. J.* 9: 67–72.

Estes, R.D. (1969). Territorial behaviour of the wildebeest (*Connochaetes taurinus* Burchell, 1823). *Z. f. Tierpsychol.* 26: 284–370.

Estes, R.D. (1972). The role of the vomeronasal organ in mammalian reproduction. *Mammalia* 36: 315–341.

Estes, R.D. (1974). Social organisation of the African *Bovidae*. *In* "The Behaviour of

Ungulates and its Relation to Management" (eds V. Geist and F. Walther), pp. 166–205. Morges: IUCN.

Ewer, R.F. (1968). "Ethology of Mammals." London: Logos Press.

Fay, L.D. (1972). "Wildlife Disease Research." Report to the Government of Kenya. Rome: FAO.

Field, C.R. (1966). Progress Report. Nuffield Unit of Tropical Animal Ecology, Uganda, mimeo.

Field, C.R. (1967). Progress Report. Nuffield Unit of Tropical Animal Ecology, Uganda, mimeo.

Field, C.R. (1968). Methods of studying the food habits of some wild ungulates in Uganda. *Proc. Nutr. Soc.* 27: 172–177.

Field, C.R. (1971). Elephant ecology in the Queen Elizabeth National Park, Uganda. *E. Afr. Wildl. J.* 9: 99–123.

Field, C.R. (1972). The food habits of wild ungulates in Uganda by analysis of stomach contents. *E. Afr. Wildl. J.* 10: 17–42.

Field, C.R. and Laws, R.M. (1970). The distribution of the larger herbivores in the Queen Elizabeth National Park, Uganda. *J. appl. Ecol.* 7: 273–294.

Floody, O.R. and Arnold, A.P. (1975). Uganda kob (*Adenota kob thomasi*): Territoriality and the Spatial Distributions of Sexual and Agonistic Behaviours at a Territorial Ground. *Z. Tierpsychol.* 37: 192–215.

Flower, S.S. (1931). Contributions to our knowledge of the duration of life in vertebrate animals. IV. Mammals. *Proc. zool. Soc. Lond.* 1931: 145–234.

Flower, W.H. and Lydekker, R. (1891). "An Introduction to the Study of Mammals, Living and Extinct." London: Adam and Charles Black.

Foster, J.B. (1968). The biomass of game animals in Nairobi National Park, 1960–1966. *J. Zool. Lond.* 155: 413–425.

Foster, J.B. and Kearney, D. (1967). Nairobi National Park census, 1966. *E. Afr. Wildl. J.* 5: 112–120.

Frechkop, S. (1954). Exploration du Parc National de l'Upemba. *Fasc.* 14. Mammifères. Brussels: IPNCB.

Geist, V. (1965). On the rutting behaviour of the mountain goat. *J. Mammal.* 45: 551–568.

Geist, V. (1971). "Mountain Sheep. A Study in Behaviour and Evolution." Chicago: Chicago University Press.

Geist, V. (1974). On the relationship of ecology and behaviour in the evolution of ungulates: theoretical considerations. *In* "The Behaviour of Ungulates and its Relation to Management" (eds V. Geist and F. Walther), pp. 235–246. Morges: IUCN.

Gentry, A.W. (1978). Bovidae. *In* "Evolution of African Mammals" (eds V.J. Maglio and H.B.S. Cooke), pp. 540–572. Massachusetts: Harvard University Press.

Gerhardt, U. (1933). V. Kloake und Begattungsorgane. *In* "Handb. Vergl. Anat. Wirbeltiere" (eds L. Bolk, E. Göppert, E. Kallius and W. Lubosch), pp. 267–350. Berlin: Urban und Schwarsenberg.

Glasgow, J.P. (1963). "The Distribution and Abundance of Tsetse." New York: Pergamon Press.

Gosling, L.M. (1969). Parturition and related behaviour in Coke's hartebeest, *Alcelaphus buselaphus cokei* Günther. *J. Reprod. Fert., Suppl.* 6: 265–286.

Gosling, L.M. (1972). The construction of antorbital gland marking sites by male oribi (*Ourebia ourebia*, Zimmerman, 1783). *Z. Tierpsychol.* 30: 271–276.

Gosling, L.M. (1974). The social behaviour of Coke's hartebeest (*Alcelaphus buselaphus cokei*). *In* "The Behaviour of Ungulates and its Relation to Management" (eds V. Geist and F. Walther), pp. 488–511. Morges: IUCN.

Graf, W. (1956). Territorialism in deer. *J. Mammal.* **37**: 165–170.

Grimsdell, J.J.R. (1969). Ecology of the buffalo, *Syncerus caffer*, in western Uganda. Ph.D. thesis, Cambridge.

Grimsdell, J.J.R. (1973). Age determination of the African buffalo. *E. Afr. Wildl. J.* **11**: 31–53.

Grimsdell, J.J.R. (1973a). Reproduction in the African buffalo, *Syncerus caffer*, in western Uganda. *J. Reprod. Fert., Suppl.* **19**: 303–318.

Grimsdell, J.J.R. and Bell, R.H.V. (1972). Population growth of red lechwe, *Kobus leche leche* Gray, in the Busanga Plain, Zambia. *E. Afr. Wildl. J.* **10**: 117–122.

Grimsdell, J.J.R. and Bell, R.H.V. (1975). Black Lechwe Research Project Final Report. Lusaka: NCSR.

Guinness, F.E., Albon, S.D. and Clutton-Brock, T.H. (1978). Factors affecting reproduction in red deer (*Cervus elaphus*) hinds on Rhum. *J. Reprod. Fert.* **54**: 325–334.

Guinness, F.E., Clutton-Brock, T.H. and Albon, S.D. (1978). Factors affecting calf mortality in red deer (*Cervus elaphus*). *J. Anim. Ecol.* **47**: 817–832.

Gwynne, M.D. and Bell, R.H.V. (1968). Selection of vegetation components by grazing ungulates in the Serengeti National Park. *Nature* **220**: 390–393.

Haagner, A. (1920). "South African Mammals." London: Witherby.

Hale-Carpenter, G.D. (1921). Report of the S.M.O.S.S. 10.1.21. Entebbe: Uganda Government Medical Department Archives. Unpublished.

Hale-Carpenter, G.D. (1922). Report of the S.M.O.S.S. 3.10.22. Entebbe: Uganda Government Medical Department Archives. Unpublished.

Haltenorth, Th. (1963). Klassifikation der Säugetiere: Artiodactyla 1 (18). *In* "Handbuch der Zoologie, 8 (32)" (eds J.-G. Helmcke, H. v. Lengerken, D. Starck and H. Wermuth), pp. 1–167. Berlin: Walter de Gruyter.

Hammond, J. (1927). "The Physiology of Reproduction in the Cow". Cambridge: CUP.

Hancock, G.R. (1933). Report to the S.M.O. Entebbe: Uganda Government Medical Department Archives. Unpublished.

Hanks, J. (1967). The use of M99 for the immobilisation of the defassa waterbuck (*Kobus defassa penricei*). *E. Afr. Wildl. J.* **5**: 96–105.

Hanks, J., Stanley-Price, M. and Wrangham, R.W. (1969). Some aspects of the ecology and behaviour of the defassa waterbuck (*Kobus defassa*) in Zambia. *Mammalia* **33**: 471–494.

Harrop, J.F. (1960). The soils of the Western Province of Uganda. *Uganda Dept. Agric. Memoirs. Ser.* **1** (**6**). Entebbe: The Government Printer.

Heape, W. (1901). The "sexual season" of mammals and the relation of "pro-oestrum" to menstruation. *Quart. J. micr. Sci.* **44**: 1–70.

Herbert, H.J. (1972). The population dynamics of the waterbuck *Kobus ellipsiprymnus* (Ogilby, 1833) in the Sabi-Sand Wildtuin. Hamburg: Paul Parey.

Heyden, K. (1969). Studien zur Systematik von *Cephalophinae* Brooke, 1876: Reduncini Simpson, 1945 und Peleini Sokolov, 1953 (-*Antilopinae* Baird, 1857). *Z. wiss. Zool.* **178**: 348–441.

Hindle, E. (1951). Abstract. *Proc. zool. Soc. Lond.* **121** (**2**): 203.

Hirst, S.M. (1969). Predation as a regulating factor of wild ungulate populations in a Transvaal lowveld nature reserve. *Zoologica Africana* **4**: 199–230.

Hirst, S.M. (1975). Ungulate-habitat relationships in a South African woodland/savanna ecosystem. *Wildlife Monographs. No.* **44**. The Wildlife Society.

Hoier, R. (1950). "A Travers Plaines et Volcans au Parc National Albert." Brussels: IPNCB.

Holsworth, W.N. (1972). Reedbuck concentrations in the Dinder National Park, Sudan. *E. Afr. Wildl. J.* **10**: 307–308.

Howard, H.E. (1920). "Territory in Bird Life." London: John Murray.

Huggett, A. St. G. and Widdas, W.F. (1951). The relationship between mammalian foetal weight and conception age. *J. Physiol.* **114**: 306–317.

Hutchins, E. (1917). Report of the Chief Veterinary Officer. Annual Report Department Agriculture, Uganda. Entebbe: The Government Printer.

Hutchinson, G.E. (1978). "An Introduction to Population Ecology." New Haven: Yale University Press.

Jackson, F.L. (1926). *In* "The Game Animals of Africa" (ed. R. Lydekker). London: Rowland Ward.

Jarman, P.J. (1972). The development of a dermal shield in impala. *J. Zool. Lond.* **166**: 349–356.

Jarman, P.J. (1974). The social organisation of antelope in relation to their ecology. *Behaviour* **48**: 215–267.

Jenkins, T.A. (1929). Horned female waterbuck. *The Field* **153**: 395.

Jewell, P.A. 1966. The concept of home range in mammals. *Symp. Zool. Soc. Lond. No.* **18**: 85–109.

Jewell, P.A. (1972). Social organisation and movements of topi (*Damaliscus korrigum*) during the rut, at Ishasha, Queen Elizabeth Park, Uganda. *Zoologica Africana* **7**: 233–255.

Johnson, C. (1964). The evolution of territoriality in the Odonata. *Evolution* **18**: 89–92.

Jungius, H. (1971). "The Biology and Behaviour of the Reedbuck (*Redunca arundinum* Boddaert, 1785) in the Kruger National Park. Hamburg: Verlag Paul Parey.

Kellas, L. (1954). Observations on the reproductive activities, measurements, and growth rate of the Dikdik (*Rhynchotragus kirkii thomasi* Neumann). *Proc. zool. Soc. Lond.* **124**: 751–784.

Kenneth, J.H. and Ritchie, G.R. (1953). Gestation Periods. A table and bibliography. *Tech. Comm.* **5**. Edinburgh: Comm. Bur. Animal Breeding and Genetics.

Kiley-Worthington, M. (1965). The waterbuck (*Kobus defassa* Rüppel 1835 and *Kobus ellipsiprymnus* Ogilby, 1833) in East Africa: Spatial distribution. A study of the sexual behaviour. *Mammalia* **29**: 177–204.

Kiley-Worthington, M. (1966). A preliminary investigation into the feeding habits of the waterbuck by faecal analysis. *E. Afr. Wildl. J.* **4**: 153–157.

Kleiber, M. (1961). "The Fire of Life. An Introduction to Animal Energetics." New York: John Wiley.

Klingel, H. (1974). A comparison of the social behaviour of the *Equidae. In* "The Behaviour of Ungulates and its Relation to Management" (eds V. Geist and F. Walther), pp. 124–132. Morges: IUCN.

Kutilek, M.J. (1974). The density and biomass of large mammals in Lake Nakuru National Park. *E. Afr. Wildl. J.* **12**: 201–212.

Lack, D. (1946). "The Life of the Robin." Revised edn. London: Witherby.

Lack, D. (1954). "The Natural Regulation of Animal Numbers." Oxford: OUP.

Lack, D. (1966). "Population Studies of Birds." Oxford: OUP.

Lamprey, H.F. (1963). Ecological separation of the large mammal species in the Tarangire Game Reserve. *E. Afr. Wildl. J.* **1**: 63–92.

Laws, R.M. and Clough, G. (1966). Observations on reproduction in the hippopotamus (*Hippopotamus amphibius* Linn.). *Symp. Zool. Soc. Lond. No.* **15**: 117–140.

Laws, R.M., Parker, I.S.C. and Johnstone, R.B. (1975). "Elephants and their Habitats. The Ecology of Elephants in North Bunyoro, Uganda." Oxford: OUP.

Ledger, H.P. (1959). A possible explanation for part of the difference in heat tolerance exhibited by *Bos taurus* and *Bos indicus* beef cattle. *Nature* **184**: 1405.

Ledger, H.P. (1968). Body composition as a basis for a comparative study of some East African mammals. *Symp. Zool. Soc. Lond. No.* **21**: 289-310.

Ledger, H.P., Payne, W.J.A., Talbot, L.M. and Zaphiro, D. (1961). The use of carcase analysis techniques for investigating the meat production potential of game and domesticated animals in semi-arid areas. *In* "Conservation of Nature and Natural Resources in Modern African States" (ed. G.G. Watterson), pp. 268-283. Morges: IUCN.

Ledger, H.P., Sachs, R. and Smith, N.S. (1967). Wildlife and food production. *World Rev. Animal Prod.* **3** (**11**): 13-36.

Lent, P. (1974). Mother-infant relationships in ungulates. *In* "The Behaviour of Ungulates and its Relation to Management" (eds V. Geist and F. Walther), pp. 14-55. Morges: IUCN.

Leslie, P.H. and Ranson, R.M. (1940). The mortality, fertility and rate of natural increase of the vole (*Microtus agrestis*) as observed in the laboratory. *J. anim. Ecol.* **9**: 27-52.

Leslie, P.H. and Ranson, R.M. (1952). The fertility and population structure of the brown rat (*Rattus norvegicus*) in corn ricks and some other habitats. *Proc. zool. Soc. Lond.* **122**: 187-238.

Leuthold, W. (1966). "Variations in Territorial Behaviour of Uganda Kob *Adenota kob Thomasi* (Neumann, 1896)". Leiden: E.J. Brill.

Leuthold, W. (1974). Observations on home range and social organisation of lesser kudu *Tragelaphus imberbis*. *In* "The Behaviour of Ungulates and its Relation to Management" (eds V. Geist and F. Walther), pp. 206-234. Morges: IUCN.

Lock, J.M. (1964). Unpublished report. Nuffield Unit of Tropical Animal Ecology, Uganda.

Lofts, B. (1970). "Animal Photoperiodism." London: Edward Arnold.

Lorenz, K. (1966). "On Aggression." London: Methuen.

Lugard, F.D. (1890). See Perham, M. and Bull, M. 1959.

Lugard, F.D. (1893). "The Rise of our East African Empire", 2 vols. Edinburgh: Blackwood.

MacFarlane, W.V. and Howard, B. (1972). Comparative water and energy economy of wild and domestic mammals. *Symp. Zool. Soc. Lond. No.* **31**: 261-296.

Malthus, T.R. (1798). An Essay on the Principle of Population, as it affects the future improvement of society, with remarks on the speculations of Mr. Godwin, M. Condorcet and other writers. London: J. Johnson.

Matthews, L.H. (1964). Overt fighting in mammals. *Inst. Biol. Symp.* **13**: 23-32.

Mayr, E. (1942). "Systematics and the Origin of Species." Columbia: Columbia University Press.

McKay, A.D. (1961). Unpublished report. Nuffield Unit of Tropical Animal Ecology, Uganda.

McNab, B.K. (1963). Bioenergetics and the determination of home range size. *Am. Nat.* **97**: 133-140.

Mentis, M.T. (1970). Estimates of natural biomasses of large herbivores in the Umfolozi Game Reserve area. *Mammalia* **34**: 363-393.

Mettam, R.W.M. (1936). Annual Report of the Veterinary Department, Uganda. Entebbe: Government Printer.

Mettam, R.W.M. (1937). A short history of rinderpest with special reference to Africa. *Uganda J.* **5**: 22-26.

Middleton, D. (1969). "The Diary of A.J. Mounteney Jephson, Emin Pasha Relief Expedition 1887-1889." Cambridge: The Hakluyt Society.

Mitchell, B.L., Shenton, J.B. and Uys, J.C.M. (1965). Predation on large mammals in the Kafue National Park, Zambia. *Zoologica Africana* 1: 297–318.

Modha, K.L. and Eltringham, S.K. (1976). Population ecology of the Uganda kob (*Adenota kob Thomasi* Neumann) in relation to the territorial system in the Rwenzori National Park, Uganda. *J. Appl. Ecol.* 13: 453–473.

Montfort, A. and Montfort, N. (1974). Notes sur l'écologie et le comportement des oribis (*Ourebia ourebi. Z*). *Terre et Vie* 28: 169–208.

Montfort, N. (1974). Quelques exemples de structures sociales chez les ongulés Africains. *In* "Zoologie et Assistance Technique" (ed. J.C. Ruwet), pp. 53–76. Liège: FULREAC.

Morris, P. (1978). The use of teeth for estimating the ages of wild animals. *In* "Development, Function and Evolution of Teeth" (eds P.M. Butler and K.A. Joysey), pp. 483–494. London and New York: Academic Press.

Morrison, J.A. (1971). Morphology of corpora lutea in the Uganda kob antelope, *Adenota kob Thomasi* (Neumann). *J. Reprod. Fert.* 26: 297–305.

Morrison, J.A. and Buechner, H.K. (1971). Reproductive phenomena during the *post partum*–preconception interval in the Uganda kob. *J. Reprod. Fert.* 26: 307–317.

Mossman, A.S. and Mossman, H.W. (1962). Ovulation, implantation and sex ratio in impala. *Science* 137: 869.

Murie, A. (1944). The wolves of Mt. McKinley, *Fauna Series, No. 5*. Washington, D.C.: US. Fauna National Parks.

Murie, J. (1867). Remarks on antelopes from the White Nile, allied to or identical with the Kobus sing-sing of Gray. *Proc. zool. Soc. Lond.* May (5): 3–8.

Mykytowycz, R. (1973). Reproduction of mammals in relation to environmental odours. *J. Reprod. Fert., Suppl.* 19: 433–446.

Nelder, J.A. (1961). The fitting of a generalisation of the logistic curve. *Biometrics* 17: 89–110.

Nice, M.M. (1941). The role of territory in bird life. *Amer. Midl. Nat.* 26: 441–487.

Noble, G.K. (1939). The role of dominance in the life of birds. *Auk* 56: 263–273.

Norton-Griffiths, M. (1975). The numbers and distribution of large mammals in Ruaha National Park, Tanzania. *E. Afr. Wildl. J.* 13: 121–140.

Ogilby, W. (1833). Characters of a new species of antelope. *Proc. zool. Soc. Lond.* 1833 (1): 47.

Osmaston, H. (1965). *In* "Uganda National Parks Handbook." Kampala: Trustees of the Uganda National Parks.

Owen-Smith, N. (1977). On territoriality in ungulates and an evolutionary model. *Quart. Rev. Biol.* 52: 1–38.

Parke, T.H. (1891). "My Personal Experiences in Equatorial Africa." London: Sampson Low.

Payne, W.J.A. (1964). Specific problems of semi-arid environments. *Proc. 6th. Int. Congr. Nutr. (Edinb.):* 213–226.

Perham, M. and Bull, M. (1959). "The Diaries of Lord Lugard", 3 vols. London: Faber and Faber.

Perry, R. (1952). "The Watcher and the Red Deer." London: Scientific Book Club.

Phelps, G. (1974). W. Lithgow. "Rare Adventures and Painful Peregrinations". London: The Folio Society.

Pienaar, U. de V. (1961). A second outbreak of anthrax amongst game animals in the Kruger National Park, 5th June to 11th October 1960. *Koedoe* 4: 4–17.

Pienaar, U. de V. (1966). "Annual Report of the Biologist of the Kruger National Park." Pretoria: Board of Trustees.

Pienaar, U. de V. (1967). Epidemiology of anthrax in wild animals and the control of anthrax epizootics in the Kruger National Park, South Africa. *Federation Proceedings* **26** (5): 1496–1502.

Pienaar, U. de V. (1969). Predator-prey relationship amongst the larger mammals of the Kruger National Park. *Koedoe* **12**: 108–176.

Pitman, C.R.S. (1948–1952). "Annual Reports of the Uganda Game Department." Entebbe: The Government Printer.

Plowright, W. (1965). Malignant cataarhal fever in East Africa. I. Behaviour of the virus in free-living populations of blue wildebeest (*Gorgon taurinus taurinus* Burchell). *Res. Vet. Sci.* **6**: 56–68.

Plowright, W., Parker, J. and Pierce, M.A. (1969). The epizootiology of African swine fever in Africa. *Vet. Rec.* **85**: 668–674.

Pocock, R.I. (1904). Proceeding of the General Meetings. *Proc. zool. Soc. Lond.* **1904**: 3–4.

Pocock, R.I. (1910). On the specialised cutaneous glands of ruminants. *Proc. zool. Soc. Lond.* **1910**: 840–986.

Posnansky, M. (1965). *In* "Uganda National Parks Handbook." Kampala: Trustees of the Uganda National Parks.

Potter, D.A. and Johnston, D.E. (1978). Raillietia whartoni sp. n. (Acari-Mesostigmata) from the Uganda kob. *J. Parasitol.* **64**: 139–142.

Radford, C.D. (1938). Notes on some new species of parasitic mites. *Parasitology* **30**: 427–440.

Riffenburgh, R.H. (1966). Growth curve estimation using parametric extremes. *Biometrics* **22**: 162–178.

Riney, T. and Kettlitz, W.L. (1964). Management of large mammals in the Transvaal. *Mammalia* **28**: 189–248.

Roberts, A. (1951). "The Mammals of South Africa." Johannesburg: Trustees, "The Mammals of South Africa" Book Fund.

Robinette, W.L. and Child, G.F.T. (1964). Notes on the biology of the lechwe (*Kobus leche*). *Puku* **2**: 84–117.

Robinson, T.J. (1951). Reproduction in the ewe. *Biol. Rev.* **26**: 121–157.

Rogerson, A. (1966). The utilisation of metabolisable energy by a wildebeest. *E. Afr. Wildl. J.* **4**: 149.

Rogerson, A. (1968). Energy utilisation by the eland and the wildebeest' *Symp. Zool. Soc. Lond. No.* **21**: 153–161.

Rollinson, D.R.I., Harker, K.W. and Taylor, J.I. (1956). Studies on the habits of zebu cattle. IV. Errors associated with recording technique. *J. agric. Sci.* **47**: 1–5.

Roosevelt, T. and Heller, E. (1914). "Life Histories of African Game Animals." 2 vols. London: John Murray.

Rüppell, E. (1835). "Neue Wirbelthiere zu der Fauna von Abyssinien Gehörig." Frankfurt-am-Main.

Sachs, R. (1979). *In* Gainer, R.S. "The Role of Anthrax in the Population Biology of Wildebeest in the Selous Game Reserve." M.Sc. thesis. University of British Columbia.

Sadleir, R.M.F. (1969). "The Ecology of Reproduction in Wild and Domestic Mammals." London: Methuen.

Sayer, J.A. and van Lavieren, L.P. (1975). The ecology of the Kafue lechwe population of Zambia before the operation of hydro-electric dams on the Kafue River. *E. Afr. Wildl. J.* **13**: 9–38.

Schenkel, R. (1966). Zum Problem der Territorialität und des Markierens bei Säugern an Beispiel des Schwarzen Nashorns und des Löwens. *Z. f. Tierpsychol.* **23**: 593–626.

Schloeth, R. (1961). Das Sozialleben des Camargue-Rindes. *Z. f. Tierpsychol.* **18**: 574–627.

Schoen, A. (1971). The effect of heat stress and water deprivation on the environmental physiology of the bushbuck, the reedbuck and the Uganda kob. *E. Afr. agric. For. J.* **37**: 1–7.

Schuster, R.H. (1976). Lekking behaviour in Kafue lechwe. *Science* **192**: 1240–1242.

Schuster, R.H. (1976a). Reproductive social organisation of Kafue lechwe: Implications for management and survival. "Proc. 4th Regional Wildl. Conf. for Eastern and Central Africa", pp. 163–183. Lusaka: The Government Printer.

Short, R.V. and Spinage, C.A. (1967). Drug immobilisation of the defassa waterbuck. *Vet. Rec.* **81**: 336–340.

Simpson, G.G. (1945). The principles of classification and a classification of mammals. *Bull. Amer. Mus. Nat. Hist.* **85**: i–xvi, 1–350.

Sinclair, A.R.E. (1973). Population increases of buffalo and wildebeest in the Serengeti. *E. Afr. Wildl. J.* **11**: 93–107.

Sinclair, A.R.E. (1974). The natural regulation of buffalo populations in East Africa. IV. The food supply as a regulating factor, and competition. *E. Afr. Wildl. J.* **12**: 291–311.

Sinclair, A.R.E. (1977). "The African Buffalo." Chicago: Chicago University Press.

Smith, A. (1840). "Illustrated Zoology of South Africa" (12), pp. 28–29. London: Smith, Elder and Co.

Smith, N.S. (1970). Appraisal of condition estimation methods for East African ungulates. *E. Afr. Wildl. J.* **8**: 123–130.

Smith, N.S. and Ledger, H.P. (1965). A method of predicting live weight from dissected leg weight. *J. Wildl. Mgmt* **29**: 504–511.

Spector, W.S. (1956). Ed. "Handbook of Biological Data." Philadelphia: W.B. Saunders.

Spinage, C.A. (1962). "Animals of East Africa." London: Collins.

Spinage, C.A. (1968). Method for deriving a survival curve of young calves in wild ungulates. *Nature* **217**: 480.

Spinage, C.A. (1968a). Naturalistic observations on the reproductive and maternal behaviour of the Uganda defassa waterbuck *Kobus defassa ugandae* Neumann. *Z. f. Tierpsychol.* **26**: 39–47.

Spinage, C.A. (1968b). A quantitative study of the daily activity of the Uganda defassa waterbuck. *E. Afr. Wildl. J.* **6**: 89–93.

Spinage, C.A. (1969). Waterbuck management data. *E. Afr. agric. For. J.* **34**: 327–335.

Spinage, C.A. (1969a). Reproduction in the Uganda defassa waterbuck *Kobus defassa ugandae* Neumann. *J. Reprod. Fert.* **18**: 445–457.

Spinage, C.A. (1969b). Quantitative assessment of ectoparasites. *E. Afr. Wildl. J.* **7**: 169–171.

Spinage, C.A. (1969c). Territoriality and social organisation of the Uganda defassa waterbuck *Kobus defassa ugandae.* *J. Zool. Lond.* **159**: 329–361.

Spinage, C.A. (1970). Population dynamics of the Uganda defassa waterbuck (*Kobus defassa ugandae* Neumann) in the Queen Elizabeth Park, Uganda. *J. Anim. Ecol.* **39**: 51–78.

Spinage, C.A. (1971). Two records of pathological conditions in the Impala (*Aepyceros melampus*). *J. Zool. Lond.* **164**: 269–270.

Spinage, C.A. (1972). African ungulate life tables. *Ecology* **53**: 645–652.

Spinage, C.A. (1973). A review of the age determination of mammals by means of teeth, with special reference to Africa. *E. Afr. Wildl. J.* **11**: 165–187.

Spinage, C.A. (1973a). The role of photoperiodism in the seasonal breeding of tropical African ungulates. *Mammal Review* **3**: 71–84.

Spinage, C.A. (1974). Territoriality and population regulation in the Uganda defassa waterbuck. *In* "The Behaviour of Ungulates and its Relation to Management" (eds V. Geist and F. Walther), pp. 635–643. Morges: IUCN.

Spinage, C.A. (1976). Incremental cementum lines in the teeth of tropical African mammals. *J. Zool. Lond.* **178**: 117–131.

Spinage, C.A. (1976a). Age determination of the female Grant's gazelle. *E. Afr. Wildl. J.* **14**: 121–134.

Spinage, C.A. and Guinness, F.E. (1972). Effects of fire in the Akagera National Park and Mutara Hunting Reserve, Rwanda. *Rev. Zoo. Bot. Afr.* **86**: 302–336.

Spinage, C.A., Guinness, Fiona, Eltringham, S.K. and Woodford, M.H. (1972). Estimation of large mammal numbers in the Akagera National Park and Mutara Hunting Reserve, Rwanda. *Terre et Vie* **4**: 561–570.

Staines, B. (1978). The dynamics and performance of a declining population of Red deer (*Cervus elaphus*). *J. Zool. Lond.* **184**: 403–420.

Stanley, H.M. (1890). *In* "Darkest Africa", 2 vols. London: Sampson Low.

Stevenson-Hamilton, J. (1912). "Animal Life in Africa." London: Heinemann.

Stevenson-Hamilton, J. (1929). "The Low-veld: Its Wildlife and its People." London: Cassell and Co.

Stewart, D. (1964). Rinderpest among wild animals in Kenya, 1960–1962. *Bull. epiz. Dis. Afr.* **12**: 39–42.

Stewart, D. and Stewart, J. (1963). The distribution of some large mammals in Kenya. *J. E. Afr. nat. Hist. Soc.* **24**: 1–52.

Swayne, H.G.C. (1895). "Seventeen Trips through Somaliland." London: Rowland Ward.

Talbot, L.M. and Talbot, M.H. (1963). The wildebeest in Western Masailand, East Africa. *Wildlife Monographs No.* **12**: 1–88. The Wildlife Society.

Talbot, L.M. and McCulloch, J.S.G. (1965). Weight estimations for East African mammals from body measurements. *J. Wildl. Mgmt* **29**: 84–89.

Talbot, L.M., Ledger, H.P. and Payne, W.J.A. (1961). The possibility of using wild animals for animal production in the semi-arid tropics of East Africa. Fetschrift sum VIII. Intern. Tierzuchtkongres Hamburg: 205–210.

Taylor, C.R. (1968). The minimum water requirements of some East African bovids. *Symp. Zool. Soc. Lond. No.* **21**: 195–206.

Taylor, C.R., Spinage, C.A. and Lyman, C.P. (1969). Water relations of the waterbuck, an East African antelope. *Amer. J. Phys.* **217**: 630–634.

Thal, J.A. (1972). "Enquête sur la Peste Bovine et les Maladies Similaires." Rapport Technique. Rome: FAO.

Thornthwaite, C.W. (1948). An approach towards a rational classification of climate. *Geogr. Rev.* **38**: 59–94.

Thornton, D.D. (1959). Unpublished Report. Nuffield Unit of Tropical Animal Ecology, Uganda.

Thornton, D.D. (1962). Unpublished Report. Nuffield Unit of Tropical Animal Ecology, Uganda.

Thornton, D.D. (1971). The effect of complete removal of hippopotamus on grassland in the Queen Elizabeth National Park, Uganda. *E. Afr. Wildl. J.* **9**: 47–55.

Tinbergen, N. (1969). On war and peace in animals and man. *Science* **160**: 1411–1418.

Tomlinson, D.N.S. (1979). The feeding behaviour of waterbuck in the Lake McIlwaine Game Enclosure. *Rhod. Sci. News* **13**: 11–14.

Tomlinson, D.N.S. (1980). Seasonal food selection by waterbuck *Kobus ellipsiprymnus* in a Rhodesian game park. *S. Afr. J. Wildl. Res.* **10**: 22–28.

Verheyen, R. (1955). Contribution à l'éthologie du waterbuck, *Kobus defassa,* et de l'antilope harnachée, *Tragulus scriptus. Mammalia* **19**: 303–319.

Von Bertalanffy, L. (1938). A quantitative theory of organic growth. *Hum. Biol.* **10**: 181–213.

Von Richter, W. (1972). Territorial behaviour of the black wildebeest *Connochaetes gnou. Zoologica Africana* 7: 207–231.

Von Richter, W. and Osterberg, R. (1977). The nutritive values of some major food plants of lechwe, puku and waterbuck along the Chobe River, Botswana. *E. Afr. Wildl. J.* 15: 91–97.

Walker, J.B. (1974). "The Ixodid Ticks of Kenya." London: Comm. Inst. Entomology.

Walther, F. (1964). Verhaltensstudien an der Gattung Tragelaphus de Blainville (1816) in Gefangenschaft unter besonderer Derücksichtigung des Sozialverhaltens. *Z. f. Tierpsychol.* 21: 393–467.

Walther, F. (1966). "Mit Huf und Horn." Berlin: P. Parey.

Watson, A. (1970). Territorial and reproductive behaviour of red grouse. *J. Reprod. Fert., Suppl.* 11: 3–14.

Watson, R.M. (1969). Reproduction of the wildebeest *Connochaetes taurinus albojubatus* Thomas, in the Serengeti region, and its significance to conservation. *J. Reprod. Fert., Suppl.* 6: 287–310.

Waser, P. (1975). Diurnal and nocturnal strategies of the bushbuck *Tragelaphus scriptus* (Pallas). *E. Afr. Wildl. J.* 13: 49–64.

Wenyon, C.M. (1926). "Protozoology", 2 vols. London: Ballière, Tindall and Cox.

Wilson, D.E. and Hirst, S.M. (1977). Ecology and factors limiting roan and sable antelope populations in South Africa. *Wildl. Monogr. No.* 54. The Wildlife Society.

Wilson, E.O. (1975). "Sociobiology." Massachusetts: Belknap Press.

Wing, L.D. and Buss, I.O. (1970). Elephants and forests. *Wildl. Monogr. No.* 19. The Wildlife Society.

Wirtz, P. (1980). Co-operative satellite males in the waterbuck. *In litt.*

Wodzicka-Tomaszewicka, M., Hutchison, J.C.D. and Bennett, N.W. (1967). Control of the annual rhythm of breeding ewes: effect of an equatorial daylength with reversed thermal seasons. *J. agric. Sci., Camb.* 68: 61–67.

Worthington, S. and Worthington, E.B. (1933). "Inland Waters of Africa." London: Macmillan.

Wright, S. (1931). Evolution in Mendelian populations. *Genetics* 16: 97–159.

Wright, S. (1960). Physiological genetics, ecology and natural selection. *In* "The Evolution of Life" (ed. S. Tax), pp. 429–475. Chicago: University of Chicago Press.

Wynne-Edwards, V.C. (1962). "Animal Dispersion in Relation to Social Behaviour." Edinburgh: Oliver and Boyd.

Young. J.Z. (1957). "The Life of Mammals." Oxford: OUP.

Subject Index